Advances in Experimental Medicine and Biology

Volume 868

More information about this series at http://www.springer.com/series/5584

Richard Bronson
Editor

The Male Role in Pregnancy Loss and Embryo Implantation Failure

Springer

Editor
Richard Bronson
Obstetrics, Gynecology and Reproductive Medicine
Stony Brook University Medical Center
Stony Brook, NY, USA

ISSN 0065-2598 ISSN 2214-8019 (electronic)
Advances in Experimental Medicine and Biology
ISBN 978-3-319-18880-5 ISBN 978-3-319-18881-2 (eBook)
DOI 10.1007/978-3-319-18881-2

Library of Congress Control Number: 2015943698

Springer Cham Heidelberg New York Dordrecht London

Printed on acid-free paper

Springer International Publishing AG Switzerland is part of Springer Science+Business Media (www.springer.com)

Preface

Only recently have we come to ask what role the male might play in pregnancy loss and failure of embryo implantation. This volume of *Advances in Experimental Biology and Medicine* contains a series of articles that reviews what is currently known about this rapidly evolving area, placing information in a potentially clinically relevant context. There are two overarching themes: (1) The fertilizing spermatozoon might lead to an impaired chance of a successful pregnancy, through abnormalities in its chromosomes, its epigenetic status, or via noncoding RNAs transported to the oocyte; (2) The composition of seminal plasma is highly complex. The female reproductive tract is exposed at coitus to these noncellular components of semen that could play important roles permissive of embryo implantation and trophectoderm outgrowth, leading to successful pregnancy, or might if perturbed be harmful.

Dimitrios Ioannou and Helen Tempest review the increasing evidence suggesting that spermatozoa of certain men have significantly high levels of aneuploidy, which is mirrored in their embryos and offspring. They provide insights into the origin and clinical relevance of paternally derived aneuploidy and review the general mechanisms through which aneuploidy arises and how numerical and sub-microscopic chromosome aberrations may result in pregnancy loss. Human spermatozoa are vulnerable to free radical attack and the generation of oxidative DNA damage. Disruptions of their genetic integrity might play a major role in determining the development of the embryo following fertilization. Dan Gavriliouk and John Aitken examine the links between DNA damage in human sperm and the appearance of mutations in the progeny leading to a variety of clinical conditions. Jason Ross and his associates focus on the role of posttranscriptional gene regulation via noncoding ribonucleic acid in the regulation of reproductive function. Douglas Carrell and his coauthors discuss the effect of paternal aging on the epigenome and its potential impact on fertility, embryonic development, and the health of offspring.

Under the second general theme of this volume of *Advances*, Judith Bulmer and Gendie Lash characterize the variation of the leukocyte population within the human endometrium throughout the menstrual cycle, particularly focusing on the ontogeny and role of uterine natural killer (NK) cells in normal and abnormal pregnancy. John Schjenken and Sarah Robertson review studies demonstrating that seminal fluid entering the female reproductive tract during coitus initiates immune adaptive processes promoting tolerance to male transplantation antigens and helps to shape subsequent fetal development. The factors involved and mechanisms that mediate these effects are discussed. Deborah Anderson and Joseph Politch review the origins and concentrations of immunomodulatory/proinflammatory factors in human seminal plasma in health and disease. Their effects on immune defense in the female genital tract that may promote fertility and fetal well-being or may contribute to female reproductive failure are discussed. Thomas Hviid reviews recent studies demonstrating the presence of immunoregulatory and tolerance-inducible human leukocyte antigen (HLA)-G in the male reproductive organs as well as its distribution in the female and male reproductive tracts. He discusses the highly variable amounts of soluble HLA-G observed in seminal fluid of different men, the genetics of its expression, and whether these differences may affect the chance of pregnancy. Gunnar Ronquist reviews the origins of prostasomes in semen and their biochemical characterization, presenting evidence that they may serve an immunomodulatory function within the female reproductive tract, perhaps through the mediation of small noncoding RNAs. Anne Schumacher and Ana Zenclussen address the contribution of paternal-derived factors in semen to fetal tolerance induction during pregnancy, with a special focus on T regulatory cell (T_{reg}) biology.

Stony Brook, NY, USA Richard Bronson

Contents

Contributors

Robert John Aitken Priority Research Centre in Reproductive Science, University of Newcastle, Callaghan, NSW, Australia

Faculty of Health and Medicine, Hunter Medical Research Institute, University of Newcastle, Callaghan, NSW, Australia

School of Environmental and Life Sciences, Discipline of Biological Sciences, University of Newcastle, University Drive, Callaghan, NSW, Australia

Deborah J. Anderson Department of Obstetrics and Gynecology, Boston University School of Medicine, Boston, MA, USA

Department of Microbiology, Boston University School of Medicine, Boston, MA, USA

Department of Medicine, Boston University School of Medicine, Boston, MA, USA

Kenneth I. Aston Andrology and IVF Laboratories, Department of Surgery, University of Utah School of Medicine, Salt Lake City, UT, USA

Judith N. Bulmer Reproductive and Vascular Biology Group, Institute of Cellular Medicine, Newcastle University, Newcastle upon Tyne, Tyne and Wear, UK

Douglas T. Carrell Andrology and IVF Laboratories, Department of Surgery, University of Utah School of Medicine, Salt Lake City, UT, USA

Department of Obstetrics and Gynecology, University of Utah School of Medicine, Salt Lake City, UT, USA

Department of Genetics, University of Utah School of Medicine, Salt Lake City, UT, USA

Dan Gavriliouk Priority Research Centre in Reproductive Science, University of Newcastle, Callaghan, NSW, Australia

Faculty of Health and Medicine, Hunter Medical Research Institute, University of Newcastle, Callaghan, NSW, Australia

Benjamin J. Hale Department of Animal Science, Iowa State University, Ames, IA, USA

Thomas Vauvert F. Hviid Department of Clinical Biochemistry, Centre for Immune Regulation and Reproductive Immunology (CIRRI), Roskilde Hospital, Copenhagen University Hospital (Roskilde), University of Copenhagen, Roskilde, Denmark

Dimitrios Ioannou Department of Human and Molecular Genetics, Herbert Wertheim College of Medicine, Florida International University, Miami, FL, USA

Timothy G. Jenkins Andrology and IVF Laboratories, Department of Surgery, University of Utah School of Medicine, Salt Lake City, UT, USA

Aileen F. Keating Department of Animal Science, Iowa State University, Ames, IA, USA

Gendie E. Lash Reproductive and Vascular Biology Group, Institute of Cellular Medicine, Newcastle University, Newcastle upon Tyne, UK

Tyson Meyer Andrology and IVF Laboratories, Department of Surgery, University of Utah School of Medicine, Salt Lake City, UT, USA

Joseph A. Politch Department of Obstetrics and Gynecology, Boston University School of Medicine, Boston, MA, USA

Sarah A. Robertson Robinson Research Institute, University of Adelaide, Adelaide, SA, Australia

School of Paediatrics and Reproductive Health, University of Adelaide, Adelaide, SA, Australia

Gunnar Ronquist Department of Medical Sciences, Clinical Chemistry, The University of Uppsala, Uppsala, Sweden

Jason W. Ross Department of Animal Science, Iowa State University, Ames, IA, USA

John E. Schjenken Robinson Research Institute, University of Adelaide, Adelaide, SA, Australia

School of Paediatrics and Reproductive Health, University of Adelaide, Adelaide, SA, Australia

Anne Schumacher Department of Experimental Obstetrics and Gynecology, Medical Faculty, Otto-von-Guericke University, Magdeburg, Germany

Helen G. Tempest Department of Human and Molecular Genetics, Herbert Wertheim College of Medicine, Florida International University, Miami, FL, USA

Biomolecular Sciences Institute, Florida International University, Miami, FL, USA

Cai-Xia Yang Department of Animal Science, Iowa State University, Ames, IA, USA

Ana Claudia Zenclussen Department of Experimental Obstetrics and Gynecology, Medical Faculty, Otto-von-Guericke University, Magdeburg, Germany

Chapter 1
Meiotic Nondisjunction: Insights into the Origin and Significance of Aneuploidy in Human Spermatozoa

Dimitrios Ioannou and Helen G. Tempest

Abstract Chromosome aneuploidy refers to changes in the chromosome complement of a genome and can include gain or loss of genetic material. The human genome is delicately balanced, and for the most part perturbations in the chromosome complement are often incompatible with embryonic development. The importance and clinical relevance of paternally derived aneuploidy is often overshadowed by the large maternal contribution; as a result, the paternal contribution to pregnancy loss due to chromosome aneuploidy is rarely considered within the clinic. However, there is increasing evidence to suggest that certain men have significantly higher levels of sperm aneuploidy, which is mirrored by an increase in aneuploidy within their embryos and offspring. Therefore, the paternal contribution to aneuploidy at least for some individuals may have greater clinical significance than is currently perceived. Thus, the main focus of this chapter is to provide insights into the origin and clinical relevance of paternally derived aneuploidy. Furthermore, this section will review the general mechanisms through which aneuploidy arises during spermatogenesis and how numerical (whole chromosome) and structural chromosome aberrations (cytogenetically visible or submicroscopic) may lead to clinically relevant aneuploidy potentially resulting in pregnancy loss, congenital malformations, and cognitive impairment.

Keywords Spermatozoa • Chromosome aneuploidy • Nondisjunction • Meiosis • Y microdeletions • Oligospermia • Chromosome disomy

---

D. Ioannou
Department of Human and Molecular Genetics, Herbert Wertheim College of Medicine, Florida International University, Miami, FL, USA

H.G. Tempest (✉)
Department of Human and Molecular Genetics, Herbert Wertheim College of Medicine, Florida International University, Miami, FL, USA

Biomolecular Sciences Institute, Florida International University, Miami, FL, USA
e-mail: htempest@fiu.edu

R. Bronson (ed.), *The Male Role in Pregnancy Loss and Embryo Implantation Failure*, Advances in Experimental Medicine and Biology 868,
DOI 10.1007/978-3-319-18881-2_1

1.1 Introduction

Within recent years the paternal influence on reproductive success has become an area of fervent research. Although it is clear that the sperm cell is required for fertilization and embryogenesis, the importance and contribution of the sperm cell in this process remains poorly understood and is frequently overshadowed by the maternal oocyte. The oocyte plays critical roles in remodeling and repairing the paternal genome upon fertilization, replacing the paternal DNA-bound proteins (protamines) with maternally derived histones and by repairing DNA damage present in the paternal genome (McLay and Clarke 2003). Additionally, the driving force for embryogenesis is often considered to be the maternal oocyte, given that it contributes the environment, energy source, enzymes, and the vast majority of support organelles required for the initial embryonic divisions (Carrell 2013). So the question remains how important is the sperm cell and the paternal genome contained within, for the process of fertilization and embryogenesis? Given the aforementioned reasons, the sperm cell is often perceived to be a silent vessel whose sole function is to safely deliver the haploid paternal genome to the maternal oocyte. However, recent studies challenge this widely held viewpoint suggesting that the sperm cell is far from a silent vessel, with the paternal genome playing an important role in reproductive outcomes.

1.2 The Paternal Genome and Its Association with Infertility

Infertility affects approximately one in six couples wishing to start a family (de Kretser 1997). Frequently, the male contribution to infertility is overlooked with much of the diagnostic/clinical workup focusing on females. However, approximately 50 % of infertility is estimated to be due to male factors solely or in combination with female factors (Chandra et al. 2005). Typically, a routine fertility workup in males is limited to family history, physical examination, and assessment of the semen parameters as per the WHO guidelines (World Health Organization 2010). The results of the semen parameter assessment in conjunction with family and reproductive history may indicate additionally genetic testing to determine whether the etiology of the perturbation in semen parameters can be identified. Such genetic tests may include Y chromosome microdeletion testing, single gene mutation testing, and karyotyping.

Prior to the development of intracytoplasmic sperm injection (ICSI) and its rapid clinical implementation of ICSI worldwide in the early 1990s, men with male factor infertility were rarely able to have biological offspring. However, the utilization of ICSI in the treatment of male factor infertility has raised concerns, given that this technique may increase the risk of pregnancy loss and transmitting genetic defects to resultant embryos. Male factor infertility can be caused by perturbations in the paternal genome [e.g., chromosome aberrations (structural or numerical), DNA damage, or single gene mutations]. If these aberrations are present in the sperm cell utilized for ICSI, they will be transmitted to the resultant embryo and may impact the process of embryogenesis and development. Prior to the development of ICSI, perturbations within the paternal genome were unlikely to be transmitted to future offspring if they resulted in infertility. However, ICSI bypasses the natural barriers in place enabling infertile men to reproduce. Given that a sperm cell carries a single copy of the paternal genome, any genetic alterations contained within a sperm that fertilizes an oocyte, be they de novo or inherited, will be directly transferred to the resulting embryo. To date, a handful of studies have associated perturbations within the paternal genome with an increased risk of failed fertilization, poor and/or arrested embryo development, fetal demise, and the birth of genetically abnormal offspring.

1.3 Karyotyping, Indications, and Prevalence of Chromosomal Aberrations

Karyotyping is frequently offered to patients with multiple spontaneous abortions and men with oligozoospermia to identify whether these findings are due to the presence of numerical or structural chromosome aberrations. Chromosome abnormalities are relatively rare in the general population arising in approximately 0.6 % of newborns (Berger 1975). However, they are among the most commonly identified cause of male infertility, with a higher incidence of chromosome aberrations observed in infertile men compared to the general population (McLachlan and O'Bryan 2010). Chromosome aberrations occur in 2 % in males presenting with infertility; however, the frequency of chromosome aberrations increases with the severity of the infertility with frequencies of 6 % and 14 % occurring in oligozoospermic and nonobstructive azoospermic males, respectively (Shi and Martin 2000). These estimates are potentially much higher, given that not all infertile males are karyotyped and the size limit of aberrations detected by karyotyping (>3–5 Mb in size). It is important to note that chromosome aberrations are a clinically significant cause of pregnancy loss, congenital malformations, and cognitive impairment in humans.

Taken collectively, additional genetic tests (karyotyping, Y chromosome microdeletion, and single gene mutation testing) reveal the etiology of the male factor infertility in an estimated 20 % of cases. This is further compounded by the fact that semen parameters are not absolutely predictive of fertility, rather they are indicative of male factor infertility. It is important to note that the WHO semen parameter reference values should not be interpreted as the minimum values required for conception. Moreover, despite widespread usage of standardized WHO guidelines, it remains a relatively subjective analysis (Lewis 2007). Therefore, despite our ability to measure altered semen parameters and extensive research to establish the genetic causes and identify clinically relevant biomarkers, the underlying etiology of the vast majority of male factor infertility (up to 80 %) is classified as idiopathic (Hotaling and Carrell 2014). Nevertheless, it seems plausible that the majority of idiopathic male factor infertility will be associated with as of yet unknown genetic factors; therefore, there is a requirement to identify these factors and develop clinically relevant tests.

1.4 Additional Genetic Tests That Could Identify Clinically Relevant Perturbations in the Paternal Genome That May Impact the Process of Embryogenesis

Within recent years, several important developments in the field have demonstrated that DNA fragmentation, alterations in the sperm epigenome, and organization of the paternal genome may play vital roles in fertilization and embryogenesis. Given the inability of the mature sperm cell to repair DNA damage, patients with very high levels of sperm DNA fragmentation may overwhelm the maternal DNA repair process leading to failed fertilization or poor embryogenesis. Furthermore, the sperm cell delivers a unique epigenetically poised genome to the oocyte, perturbations in which may lead to failed fertilization and perturbed embryogenesis (Carrell 2012; Hammoud et al. 2010; Jenkins and Carrell 2012; Kumar et al. 2013; Schagdarsurengin et al. 2012). Additionally, the paternal genome has a unique organization within the human sperm cell (Millan et al. 2012; Zalenskaya and Zalensky 2004; Zalensky and Zalenskaya 2007) which may function as an additional layer of epigenetic regulation (Millan et al. 2012) and be critical for fertilization and embryogenesis. Currently, this research field is in its infancy, but if specific epigenetic patterns and genome organization are critical for embryogenesis, identification of perturbations may lead to the development of clinically relevant tests.

1.5 The Clinical Consequence of Chromosome Aneuploidy

Chromosome aneuploidy (the presence of extra or missing chromosomes—trisomies and monosomies, respectively) is the leading cause of fetal demise and cognitive impairment in humans and is estimated to occur in up to 50–80 % of conceptions (Hassold and Hunt 2001; Hassold et al. 1993; Munne et al. 2004; Vanneste et al. 2009) and 5 % of clinically recognized pregnancies (Hassold and Hunt 2001). The vast majority of aneuploid conceptuses are spontaneously aborted prior to establishment of a clinically recognized pregnancy. Full-blown (non-mosaic) aneuploidies are only viable in a monosomic state for the X chromosome (Turner syndrome) and for a handful of chromosomes in a trisomic state, chromosomes 13, 18, and 21 (Patau, Edwards, and Down syndrome, respectively), and the sex chromosomes (e.g., XYY, XXX, and XXY [Klinefelter syndrome]). Although these aneuploidies are clinically viable, it should be noted that the vast majority of these will be spontaneously aborted often prior to establishment of a clinically recognized pregnancy. It is well established that the largest risk factor for chromosome aneuploidy is advancing maternal age (Hassold et al. 1993, 2007; Hassold and Hunt 2001, 2009). The vast majority of chromosome aneuploidy involving chromosomes 1–22 (autosomes) and trisomy X in the early embryo is predominantly maternal in origin (70–100 %) (Hassold et al. 2007); as a consequence the paternal contribution to chromosome aneuploidy is frequently overlooked. Nevertheless, the paternal contribution for the sex chromosome aneuploidies is considerably higher accounting for between 50 and 100 % of cases (Hassold et al. 1993). This is of clinical relevance given that aneuploidies involving the sex chromosomes are clinically viable and more common than autosomal aneuploidies in both spontaneous abortions and neonates (Templado et al. 2013).

The most common cause of chromosome aneuploidy is perturbed segregation of chromosomes during maternal or paternal meiosis. For the most part these are typically de novo events that arise in the gametes of individuals with a normal somatic karyotype. Individuals with an abnormal karyotype, be it numerical or structural in nature, may be at an increased risk to transmit chromosome aberrations to their future offspring and will be considered in subsequent sections. Given a normal karyotype, meiosis in the male is initiated within primary spermatocytes; chromosomes pair and undergo meiotic recombination, followed by two successive rounds of cell division leading to the formation of four genetically unique haploid spermatids that undergo cytodifferentiation to form mature spermatozoa. Critical to ensuring the appropriate segregation of chromosomes in meiosis is the pairing and synapsis of homologous chromosomes and initiation of meiotic recombination prior to the first meiotic division (meiosis I) (Fig. 1.1). In the normal situation the first meiotic division results in homologous chromosomes segregating to the opposite poles to produce two daughter cells (each homologous chromosome consists of a pair of sister chromatids). In the second meiotic division (meiosis II), the two sister chromatids segregate to opposite poles to create four genetically diverse daughter cells (Fig. 1.2). Chromosome aneuploidy can arise following inappropriate segregation of homologous chromosomes or chromatids (nondisjunction), anaphase lag, or due to an ineffective meiotic checkpoint control (Ramasamy et al. 2014). The main etiology through which chromosome aneuploidy is considered to arise is through nondisjunction (Griffin 1996; Hassold et al. 1993; Marquez et al. 1996). There are several ways in which nondisjunction can occur: (1) both homologous chromosomes may travel to the same pole in meiosis I; (2) both sister chromatids may travel to the same pole in meiosis II; and (3) premature separation of the sister chromatids can occur in meiosis I or II potentially resulting in inappropriate segregation to the same pole. The net effect of these various nondisjunction events will be that of the four sperms produced; some will contain the normal haploid complement of chromosomes, whereas some will be missing a chromosome (nullisomic) or possess an additional chromosome (disomy) (Fig. 1.2). If a nullisomic or disomic sperm were to fertilize a normal haploid oocyte, it would give rise to an embryo

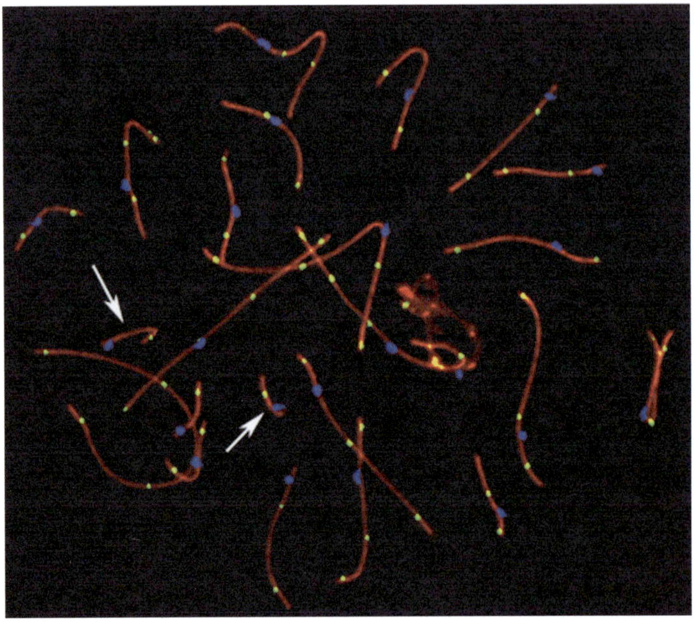

Fig. 1.1 Human pachytene spermatocyte demonstrating synapsis of homologous chromosome pairs and sites of meiotic recombination (chiasmata). The protein complex (synaptonemal complex) that forms the physical connection between homologous chromosomes is labeled in *red*, chromosome centromeres in *blue*, and sites of recombination in *yellow*. Note, all bivalents possess multiple chiasmata, with the exception of chromosomes 21 and 22 (*arrowed*) and the sex body which only possess a single chiasma; thus, these particular bivalents have an increased risk of chromosome nondisjunction if recombination does not take place (achiasmate bivalents)

that is monosomic or trisomic, respectively, for the involved chromosome(s). The process of synapsis and meiotic recombination has been shown to be extremely important, not only to create genetic diversity but also to ensure the correct segregation of chromosomes throughout meiosis. Perturbations in chromosome synapsis and meiotic recombination (reduced/absent recombination and/or altered placement of recombination) have been associated with defective spermatogenesis (reduced sperm counts) and increased levels of aneuploidy within sperm (Ferguson et al. 2007; Martin 2008; Sun et al. 2008; Tempest 2011). In addition to meiotic nondisjunction, chromosome nondisjunction can also occur postzygotically during mitosis. If postzygotic mitotic nondisjunction occurs, the embryo will be mosaic, possessing a mixture of diploid and aneuploid cells; the proportion of aneuploid cells will depend on the stage of embryonic development the nondisjunction event occurred. An early nondisjunction event will affect a greater proportion of cells. A mosaic karyotype can also arise following a meiotic nondisjunction event, if one of the additional chromosomes fails to be incorporated in the daughter cell as the result of trisomy rescue or anaphase lag during the postzygotic mitotic cell divisions

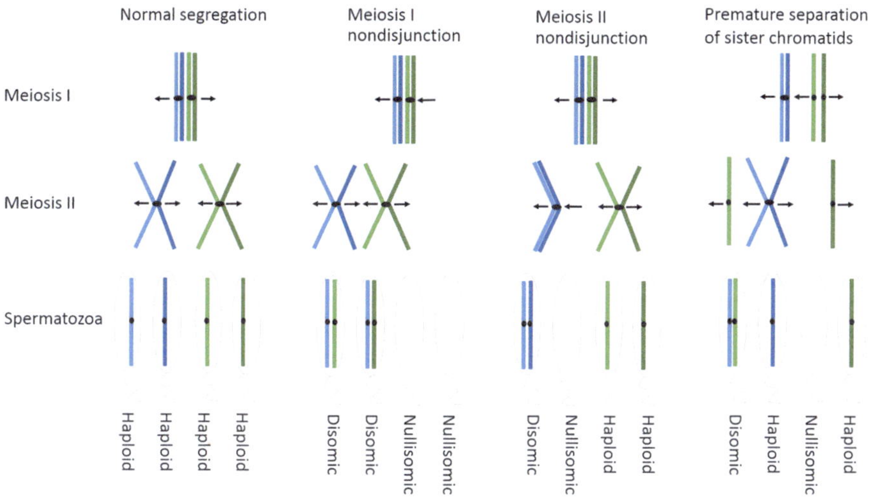

Fig. 1.2 Examples of nondisjunction mechanisms. The image on the *left* illustrates the normal segregation of homologous chromosomes in meiosis I and sister chromatids in meiosis II and the resulting spermatozoa. Following which, nondisjunction events occurring in meiosis I or meiosis II are shown, in which homologues (meiosis I) or sister chromatids (meiosis II) travel to the same pole. Note, in meiosis I the nondisjunction event may be due to a "true" nondisjunction event or the result of an achiasmate bivalent. The segregation pattern shown on the *right* illustrates how nondisjunction can arise as the result of premature separation of the sister chromatids (depicted in meiosis I, this can also arise in meiosis II). The outcome of these various segregation patterns will result in spermatozoa that will be either haploid (correct copy number), disomic (gain of a chromosome), or nullisomic (loss of a chromosome) for the chromosome(s) involved

of a trisomic embryo (Conlin et al. 2010). In this situation, the earlier this occurs during development results in a lower proportion of trisomic cells. For the purposes of this chapter, we will only consider further meiotic nondisjunction events and how chromosome aberrations may impact the process of meiosis and hence spermatogenesis.

1.6 Assessment of the Levels of Chromosome Aneuploidy Within Sperm

Given that the vast majority of chromosome aneuploidies perish in utero, often prior to a clinical pregnancy being established, an accurate assessment of the frequency can only be obtained by studying its frequency in the gametes (Templado et al. 2013). Since the advent of fluorescence in situ hybridization (FISH), chromosome aneuploidy has been relatively straightforward to assess within sperm (Fig. 1.3). FISH has been routinely utilized for decades in a wide range of clinical application; however, the majority of data on sperm aneuploidy is derived from research rather

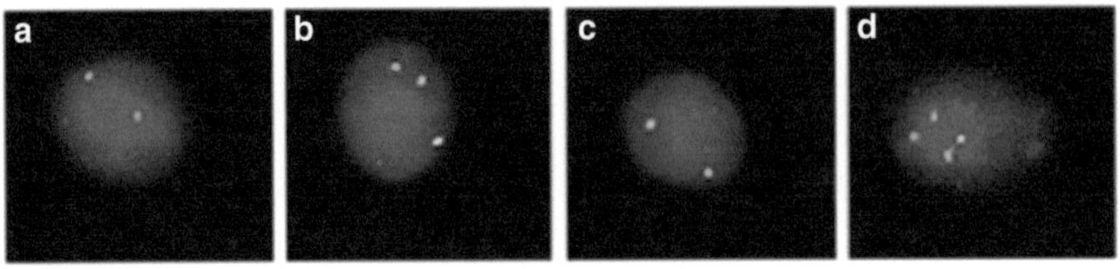

Fig. 1.3 Fluorescence in situ hybridization (FISH) for chromosomes 18, 21, X, and Y in sperm nuclei. The sperm nuclei are counterstained with DAPI (pseudo-colored *gray*) with chromosomes 18, 21, X, and Y probed in *aqua*, *green*, *gold*, and *red* fluorochromes, respectively. Panels (**a**–**d**) provide examples of normal and abnormal sperm FISH results for the investigated chromosomes: Panel (**a**) normal haploid Y-bearing sperm, Panel (**b**) XY disomy, Panel (**c**) YY disomy, and Panel (**d**) XY disomy and disomy 21

than clinical studies. To date, there are over 50 published studies examining the levels of sperm aneuploidy predominantly in normozoospermic and infertile men. These studies have revealed several important findings: (1) all men have a proportion of aneuploid sperm within their ejaculate, and (2) virtually all studies report significantly higher levels of aneuploidy in infertile men compared to individuals with normal semen parameters (Hann et al. 2011; Harton and Tempest 2012; Shi and Martin 2000; Tempest 2011; Tempest and Griffin 2004; Templado et al. 2011). From the published FISH studies, the lower estimates of sperm aneuploidy frequencies in males with normal semen parameters have been estimated to be around 3–5 % (Pang et al. 1999; Templado et al. 2011; Chatziparasidou et al. 2014). This aneuploidy estimate is based on extrapolated data given that only 2–5 chromosomes can be reliably scored in a single sperm cell due to the limited number of fluorochromes available. In addition, most studies have selected chromosomes that are clinically significant, namely, those chromosomes that are viable in an aneuploid state (chromosomes 13, 18, 21, X, and Y). For the most part, studies report similar levels of disomy with an average of approximately 0.1 % for each chromosome (Shi and Martin 2000; Tempest and Griffin 2004; Templado et al. 2011). These estimates are extrapolated from the data on investigated chromosomes; it is important to note that there are a number of chromosomes for which there is little or no aneuploidy FISH data available (Shi and Martin 2000; Tempest and Griffin 2004; Templado et al. 2011). However, the vast majority of studies have observed that specific chromosomes are more prone to chromosome nondisjunction with much higher frequencies of disomy found within sperm. Aneuploidy for chromosomes 21 and 22 and the sex chromosomes is frequently reported to be approximately threefold higher than the other chromosomes (Shi and Martin 2000; Tempest and Griffin 2004; Templado et al. 2011). In males, there is an average of 50 sites of recombination (chiasmata) in each pachytene spermatocyte; the chiasmata are distributed across chromosomes, and in the normal situation there is at least one chiasma located on the long and short arm of the chromosome (Hassold et al. 2000). Longer chromosomes such as chromosome 1 have a greater number of chiasmata (~5), whereas the smallest chromosomes 21 and 22 and the sex chromosomes usually only possess a single chiasma (Martin 2008; Sun et al. 2005, 2008). As discussed previously, meiotic recombination appears to play an important role in ensuring chromosomes disjoin correctly during meiosis. Therefore, the presence of multiple chiasmata along the length of a chromosome may function in part, as an insurance policy to prevent nondisjunction. Thus, smaller chromosomes possessing a single chiasma (e.g., chromosomes 21, 22, X, and Y) are more prone to be achiasmate (lacking a chiasma) (Sun et al. 2008), losing the insurance policy of additional chiasmata to reduce the risk of nondisjunction. Studies have demonstrated that some individuals have a higher proportion of achiasmate bivalents and that this translates to a significantly higher level of sperm aneuploidy (Ferguson et al. 2007; Sun et al. 2008).

The vast majority of sperm FISH studies have analyzed and compared the frequencies of sperm aneuploidy between infertile men and either normozoospermic men, or men with proven fertility. All but a handful of these studies have identified a significant increase in sperm aneuploidy levels for at least one investigated chromosome in men with reduced semen parameters compared to normozoospermic men (Shi and Martin 2000; Tempest et al. 2004; Templado et al. 2011; Chatziparasidou et al. 2014). Collectively these studies provide strong evidence that infertile men have a significantly higher proportion of aneuploid sperm (approximately threefold), with severe oligoasthenoteratozoospermic and nonobstructive azoospermic individuals often having the highest levels (Shi and Martin 2000; Tempest et al. 2004; Templado et al. 2011; Chatziparasidou et al. 2014). From the published data, it is also clear that there are notable differences in the frequencies reported between studies; thus, baseline levels are difficult to establish. Differences between studies are likely due to technical differences including FISH probes utilized, number of cells scored, differences in scoring criteria, and subjective differences between scorers. In addition, the differences could be due to interindividual differences in sperm aneuploidy frequencies. It would seem plausible that some individuals may be more prone to chromosome nondisjunction and anaphase lag and/or have a less efficient meiotic checkpoint than others that may lead to increased levels of sperm aneuploidy. Furthermore, exogenous factors may have the ability to increase or decrease sperm aneuploidy levels. To date, several studies have examined whether sperm aneuploidy levels are consistent over time and if interindividual differences in sperm aneuploidy exist. The general consensus of these studies reveals that sperm aneuploidy levels remain remarkably consistent over time within individuals; nevertheless, there are stable variants who consistently produce higher levels of sperm aneuploidy (Rubes et al. 2005; Tempest et al. 2009). Preliminary data suggests that exogenous factors (e.g., diet, chemotherapy, or environment) may affect sperm aneuploidy levels resulting in a transient increase or potentially a decrease in aneuploidy levels (Harkonen 2005; McAuliffe et al. 2012, 2014; Tempest et al. 2005, 2009; Young et al. 2013). Despite the findings of significantly higher levels of chromosome aneuploidy within the sperm or certain subsets of men, sperm aneuploidy is rarely examined clinically, with the test only available at a handful of reference laboratories worldwide (Carrell 2008; Ramasamy et al. 2014). The lack of robust clinical studies provides us with a relatively poor understanding of the role of sperm aneuploidy in embryogenesis. Given that certain individuals produce higher levels of sperm aneuploidy, perhaps the most important question to address is whether these increased levels translate to an increased risk of paternally derived aneuploid embryos and offspring. This question is particularly difficult to address and is confounded by the significant maternal contribution to chromosome aneuploidy and that the parental origin of trisomies is rarely identified. Thus, if high sperm aneuploidy levels are identified, it is not yet clear how, or if, this should be utilized to counsel patients (Harton and Tempest 2012; Carrell 2008; Hann et al. 2011; Tempest 2011; Templado et al. 2011). To date, a handful of studies have tried to address whether sperm aneuploidy levels translate to embryo aneuploidies. Several studies have retrospectively identified increased levels of sperm aneuploidy in the fathers of paternally derived aneuploid offspring (Arnedo et al. 2006; Blanco et al. 1998; Martinez-Pasarell et al. 1999; Soares et al. 2001; Moosani et al. 1999). These studies suggest that in almost all cases these individuals had significantly higher levels of sperm aneuploidy for multiple chromosomes compared to fertile men with no history of aneuploid offspring (Harton and Tempest 2012). Other studies have provided preliminary evidence to suggest that higher levels of sperm aneuploidy are associated with recurrent ICSI failure (Nicopoullos et al. 2008; Petit et al. 2005), increased chromosome abnormalities in preimplantation embryos (Gianaroli et al. 2005), and lower pregnancy rates and live births (Nagvenkar et al. 2005). Furthermore, the approximate threefold increase in sperm aneuploidy observed in infertile men is mirrored by a threefold increase in de novo chromosome aberrations observed in children born after ICSI (Aboulghar et al. 2001; Bonduelle et al. 2002; Devroey and Van Steirteghem 2004; Van Steirteghem et al. 2002). Clearly, none of these studies provide direct evidence that sperm aneuploidy directly translates to embryo aneuploidy. However, albeit in a very small number of studies, there is compelling evidence to suggest that sperm aneuploidy may play a greater role in transmitting aneuploidy to embryos than previously perceived (Harton and Tempest 2012; Tempest 2011).

1.7 How Do Chromosome Aberrations Affect Meiosis?

1.7.1 Numerical Sex Chromosome Aneuploidies and Their Impact on Fertility and Meiosis

Sex chromosome aneuploidies are relatively common in the general population. The incidence of both Klinefelter syndrome (47,XXY) and 47,XYY syndrome is estimated to occur in 1 in 500 to 1 in 1,000 male live births (Morris et al. 2008). Individuals with Klinefelter syndrome typically present with nonobstructive azoospermia (accounting for ~11 % of nonobstructive azoospermia) or potentially oligozoospermia with a mosaic karyotype (Van Assche et al. 1996). Klinefelter syndrome is often perceived to have a classical phenotype (e.g., tall stature, gynecomastia, and hypogonadism); however, many cases have a highly variable phenotype and may not be identified until they present with infertility (Paduch et al. 2009). These men are born with spermatogonia, but during early puberty the spermatogonia undergo a massive wave of apoptosis leading to azoospermia. In around 50 % of cases, sperm can be recovered following testicular sperm extraction and can be used with ICSI to allow patients to have biological offspring (Paduch et al. 2009). Individuals with a 47,XYY karyotype also present with a variable phenotype; the majority of cases may have no phenotypic abnormalities, whereas some individuals may have a greater risk for behavioral problems, mild learning disabilities, and tall stature (Kim et al. 2013). Men with 47,XYY syndrome exhibit variable semen parameters ranging from normozoospermia to azoospermia (Kim et al. 2013) and hence, as with Klinefelter syndrome, may only be diagnosed if they present with fertility problems. In both of these cases, there is an additional sex chromosome that has to proceed and be segregated through meiosis; therefore, if sperm is present, there is a theoretical risk that 50 % of the sperm produced will be aneuploid for the sex chromosomes. Thus, in these situations there is a significant risk of fetal demise and transmission of a sex chromosome aneuploidy in future offspring. FISH has been utilized to assess the levels of sex chromosome aneuploidy in the sperm from 47,XXY and 47,XYY males. The published studies are highly variable, but report significantly lower sperm aneuploidy levels than the theoretical 50 %. In brief, the range of sperm aneuploidy for the sex chromosomes reported in Klinefelter syndrome (non-mosaic/mosaic) and 47,XYY men is between 1–25 %/0–7 % and 0.1–14 %, respectively (Ferlin et al. 2005; Sarrate et al. 2005; Tempest 2011). These studies provide additional evidence of the presence of an as-of-yet unknown meiotic checkpoint that is relatively efficient in eliminating a large proportion of aneuploid sperm cells (Harton and Tempest 2012; Burgoyne et al. 2009) and may contribute to the low sperm count observed in some 47,XYY males. Nonetheless, despite the presence of a meiotic checkpoint, there remains a considerable proportion of aneuploid sperm that are capable of completing meiosis leading to significantly higher levels of sperm aneuploidy compared to karyotypically normal men. To date, several studies have reported that the increase in sperm aneuploidy observed is also mirrored by an equivalent increase in sex chromosome aneuploidies in preimplantation embryos (Gonzalez-Merino et al. 2007; Staessen et al. 2003). Additionally, approximately 10 % of cases in the literature have resulted in aneuploid offspring (two 47,XXY conceptuses) (Friedler et al. 2001; Ron-El et al. 2000). Therefore, couples should be offered genetic counseling to inform them of their potential increased risk of spontaneous abortions and aneuploid offspring.

1.7.2 Structural Alterations in the Paternal Genome and Their Impact on Fertility and Meiosis

Structural chromosome aberrations can be classified as cytogenetic aberrations or genomic structural variants and can include translocations, inversions, insertions, or deletions. Chromosome aberrations are considered balanced when all the DNA is present, but its order or location on a chromosome(s) has been rearranged or unbalanced if there is gain or loss of genetic material. Cytogenetically visible aberrations can be readily detected by karyotyping and are relatively large in size (>3–5 Mb) (Shaffer and Bejjani 2004), whereas genomic structural variants are not microscopically visible (typically, >1 kb) (Freeman et al. 2006).

1.7.3 Karyotype Aberrations and Their Impact on Fertility and Meiosis

The incidence of cytogenetically visible chromosome aberrations is higher in males with fertility problems than that of the general population (Harton and Tempest 2012). Carriers of balanced structural chromosome rearrangements (e.g., translocations and inversions) are usually phenotypically normal; however, they often present with infertility. Structural chromosome rearrangements pose problems during meiosis, requiring unique pairing structures to form and to facilitate homologous chromosome pairing (e.g., quadrivalents, trivalents and inversion loops for reciprocal translocations, Robertsonian translocations and inversions, respectively). The formation of these unique pairing structures imposes time constraints and/or may lead to failure of chromosome pairing, which could result in the activation of meiotic checkpoints potentially eliminating these cells (Shah et al. 2003). However, some cells are able to successfully form these unique pairing structures and complete meiosis. In this situation, problems can arise when segregating chromosomes to daughter cells.

1.7.4 The Impact of Chromosome Translocations on Fertility and Meiosis

In the case of chromosome translocations, several outcomes are possible: gametes may (1) contain the normal chromosome complement, (2) be balanced carrying the chromosome translocation, or (3) be unbalanced (containing segments that may be monosomic and/or trisomic for the chromosome regions involved in the translocation). Sperm cells that are chromosomally normal or balanced are unlikely to affect embryogenesis or development. However, a significant proportion of sperm will be unbalanced. The phenotypic consequences of unbalanced segregation in chromosome translocations are often difficult to predict, given that most chromosomes translocations are unique to individual families. In these cases a careful review of the family history and assessment of the chromosomes involved and the size of the translocation can assist with counseling for the various possible outcomes. The vast majority of unbalanced sperm would result in early embryonic arrest or spontaneous abortions due to the incompatibility of partial trisomies and monosomies with embryogenesis. However, it is important to note that some unbalanced segregation products may be clinically viable, potentially resulting in congenital malformations and cognitive impairment. This has an increased likelihood if the chromosome translocation involves gene-poor regions of the genome, is small in size, and/or is known to be tolerated in a monosomic (e.g., chromosome X) or trisomic state (e.g., chromosomes 13, 18, 21, X, and Y). Using FISH probes for the specific chromosomes involved in the translocation, it is possible to evaluate the proportion of unbalanced sperm produced. The published literature reports extremely variable levels of sperm aneuploidy, which likely reflects patient-specific translocations. Studies evaluating the frequencies of unbalanced sperm from 30 reciprocal translocation carriers report partial aneuploidy levels of 29–81 % (Ferlin et al. 2007; Sarrate et al. 2005; Tempest 2011), whereas studies from 20 Robertsonian translocation carriers report aneuploidy levels of between 3 and 36 % (Ferlin et al. 2007; Sarrate et al. 2005; Tempest 2011).

1.7.5 The Impact of Chromosome Inversions on Fertility and Meiosis

Meiotic segregation in carriers of balanced inversions usually results in normal or balanced gametes, unless meiotic recombination takes place within the inversion. If recombination takes place within the inversion, resulting gametes may be monosomic and/or trisomic for the regions involved and potentially acentric or dicentric if the inversion involves the centromere (paracentric inversion). The clinical viability of unbalanced gametes is essentially the same as for translocations, in that, the vast majority will be spontaneously aborted often prior to a clinical pregnancy being established or potentially viable with risks of congenital abnormalities and/or cognitive impairment depending on the chromosome region involved and size of the unbalanced segments. In the case of inversions, larger inversions have a higher likelihood of resulting in unbalanced gametes as there is an increased risk of recombination occurring within a larger segment. To date, a handful of studies have examined the proportion of unbalanced sperm using FISH. As with translocations, the percentage of unbalanced sperm produced in inversion carriers varies widely between studies (1–54 %) (Anton et al. 2002, 2005; Jaarola et al. 1998; Mikhaail-Philips et al. 2004, 2005; Yakut et al. 2003), with the highest levels most likely reflecting larger inversions.

In the case of translocations and inversions, the estimates of sperm aneuploidy should be used with caution and are not generally applicable due to the fact that the vast majority of aberrations are unique. Therefore, individual assessment using personalized FISH probes for the chromosomal rearrangement can be readily used to obtain patient-specific risks in conjunction with family history to determine frequency of unbalanced sperm and whether unbalanced products may be clinically viable.

1.7.6 Genomic Structural Aberrations

It is clear that cytogenetically visible chromosome aberrations, numerical or structural in nature, can have a tremendous impact on embryogenesis resulting in embryonic arrest, fetal demise, and potentially live-born offspring with wide-ranging clinical phenotypes. Despite the clear association of abnormal karyotypes and embryogenesis, it is noteworthy that a significant proportion of early failed pregnancies with developmental defects are karyotypically normal. Studies suggest that up to ~20 % of morphologically abnormal embryos possess a normal karyotype (Philipp et al. 2003; Rajcan-Separovic et al. 2010). It seems plausible that many of these seemingly euploid developmentally abnormal embryos may be the result of submicroscopic perturbations that are not detectable by routine karyotyping. Recent technological advances that include chromosomal microarrays and sequencing now enable the routine detection of insertions, deletions, inversions, and duplications that are as small as 50 bp in size. These variants can arise through multiple mechanisms; however, they frequently arise as a result of errors during meiotic recombination. The paternal and maternal transmission of de novo submicroscopic chromosome aberrations following errors in meiotic recombination is likely to be equal, given that both male and female gametes only progress through meiosis on a single occasion. De novo submicroscopic chromosome aberrations can be benign or pathogenic in nature depending on their location and genetic content (Wapner et al. 2012). The clinical utilization of chromosome microarrays has been extremely beneficial in the evaluation of children with neurocognitive developmental delays and congenital structural malformations (Wapner et al. 2012). However, the consequence of these submicroscopic aberrations on fertilization, embryogenesis, and fetal development remains poorly understood. To date, a handful of studies have reported that between 4 and 13 % of miscarriages may possess submicroscopic chromosome aberrations that would not be detectable by routine karyotyping (Rajcan-Separovic et al. 2010; Shaffer et al. 2008; Shimokawa et al. 2006). It is important to note that the majority of studies did not determine whether these submicroscopic alterations were de novo, benign, or potentially pathogenic. A recent large study compared the use of chromosomal microarray and karyotyping for prenatal diagnosis enrolling over 4,400 women with indications for prenatal diagnosis including advanced maternal age, abnormal Down syndrome screening result, and structural abnormalities identified on ultrasound. This study reported that 6 % of samples were found to have a submicroscopic clinically significant unbalanced chromosome aberration (Wapner et al. 2012). The occurrence of submicroscopic aberrations in the prenatal diagnosis study is on the lower end compared to that reported in the miscarriage studies; this is most likely due to the fact that the prenatal diagnosis study evaluated gestationally older pregnancies than the miscarriage studies and only included potentially pathogenic alterations. The gestational age is critical when considering chromosome aberrations. As embryos develop a lower proportion of chromosomally abnormal embryos will be identified as the vast majority of aberrations result in spontaneous abortions, thus the prevalence of chromosome aberrations decreases with gestational age. Another critical point to note is the inability of chromosome microarrays used by these studies to detect balanced chromosome aberrations or triploidy; therefore, prevalence of the chromosomal aberrations reported could be conceivably higher (Wapner et al. 2012).

1.8 Conclusions

Chromosome aneuploidy, complete or partial in nature, can perturb embryogenesis and development. Somewhat surprisingly, the role of paternally derived chromosome aneuploidy in embryogenesis and development remains questionable in the clinic. A survey administered to fertility clinics in the UK perceived there is to be an increased risk of transmitting paternal genetic abnormalities following ICSI, but despite the concern most did not offer sperm aneuploidy screening (Griffin et al. 2003). Published studies suggest sperm aneuploidy has the potential to be a clinically useful screening tool to identify individuals with an increased risk of transmitting chromosome aneuploidy to future offspring. However, the clinical implementation of widespread routine sperm aneuploidy screening has been hampered by a number of factors that include (1) large maternal contribution to aneuploidy, (2) various technical challenges (e.g., wide variations between studies, requirement to score large numbers of cells (>5,000), limited number of chromosomes tested (3–5 chromosomes) per cell, and the inability to test the individual sperm that will be used for ICSI), (3) identification of individuals who may benefit from sperm aneuploidy scoring, and (4) what is a clinically significant level of sperm aneuploidy and how should patients with higher levels of aneuploidy be counseled. These important considerations have largely precluded the widespread clinical application of sperm aneuploidy screening, suggesting that at the moment the drawbacks outnumber the potential benefits. However, it is important to note that published studies have demonstrated chromosome aberrations (numerical or structural) can initiate and complete meiosis. Furthermore, the levels of aneuploidy observed in the sperm of infertile men or men with karyotype aberrations are mirrored and translate to increased levels of aneuploidy in preimplantation embryos and offspring, with several potential outcomes including fetal demise and congenital and cognitive impairments. Nevertheless, these findings are fundamentally based on a handful of small studies, and the true clinical ramifications of such findings will be difficult to determine until larger studies in a clinical setting are initiated. It seems unlikely that such studies will be initiated until some of the technical issues are resolved allowing rapid, cost-effective assessment of sperm aneuploidy levels. Automated capture and analysis software is currently available for FISH scoring and has been successfully applied to measure sperm aneuploidy and more commonly clinically for oncological FISH assays. However, automated approaches to assess sperm aneuploidy are not available in most clinics and have failed to significantly reduce the cost of the test (Carrell and Emery 2008; Tempest et al. 2010), primarily due to the initial outlay for capture system, software, and the relatively high cost of commercial FISH probes. Additionally, due to the lack of distinguishable fluorochromes, scoring all 24 chromosomes simultaneously in a single cell does not provide a genome-wide appreciation of the levels of aneuploidy within a single cell. The development and application of rapid multicolor FISH assays in combination with FISH reprobing and automation may allow all 24 chromosomes to be assessed in a single cell (Ioannou et al. 2011, 2012). Furthermore, it is also important to note that FISH as it is currently performed for aneuploidy assessments provides an extremely low resolution, typically only assessing a single chromosome region. Furthermore, FISH does not readily provide the possibility to screen for de novo structural aberrations or small genomic variants that may significantly contribute to spontaneous abortions. Technologically, it is possible to assess genome-wide levels of nondisjunction, unbalanced rearrangements, and smaller genomic variants in a single sperm cell using higher-resolution methodologies (e.g., sequencing, chromosome microarrays, and SNP arrays). Currently this remains prohibitively expensive as a routine clinical test given the large number of sperm cells that would need to be screened due to the relatively low proportion of aneuploidy in sperm. Currently, estimates of the proportion of sperm with de novo structural aberrations and smaller genomic perturbations are unknown, but it is reasonable to suggest that these may be similar or lower than aneuploidy levels. Thus, globally the proportion of sperm that may be perturbed may be clinically significant but still relatively low and hence not cost-effective to evaluate unless specific patient cohorts who may benefit can be identified.

It is clear that the paternal genome plays an important role in embryogenesis and development and is capable of transmitting genetic defects to embryos and hence future offspring. Studies have demonstrated that perturbations of the paternal epigenome and/or genome may have a tremendous impact on embryogenesis. While the maternal contribution to aneuploidy is considerable, there remains a clinically significant paternal contribution, particularly in infertile men and men with karyotype abnormalities. Furthermore, the paternal genome has been shown to be more mutagenic leading to an increased risk of de novo mutations and potentially at least an equal contribution to the generation of de novo genomic variants which have recently been shown to be associated with fetal demise and congenital and cognitive impairments. These factors should be considered particularly in couples who have experienced recurrent pregnancy loss, failed fertilization, and unexplained male factor infertility. Depending on the results of the semen parameter assessment and family history, additional genetic testing may be warranted including DNA fragmentation, karyotyping, Y chromosome microdeletions, single gene mutation testing, and potentially sperm aneuploidy assessment, particularly in individuals with karyotype aberrations. This information should be used to carefully counsel patients and provide them with individualized risk assessments to allow patients to make informed decisions regarding their reproductive future.

References

Aboulghar H, Aboulghar M, Mansour R, Serour G, Amin Y, Al-Inany H (2001) A prospective controlled study of karyotyping for 430 consecutive babies conceived through intracytoplasmic sperm injection. Fertil Steril 76(2):249–253

Anton E, Blanco J, Egozcue J, Vidal F (2002) Risk assessment and segregation analysis in a pericentric inversion inv6p23q25 carrier using FISH on decondensed sperm nuclei. Cytogenet Genome Res 97(3–4):149–154

Anton E, Blanco J, Egozcue J, Vidal F (2005) Sperm studies in heterozygote inversion carriers: a review. Cytogenet Genome Res 111(3–4):297–304

Arnedo N, Templado C, Sanchez-Blanque Y, Rajmil O, Nogues C (2006) Sperm aneuploidy in fathers of Klinefelter's syndrome offspring assessed by multicolour fluorescent in situ hybridization using probes for chromosomes 6, 13, 18, 21, 22, X and Y. Hum Reprod 21(2):524–528. doi:10.1093/humrep/dei321

Berger R (1975) The incidence of constitutional chromosome aberrations. J Genet Hum 23(Suppl):42–49

Blanco J, Gabau E, Gomez D, Baena N, Guitart M, Egozcue J, Vidal F (1998) Chromosome 21 disomy in the spermatozoa of the fathers of children with trisomy 21, in a population with a high prevalence of Down syndrome: increased incidence in cases of paternal origin. Am J Hum Genet 63(4):1067–1072. doi:10.1086/302058

Bonduelle M, Van Assche E, Joris H, Keymolen K, Devroey P, Van Steirteghem A, Liebaers I (2002) Prenatal testing in ICSI pregnancies: incidence of chromosomal anomalies in 1586 karyotypes and relation to sperm parameters. Hum Reprod 17(10):2600–2614

Burgoyne PS, Mahadevaiah SK, Turner JM (2009) The consequences of asynapsis for mammalian meiosis. Nat Rev Genet 10(3):207–216

Carrell DT (2008) The clinical implementation of sperm chromosome aneuploidy testing: pitfalls and promises. J Androl 29(2):124–133. doi:10.2164/jandrol.107.003699

Carrell DT (2012) Epigenetics of the male gamete. Fertil Steril 97(2):267–274

Carrell DT (ed) (2013) The reproductive fitness of the human male gamete. Paternal influences on human reproductive success. Cambridge University Press, Cambridge

Carrell DT, Emery BR (2008) Use of automated imaging and analysis technology for the detection of aneuploidy in human sperm. Fertil Steril 90(2):434–437. doi:10.1016/j.fertnstert.2007.06.095

Chandra A, Martinez GM, Mosher WD, Abma JC, Jones J (2005) Fertility, family planning, and reproductive health of U.S. women: data from the 2002 National Survey of Family Growth. Vital Health Stat 23(25):1–160

Chatziparasidou A, Christoforidis N, Samolada G, Nijs M (2014) Sperm aneuploidy in infertile male patients: a systematic review of the literature. Andrologia. doi:10.1111/and.12362

Conlin LK, Thiel BD, Bonnemann CG, Medne L, Ernst LM, Zackai EH, Deardorff MA, Krantz ID, Hakonarson H, Spinner NB (2010) Mechanisms of mosaicism, chimerism and uniparental disomy identified by single nucleotide polymorphism array analysis. Hum Mol Genet 19(7):1263–1275. doi:10.1093/hmg/ddq003

de Kretser DM (1997) Male infertility. Lancet 349(9054):787–790

Devroey P, Van Steirteghem A (2004) A review of ten years experience of ICSI. Hum Reprod Update 10(1):19–28

Ferguson KA, Wong EC, Chow V, Nigro M, Ma S (2007) Abnormal meiotic recombination in infertile men and its association with sperm aneuploidy. Hum Mol Genet 16(23):2870–2879

Ferlin A, Garolla A, Foresta C (2005) Chromosome abnormalities in sperm of individuals with constitutional sex chromosomal abnormalities. Cytogenet Genome Res 111(3–4):310–316. doi:10.1159/000086905

Ferlin A, Raicu F, Gatta V, Zuccarello D, Palka G, Foresta C (2007) Male infertility: role of genetic background. Reprod Biomed Online 14(6):734–745

Freeman JL, Perry GH, Feuk L, Redon R, McCarroll SA, Altshuler DM, Aburatani H, Jones KW, Tyler-Smith C, Hurles ME, Carter NP, Scherer SW, Lee C (2006) Copy number variation: new insights in genome diversity. Genome Res 16(8):949–961. doi:10.1101/gr.3677206

Friedler S, Raziel A, Strassburger D, Schachter M, Bern O, Ron-El R (2001) Outcome of ICSI using fresh and cryopreserved-thawed testicular spermatozoa in patients with non-mosaic Klinefelter's syndrome. Hum Reprod 16(12):2616–2620

Gianaroli L, Magli MC, Ferraretti AP (2005) Sperm and blastomere aneuploidy detection in reproductive genetics and medicine. J Histochem Cytochem 53(3):261–267. doi:10.1369/jhc.4B6434.2005

Gonzalez-Merino E, Hans C, Abramowicz M, Englert Y, Emiliani S (2007) Aneuploidy study in sperm and preimplantation embryos from nonmosaic 47, XYY men. Fertil Steril 88(3):600–606. doi:10.1016/j.fertnstert.2006.12.020

Griffin DK (1996) The incidence, origin, and etiology of aneuploidy. Int Rev Cytol 167: 263–296

Griffin DK, Hyland P, Tempest HG, Homa ST (2003) Safety issues in assisted reproduction technology: should men undergoing ICSI be screened for chromosome abnormalities in their sperm? Hum Reprod 18(2):229–235

Hammoud SS, Purwar J, Pflueger C, Cairns BR, Carrell DT (2010) Alterations in sperm DNA methylation patterns at imprinted loci in two classes of infertility. Fertil Steril 94(5):1728–1733. doi:10.1016/j.fertnstert.2009.09.010

Hann MC, Lau PE, Tempest HG (2011) Meiotic recombination and male infertility: from basic science to clinical reality? Asian J Androl 13(2):212–218. doi:10.1038/aja.2011.1

Harkonen K (2005) Pesticides and the induction of aneuploidy in human sperm. Cytogenet Genome Res 111(3–4):378–383. doi:10.1159/000086915

Harton GL, Tempest HG (2012) Chromosomal disorders and male infertility. Asian J Androl 14(1):32–39. doi:10.1038/aja.2011.66

Hassold T, Hunt P (2001) To err (meiotically) is human: the genesis of human aneuploidy. Nat Rev Genet 2(4):280–291. doi:10.1038/35066065

Hassold T, Hunt P (2009) Maternal age and chromosomally abnormal pregnancies: what we know and what we wish we knew. Curr Opin Pediatr 21(6):703–708

Hassold T, Hunt PA, Sherman S (1993) Trisomy in humans: incidence, origin and etiology. Curr Opin Genet Dev 3(3):398–403

Hassold T, Sherman S, Hunt P (2000) Counting cross-overs: characterizing meiotic recombination in mammals. Hum Mol Genet 9(16):2409–2419

Hassold T, Hall H, Hunt P (2007) The origin of human aneuploidy: where we have been, where we are going. Hum Mol Genet 16(Spec No 2):R203–R208

Hotaling J, Carrell DT (2014) Clinical genetic testing for male factor infertility: current applications and future directions. Andrology 2(3):339–350. doi:10.1111/j.2047-2927.2014.00200.x

Ioannou D, Meershoek EJ, Thornhill AR, Ellis M, Griffin DK (2011) Multicolour interphase cytogenetics: 24 chromosome probes, 6 colours, 4 layers. Mol Cell Probes 25(5–6):199–205. doi:10.1016/j.mcp.2011.08.002

Ioannou D, Fonseka KG, Meershoek EJ, Thornhill AR, Abogrein A, Ellis M, Griffin DK (2012) Twenty-four chromosome FISH in human IVF embryos reveals patterns of post-zygotic chromosome segregation and nuclear organisation. Chromosome Res 20(4):447–460. doi:10.1007/s10577-012-9294-z

Jaarola M, Martin RH, Ashley T (1998) Direct evidence for suppression of recombination within two pericentric inversions in humans: a new sperm-FISH technique. Am J Hum Genet 63(1):218–224. doi:10.1086/301900

Jenkins TG, Carrell DT (2012) The sperm epigenome and potential implications for the developing embryo. Reproduction 143(6):727–734. doi:10.1530/REP-11-0450

Kim IW, Khadilkar AC, Ko EY, Sabanegh ES Jr (2013) 47, XYY syndrome and male infertility. Rev urol 15(4):188–196

Kumar M, Kumar K, Jain S, Hassan T, Dada R (2013) Novel insights into the genetic and epigenetic paternal contribution to the human embryo. Clinics 68(Suppl 1):5–14

Lewis SE (2007) Is sperm evaluation useful in predicting human fertility? Reproduction 134: 31–40
Marquez C, Egozcue J, Martorell MR, Moreno V, Templado C (1996) Colcemid increases the frequency of chromosome abnormalities in human sperm. Cytogenet Cell Genet 72(2–3):164–170
Martin RH (2008) Meiotic errors in human oogenesis and spermatogenesis. Reprod Biomed Online 16(4):523–531
Martinez-Pasarell O, Nogues C, Bosch M, Egozcue J, Templado C (1999) Analysis of sex chromosome aneuploidy in sperm from fathers of Turner syndrome patients. Hum Genet 104(4):345–349
McAuliffe ME, Williams PL, Korrick SA, Altshul LM, Perry MJ (2012) Environmental exposure to polychlorinated biphenyls and p, p′-DDE and sperm sex-chromosome disomy. Environ Health Perspect 120(4):535–540. doi:10.1289/ehp.1104017
McAuliffe ME, Williams PL, Korrick SA, Dadd R, Marchetti F, Martenies SE, Perry MJ (2014) Human sperm sex chromosome disomy and sperm DNA damage assessed by the neutral comet assay. Hum Reprod 29(10):2148–2155. doi:10.1093/humrep/deu177
McLachlan RI, O'Bryan MK (2010) Clinical review#: state of the art for genetic testing of infertile men. J Clin Endocrinol Metab 95(3):1013–1024. doi:10.1210/jc.2009-1925
McLay DW, Clarke HJ (2003) Remodelling the paternal chromatin at fertilization in mammals. Reproduction 125(5):625–633
Mikhaail-Philips MM, Ko E, Chernos J, Greene C, Rademaker A, Martin RH (2004) Analysis of chromosome segregation in sperm from a chromosome 2 inversion heterozygote and assessment of an interchromosomal effect. Am J Med Genet A 127A(2):139–143. doi:10.1002/ajmg.a.20693
Mikhaail-Philips MM, McGillivray BC, Hamilton SJ, Ko E, Chernos J, Rademaker A, Martin RH (2005) Unusual segregation products in sperm from a pericentric inversion 17 heterozygote. Hum Genet 117(4):357–365. doi:10.1007/s00439-004-1245-0
Millan NM, Lau P, Hann M, Ioannou D, Hoffman D, Barrionuevo M, Maxson W, Ory S, Tempest HG (2012) Hierarchical radial and polar organisation of chromosomes in human sperm. Chromosome Res 20:875–887. doi:10.1007/s10577-012-9323-y
Moosani N, Chernos J, Lowry RB, Martin RH, Rademaker A (1999) A 47, XXY fetus resulting from ICSI in a man with an elevated frequency of 24, XY spermatozoa. Hum Reprod 14(4):1137–1138
Morris JK, Alberman E, Scott C, Jacobs P (2008) Is the prevalence of Klinefelter syndrome increasing? Eur J Hum Genet 16(2):163–170. doi:10.1038/sj.ejhg.5201956
Munne S, Sandalinas M, Magli C, Gianaroli L, Cohen J, Warburton D (2004) Increased rate of aneuploid embryos in young women with previous aneuploid conceptions. Prenat Diagn 24(8):638–643. doi:10.1002/pd.957
Nagvenkar P, Zaveri K, Hinduja I (2005) Comparison of the sperm aneuploidy rate in severe oligozoospermic and oligozoospermic men and its relation to intracytoplasmic sperm injection outcome. Fertil Steril 84(4):925–931. doi:10.1016/j.fertnstert.2005.04.048
Nicopoullos JD, Gilling-Smith C, Almeida PA, Homa S, Nice L, Tempest H, Ramsay JW (2008) The role of sperm aneuploidy as a predictor of the success of intracytoplasmic sperm injection? Hum Reprod 23(2):240–250. doi:10.1093/humrep/dem395
Paduch DA, Bolyakov A, Cohen P, Travis A (2009) Reproduction in men with Klinefelter syndrome: the past, the present, and the future. Semin Reprod Med 27(2):137–148. doi:10.1055/s-0029-1202302
Pang MG, Hoegerman SF, Cuticchia AJ, Moon SY, Doncel GF, Acosta AA, Kearns WG (1999) Detection of aneuploidy for chromosomes 4, 6, 7, 8, 9, 10, 11, 12, 13, 17, 18, 21, X and Y by fluorescence in-situ hybridization in spermatozoa from nine patients with oligoasthenoteratozoospermia undergoing intracytoplasmic sperm injection. Hum Reprod 14(5):1266–1273
Petit FM, Frydman N, Benkhalifa M, Le Du A, Aboura A, Fanchin R, Frydman R, Tachdjian G (2005) Could sperm aneuploidy rate determination be used as a predictive test before intracytoplasmic sperm injection? J Androl 26(2):235–241

Philipp T, Philipp K, Reiner A, Beer F, Kalousek DK (2003) Embryoscopic and cytogenetic analysis of 233 missed abortions: factors involved in the pathogenesis of developmental defects of early failed pregnancies. Hum Reprod 18(8):1724–1732

Rajcan-Separovic E, Qiao Y, Tyson C, Harvard C, Fawcett C, Kalousek D, Stephenson M, Philipp T (2010) Genomic changes detected by array CGH in human embryos with developmental defects. Mol Hum Reprod 16(2):125–134. doi:10.1093/molehr/gap083

Ramasamy R, Besada S, Lamb DJ (2014) Fluorescent in situ hybridization of human sperm: diagnostics, indications, and therapeutic implications. Fertil Steril 102(6):1534–1539. doi:10.1016/j.fertnstert.2014.09.013

Ron-El R, Strassburger D, Gelman-Kohan S, Friedler S, Raziel A, Appelman Z (2000) A 47, XXY fetus conceived after ICSI of spermatozoa from a patient with non-mosaic Klinefelter's syndrome: case report. Hum Reprod 15(8):1804–1806

Rubes J, Vozdova M, Oracova E, Perreault SD (2005) Individual variation in the frequency of sperm aneuploidy in humans. Cytogenet Genome Res 111(3–4):229–236

Sarrate Z, Blanco J, Anton E, Egozcue S, Egozcue J, Vidal F (2005) FISH studies of chromosome abnormalities in germ cells and its relevance in reproductive counseling. Asian J Androl 7(3):227–236. doi:10.1111/j.1745-7262.2005.00061.x

Schagdarsurengin U, Paradowska A, Steger K (2012) Analysing the sperm epigenome: roles in early embryogenesis and assisted reproduction. Nat Rev Urol 9(11):609–619. doi:10.1038/nrurol.2012.183

Shaffer LG, Bejjani BA (2004) A cytogeneticist's perspective on genomic microarrays. Hum Reprod Update 10(3):221–226. doi:10.1093/humupd/dmh022

Shaffer LG, Coppinger J, Alliman S, Torchia BA, Theisen A, Ballif BC, Bejjani BA (2008) Comparison of microarray-based detection rates for cytogenetic abnormalities in prenatal and neonatal specimens. Prenat Diagn 28(9):789–795. doi:10.1002/pd.2053

Shah K, Sivapalan G, Gibbons N, Tempest H, Griffin DK (2003) The genetic basis of infertility. Reproduction 126(1):13–25

Shi Q, Martin RH (2000) Aneuploidy in human sperm: a review of the frequency and distribution of aneuploidy, effects of donor age and lifestyle factors. Cytogenet Cell Genet 90(3–4):219–226

Shimokawa O, Harada N, Miyake N, Satoh K, Mizuguchi T, Niikawa N, Matsumoto N (2006) Array comparative genomic hybridization analysis in first-trimester spontaneous abortions with 'normal' karyotypes. Am J Med Genet A 140(18):1931–1935. doi:10.1002/ajmg.a.31421

Soares SR, Templado C, Blanco J, Egozcue J, Vidal F (2001) Numerical chromosome abnormalities in the spermatozoa of the fathers of children with trisomy 21 of paternal origin: generalised tendency to meiotic non-disjunction. Hum Genet 108(2):134–139

Staessen C, Tournaye H, Van Assche E, Michiels A, Van Landuyt L, Devroey P, Liebaers I, Van Steirteghem A (2003) PGD in 47, XXY Klinefelter's syndrome patients. Hum Reprod Update 9(4):319–330

Sun F, Trpkov K, Rademaker A, Ko E, Martin RH (2005) Variation in meiotic recombination frequencies among human males. Hum Genet 116(3):172–178

Sun F, Mikhaail-Philips M, Oliver-Bonet M, Ko E, Rademaker A, Turek P, Martin RH (2008) The relationship between meiotic recombination in human spermatocytes and aneuploidy in sperm. Hum Reprod 23(8):1691–1697

Tempest HG (2011) Meiotic recombination errors, the origin of sperm aneuploidy and clinical recommendations. Syst Biol Reprod Med 57(1–2):93–101. doi:10.3109/19396368.2010.504879

Tempest HG, Griffin DK (2004) The relationship between male infertility and increased levels of sperm disomy. Cytogenet Genome Res 107(1–2):83–94. doi:10.1159/000079575

Tempest HG, Homa ST, Dalakiouridou M, Christopikou D, Wright D, Zhai XP, Griffin DK (2004) The association between male infertility and sperm disomy: evidence for variation in disomy levels among individuals and a correlation between particular semen parameters and disomy of specific chromosome pairs. Reprod Biol Endocrinol 2:82. doi:10.1186/1477-7827-2-82

Tempest HG, Homa ST, Zhai XP, Griffin DK (2005) Significant reduction of sperm disomy in six men: effect of traditional Chinese medicine? Asian J Androl 7(4):419–425. doi:10.1111/j.1745-7262.2005.00068.x
Tempest HG, Ko E, Rademaker A, Chan P, Robaire B, Martin RH (2009) Intra-individual and inter-individual variations in sperm aneuploidy frequencies in normal men. Fertil Steril 91(1):185–192. doi:10.1016/j.fertnstert.2007.11.002
Tempest HG, Cheng SY, Gillott DJ, Handyside AH, Thornhill AR, Griffin DK (2010) Scoring of sperm chromosomal abnormalities by manual and automated approaches: qualitative and quantitative comparisons. Asian J Androl 12(2):257–262. doi:10.1038/aja.2009.85
Templado C, Vidal F, Estop A (2011) Aneuploidy in human spermatozoa. Cytogenet Genome Res 133(2–4):91–99
Templado C, Uroz L, Estop A (2013) New insights on the origin and relevance of aneuploidy in human spermatozoa. Mol Hum Reprod. doi:10.1093/molehr/gat039
Van Assche E, Bonduelle M, Tournaye H, Joris H, Verheyen G, Devroey P, Van Steirteghem A, Liebaers I (1996) Cytogenetics of infertile men. Hum Reprod 11(Suppl 4):1–24, discussion 25–26
Van Steirteghem A, Bonduelle M, Devroey P, Liebaers I (2002) Follow-up of children born after ICSI. Hum Reprod Update 8(2):111–116
Vanneste E, Voet T, Le Caignec C, Ampe M, Konings P, Melotte C, Debrock S, Amyere M, Vikkula M, Schuit F, Fryns JP, Verbeke G, D'Hooghe T, Moreau Y, Vermeesch JR (2009) Chromosome instability is common in human cleavage-stage embryos. Nat Med 15(5):577–583. doi:10.1038/nm.1924
Wapner RJ, Martin CL, Levy B, Ballif BC, Eng CM, Zachary JM, Savage M, Platt LD, Saltzman D, Grobman WA, Klugman S, Scholl T, Simpson JL, McCall K, Aggarwal VS, Bunke B, Nahum O, Patel A, Lamb AN, Thom EA, Beaudet al, Ledbetter DH, Shaffer LG, Jackson L (2012) Chromosomal microarray versus karyotyping for prenatal diagnosis. N Engl J Med 367(23):2175–2184. doi:10.1056/NEJMoa1203382
World Health Organization (2010) WHO laboratory manual for the examination and processing of human semen, 5th edn. World Health Organization, Geneva
Yakut T, Acar H, Egeli U, Kimya Y (2003) Frequency of recombinant and nonrecombinant products of pericentric inversion of chromosome 1 in sperm nuclei of carrier: by FISH technique. Mol Reprod Dev 66(1):67–71. doi:10.1002/mrd.10325
Young HA, Meeker JD, Martenies SE, Figueroa ZI, Barr DB, Perry MJ (2013) Environmental exposure to pyrethroids and sperm sex chromosome disomy: a cross-sectional study. Environ Health 12:111. doi:10.1186/1476-069X-12-111
Zalenskaya IA, Zalensky AO (2004) Non-random positioning of chromosomes in human sperm nuclei. Chromosome Res 12(2):163–173
Zalensky A, Zalenskaya I (2007) Organization of chromosomes in spermatozoa: an additional layer of epigenetic information? Biochem Soc Trans 35(Pt 3):609–611

Chapter 2
Damage to Sperm DNA Mediated by Reactive Oxygen Species: Its Impact on Human Reproduction and the Health Trajectory of Offspring

Dan Gavriliouk and Robert John Aitken

Abstract Disruptions to the genetic integrity of the mammalian spermatozoon play a major role in determining the subsequent developmental trajectory of the embryo. This chapter examines the causative links that connect DNA damage in human spermatozoa and the appearance of mutations in the progeny responsible for a variety of clinical conditions from autism to cancer. Integral to this discussion is an abundance of evidence indicating that human spermatozoa are vulnerable to free radical attack and the generation of oxidative DNA damage. The resolution of this damage appears to be initiated by the spermatozoa but is driven to completion by the oocyte in a round of DNA repair that follows fertilization. The persistence of unresolved oxidative DNA damage following zygote formation has the potential to create mutations/epimutations in the offspring that may have a profound impact on the health of the progeny. It is proposed that the creation of oxidative stress in the male germ line is a consequence of a wide variety of environmental/lifestyle factors that influence the health and well-being of the offspring as a consequence of mutational change

D. Gavriliouk
Priority Research Centre in Reproductive Science, University of Newcastle, Callaghan, NSW 2308, Australia

Faculty of Health and Medicine, Hunter Medical Research Institute, University of Newcastle, Callaghan, NSW 2308, Australia

R.J. Aitken, Sc.D., F.R.S.E., F.A.A. (✉)
Priority Research Centre in Reproductive Science, University of Newcastle, Callaghan, NSW 2308, Australia

Faculty of Health and Medicine, Hunter Medical Research Institute, University of Newcastle, Callaghan, NSW 2308, Australia

School of Environmental and Life Sciences, Discipline of Biological Sciences, University of Newcastle, University Drive, Callaghan, NSW 2308, Australia
e-mail: John.Aitken@newcastle.edu.au; Brett.Nixon@newcastle.edu.au

R. Bronson (ed.), *The Male Role in Pregnancy Loss and Embryo Implantation Failure*, Advances in Experimental Medicine and Biology 868,
DOI 10.1007/978-3-319-18881-2_2

induced by the aberrant repair of oxidative DNA damage in the zygote. Factors such as paternal age, subfertility, smoking, obesity, and exposure to a range of environmental influences, including radio-frequency electromagnetic radiation and xenobiotics, have all been implicated in this process. Identifying the contributors to oxidative stress in the germ line and resolving the mechanisms by which such stressors influence the mutational load carried by the progeny will be an important task for the future. This task is particularly pressing, given the extensive use of assisted reproductive technologies to achieve pregnancies in vitro that would have been prevented in vivo by the complex array of mechanisms that nature has put in place to ensure that only the fittest gametes participate in the generative process.

Keywords Reactive oxygen species (ROS) • Oxidative stress • Sperm mitochondria • DNA damage • Apoptosis • Histones • Protamines • Base excision repair

2.1 Introduction

The integrity of DNA in the sperm nucleus is an important determinant of semen quality since it defines not only the success of fertilization but also the normality of embryonic development and the health trajectory of the offspring. As a consequence, DNA damage in these cells is associated with the impairment of fertility, an increase in the incidence of miscarriage, and a variety of defects in the progeny ranging from neurological conditions such as autism to cancer (Evenson et al. 1999; Larson et al. 2000; Aitken and Baker 2013; Aitken et al. 2013). The backbone of the DNA helix is frequently cleaved in spermatozoa resulting in either single (SSB)- or double-strand breaks (DSB), while oxidative attack leads to the formation of base adducts particularly 8-hydroxy-2′-deoxyguanosine (8OHdG) (De Iuliis et al. 2009). The sources of this DNA damage are complex and include age, genetic disposition, lifestyle, and exposure to various external factors including ionizing radiation and a wide range of xenobiotics including chemical carcinogens (Aitken and De Iuliis 2010).

While the induction of strand breaks and progressive accumulation of base adducts can eventually give rise to mutations, there is uncertainty as to when in the reproductive process such mutagenic change occurs. In some cases, such as Apert syndrome, there is good evidence to support a model whereby the mutation causing this disease arises in spermatogonial stem cells as a function of paternal age. The consensus view of this process asserts that as men age their germ cells experience multiple rounds of premeiotic replication, and with each cellular iteration the risk of a mutation occurring as a consequence of replication error correspondingly increases. Apert syndrome involves FGFR2 (fibroblast growth factor receptor 2) mutations, which are thought to become overrepresented in the sperm population as a consequence of age-dependent clonal expansion, mutant spermatogonial stem cells having a proliferative advantage over nonmutated cells (Goriely et al. 2003). Accordingly this mutation appears in clusters within the seminiferous tubules possibly as a consequence of failures of asymmetrical division within the germ line (Shinde et al. 2013). Such a germ line selection model may also explain the origin of other dominant genetic diseases such as achondroplasia, the incidence of which is also correlated with paternal age (Crow 2000). However, in this case, there appears to be a major discrepancy between the incidence of the causative mutation in spermatozoa and the appearance of the disease in the progeny of aging males (Hurst and Ellegren 2002). In order to explain this discrepancy, we have proposed an alternative hypothesis for the origin of such mutations whereby the latter are held to arise as a result of inefficient or aberrant repair of damaged sperm DNA within the oocyte, immediately after fertilization (Aitken et al. 2004). The first cell division following fertilization is unarguably the most important. The zygote has one opportunity to repair the DNA damage brought into the oocyte by the fertilizing spermatozoon before S-phase of the first mitotic division is initiated. Any inadequacies in this repair and replication process could have major consequences for the embryo, since any infidelities in the transmission of genetic information through the first cell division will subsequently be replicated throughout embryogenesis (Fig. 2.1).

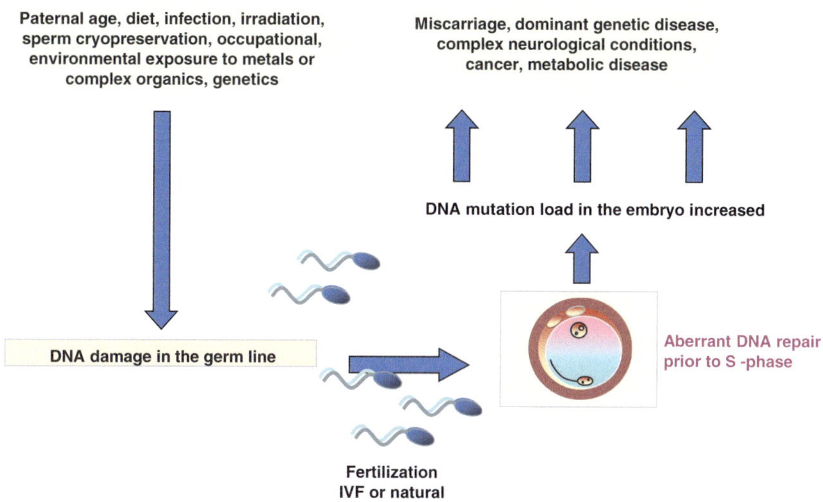

Fig. 2.1 Potential origins of genetic and epigenetic changes in the germ line. The "aberrant repair hypothesis" essentially posits that a variety of clinical, biological, lifestyle, and environmental factors including paternal age, obesity, gamete cryopreservation, exposure to radiation, transition metals, or a wide range of small molecular mass xenobiotics can trigger mitochondrial ROS generation and oxidative DNA damage in spermatozoa. These cells then fertilize the oocyte, possibly with the help of ART techniques such as ICSI (intracytoplasmic sperm injection), and it is then up to the oocyte to affect repair of the damaged paternal genome. Any inefficiencies or mistakes resulting from this round of postfertilization DNA repair have the potential to create mutations or epimutations that will impact upon the normal development and health trajectory of the offspring (Aitken and Krausz 2001)

Since DNA damage is such a common feature of defective spermatozoa generated by male-factor patients in an assisted reproductive technology (ART) setting (Lopes et al. 1998), it is inevitable that a significant number of in vitro conceptions are achieved using DNA-damaged spermatozoa that would have been excluded from this process in vivo. Although the use of ART and particularly ICSI (intracytoplasmic sperm injection) to achieve fertilization in cases of severe male infertility has been extremely successful in improving conception rates, such apparent therapeutic advances might have been achieved at a cost, in terms of the health and well-being of the offspring. Of critical importance to this discussion is a clear understanding of the oocyte's competence for DNA repair. Typically, SSB are repaired by the base excision repair (BER) and nucleotide excision repair (NER) pathways, while DSB are repaired by nonhomologous end join (NHEJ) and homologous recombination (HR). The significance of these pathways in the oocyte has not been clearly articulated, but for the reasons given below, the BER pathway is probably the single most important DNA repair strategy associated with the oocyte's central task of tackling the DNA damage brought into the zygote by the fertilizing spermatozoon.

It is the purpose of this chapter to examine the source and nature of DNA damage in human spermatozoa, to examine how such damage is repaired following fertilization, and to review how errors in this repair process impact on the health and well-being of the offspring. This discussion will begin with a consideration of the cause of DNA damage in human spermatozoa, and in this context, the spermatozoon's susceptibility in oxidative stress is fundamental (Aitken and Clarkson 1987).

2.2 Oxidative Stress and Spermatozoa

2.2.1 The Perpetual Cycle of ROS Generation in Spermatozoa

A major cause of DNA damage in spermatozoa is oxidative stress mediated by a variety of reactive oxygen species (ROS) including free radicals such as superoxide anion ($O_2^{-\bullet}$), nitric oxide ($NO^\bullet$), or the hydroxyl radical ($OH^\bullet$) as well as powerful oxidants such as hydrogen peroxide (H_2O_2) or peroxynitrite ($ONOO^-$). Spermatozoa are particularly prone to oxidative stress because their antioxidant defensive capacity is limited, due to the removal of a majority of their cytoplasm during spermatogenesis and a consequential reduction in cytoplasmic antioxidants such as catalase or superoxide dismutase. Furthermore, these cells are professional generators of ROS, with a vast majority of these reactive oxygen metabolites being generated as a consequence of electron leakage from the sperm mitochondria (Koppers et al. 2008).

The vulnerability of spermatozoa to oxidative stress also reflects the abundance of substrates these cells offer up for free radical attack. Thus, the membranous constituents of spermatozoa contain high concentrations of polyunsaturated fatty acids, particularly docosahexaenoic acid, the double bonds of which are vulnerable to attack by ROS and the initiation of lipid peroxidation cascades. The hydrogen abstraction event that initiates lipid peroxidation is promoted because the carbon–hydrogen dissociation energies are lowest at the bis-allylic methylene position, generating carbon-centered lipid radicals that then combine with oxygen to generate peroxyl (ROO•) and alkoxyl (RO•) radicals that, in order to stabilize, abstract hydrogen atoms from adjacent carbons. These chemical reactions create additional lipid radicals that then perpetuate the lipid peroxidation chain reaction, culminating in the generation of small molecular mass electrophilic lipid aldehydes such as 4-hydroxynonenal (4HNE), acrolein, and malondialdehyde.

Further to this lipid-based vulnerability to free radical attack, we have also recently demonstrated that the lipid aldehydes generated as a result of lipid peroxidation actually trigger the generation of yet more free radicals from the sperm mitochondria in a self-perpetuating cycle (Fig. 2.2). According to this scheme, lipid aldehydes such as acrolein or 4HNE form adducts with proteins in the mitochondrial electron transport chain (ECT) that perturb the normal, controlled 4-electron reduction of oxygen to water. This results in the leakage of electrons from the ECT that affect the 1-electron reduction of nature's universal electron acceptor, oxygen, to generate $O_2^{-\bullet}$, which then rapidly dismutates to H_2O_2 under the influence of mitochondrial superoxide dismutase (Aitken et al. 2013). Intriguingly, defective spermatozoa from subfertile males contain a superabundance of superoxide dismutase, possibly reflecting the retention of excess residual cytoplasm during the terminal stages of spermiogenesis (Aitken et al. 1996; Gomez et al. 1996; Sanocka et al. 1997). The fact that correlations have also been observed between superoxide dismutase and other cytoplasmic enzymes such as glucose-6-phophate dehydrogenase or creatine kinase supports this view (Aitken et al. 1996; Gomez et al. 1996). Normally the presence of excess superoxide dismutase would be considered an asset for any cell seeking to protect itself from oxidative stress. However, unless this enzyme is accompanied by a corresponding increase in the presence of enzymes that can scavenge H_2O_2 such as glutathione peroxidase or catalase (both of which are present in limited supply in human spermatozoa; Storey et al. 1998; Sanocka et al. 1997), the presence of excess superoxide dismutase simply turns a short-lived, membrane-impermeant, relatively inert free radical in the form of $O_2^{-\bullet}$ into a long-lived, membrane-permeant reactive oxidant, H_2O_2, that can attack a wide variety of substrates in spermatozoa. Such attacks not only target the polyunsaturated fats that abound in the sperm plasma membrane but also the DNA in the sperm nucleus and mitochondria.

2.2.2 *Primary Causes of Excess ROS Generation by Sperm Mitochondria*

If oxidative stress stimulates a lipid peroxidation process characterized by the generation of aldehydes that perpetuate the generation of ROS from the sperm mitochondria, then the question arises as to what triggers oxidative stress in the first instance. In this context one of the most important activators of ROS generation by sperm mitochondria is the induction of apoptosis in response to senescence or other adverse circumstances.

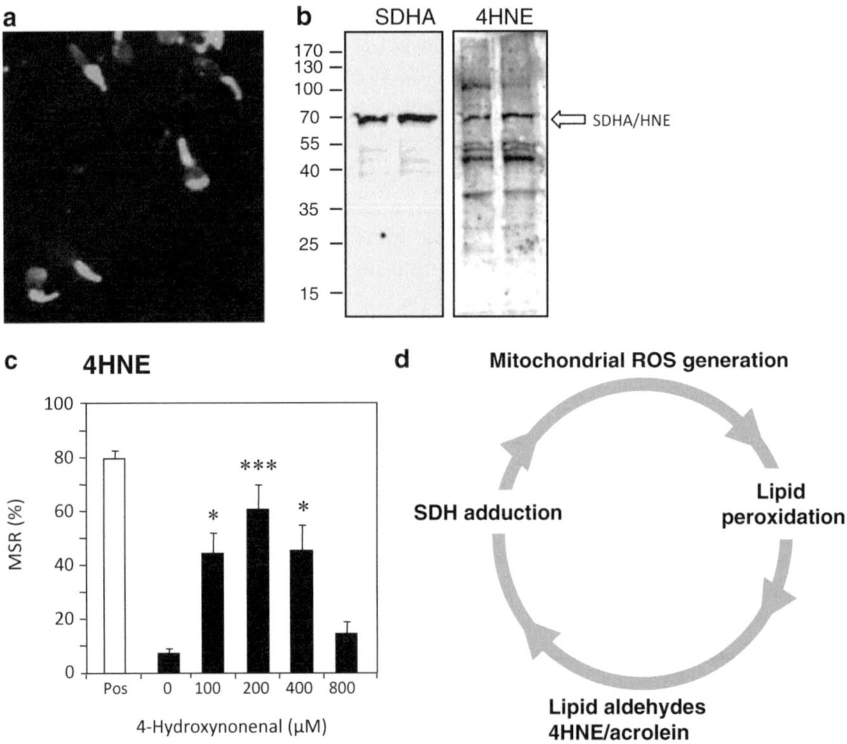

Fig. 2.2 The self-propagating nature of ROS generation by spermatozoa. (**a**) Immunocytochemical analysis with an anti-4HNE antibody reveals that this electrophilic aldehyde binds to multiple proteins in spermatozoa, particularly in the sperm midpiece where the mitochondria are located. (**b**) Western blot analysis confirms that multiple proteins are adducted by 4HNE including major proteins from the mitochondrial electron transport chain (ETC) such as succinic acid dehydrogenase (SDHA). (**c**) The binding of 4HNE to the sperm mitochondria perturbs electron flow through the ETC leading to electron leakage and the generation of ROS; MSR describes the percentage of live cells (defined using SYTOX Green®) that are generating a positive signal with MitoSOX Red ™, a probe for mitochondrial ROS generation, in a flow cytometer. The bell-shaped dose–response curve reflects the ability of 4HNE to compromise cell viability at higher doses (Aitken et al. 2012a, b). (**d**) As a result of these findings, we propose a self-perpetuating cycle of ROS-induced ROS generation wherein oxidative stress leads to the formation of lipid aldehydes that bind to proteins within the ETC and stimulate yet more ROS generation and lipid peroxidation

2.2.2.1 The Intrinsic Apoptotic Cascade in Spermatozoa

One of the most important points to make about apoptosis in spermatozoa is that a regulated, apoptotic cell death is the default position for this cell type and is largely intrinsically induced. To the authors' knowledge there are no extrinsic factors that will reliably and robustly trigger apoptosis in spermatozoa via a receptor-mediated mechanism. Although powerful bacterially derived factors such as lipopolysaccharide (LPS) have been claimed to trigger apoptosis in human spermatozoa (Hakimi et al. 2006), we have been unable to consistently replicate these findings, despite many attempts. The one condition that will consistently trigger an apoptotic response in these cells is senescence, which is itself a reflection of the oxidative stress created via the formation of lipid aldehydes as a consequence of cell metabolism. If these electrophilic aldehydes are scavenged by powerful nucleophiles such as the thiol, penicillamine (Aitken et al. 2012a, b), then the survival of spermatozoa in vitro is significantly increased. The prolonged life span of mammalian spermatozoa in vivo contrasts dramatically with their limited survival in vitro and suggests that both the male and female reproductive tracts must be capable of controlling the bioavailability of such lipid aldehydes via molecular mechanisms that are still poorly unresolved.

It is the fate of a great majority of spermatozoa to experience a senescence-induced apoptotic death in either the male or female reproductive tract. Whatever the location of this death, its manner is of critical importance. Spermatozoa are terminally differentiated cells that arise long after immunological tolerance has been generated and are potentially immunogenic in both males and females. Thus, when these cells die and are phagocytosed by neutrophils or macrophages, it is important that the phagocytic process responsible for their removal is "silent" and does not trigger a pro-inflammatory, phlogistic response. Such silent phagocytosis occurs commonly in biological systems and is mediated by the phagocyte's ability to recognize apoptotic markers such as phosphatidylserine and other "eat-me" signals on the exofacial surface of the cell being engulfed (Hochreiter-Hufford and Ravichandran 2013). In order for the massive phagocytic event that follows insemination to be truly silent, it is critical that the spermatozoa undergoing phagocytosis have undergone an apoptotic death. This apoptotic death is generally triggered by the oxidative stress that accompanies the physiological exertion needed to sustain high levels of sperm motility over prolonged periods of time.

It should also be recognized that oxidative stress is an inevitable consequence of the redox mechanisms that are needed to drive the capacitation process forward. "Capacitation" is a general term that covers a range of physiological changes that spermatozoa must undergo if they are to undergo a physiological acrosome reaction in the immediate vicinity of the oocyte. Preparation for acrosomal exocytosis involves a complex array of biochemical changes including cholesterol efflux, a quantum increase in levels of tyrosine phosphorylation, and the expression of hyperactivated movement, all of which are known to be redox regulated (de Lamirande and Gagnon 1993; Aitken et al. 1998; Brouwers et al. 2011). It is theoretically possible for a variety of ROS to be responsible for the induction of these changes associated with capacitation, but the constitutive generation of $NO^{\bullet}$ by spermatozoa suggests that the powerful oxidant, peroxynitrite ($ONOO^-$), is a major product of these cells with a proven capacity to stimulate capacitation (de Lamirande and Lamothe 2009; Rodriguez et al. 2011). If fertilization does not occur, it has been proposed that the continued generation of $ONOO^-$ by capacitating spermatozoa leads to a state of "over-capacitation" whereby the ROS generation that drives capacitation eventually overwhelms the limited antioxidant capacity of these cells, leading to a state of oxidative stress (Aitken 2011). The appearance of $ONOO^-$-mediated oxidative stress leads to a loss of sperm function (Uribe et al. 2015) and the generation of lipid aldehydes that via the mechanisms described above, stimulate yet more ROS generation, ultimately precipitating a state of apoptosis characterized by rapid motility loss, mitochondrial ROS generation, caspase activation in the cytosol, annexin V binding to the cell surface, cytoplasmic vacuolization, and oxidative DNA damage (Koppers et al. 2011). During this intrinsic apoptotic cascade, it is only after the spermatozoa have become immobilized, and therefore prone to phagocytosis, that markers of apoptosis such as phosphatidylserine externalization start to appear (Koppers et al. 2011).

Sperm senescence and apoptosis is not normally induced by an external factor, but is an intrinsic process dependent on a fall in phosphoinositide 3-kinase (PI3K) activity. The latter is highly active in mammalian spermatozoa and generates the novel phosphoinositide, PtdIns(3,4,5)P3, which binds to the PH domain of kinases such as AKT1 causing the latter to translocate to the sperm surface where it becomes activated by phosphorylation under the influence of PDK-1. This pathway leads to the promotion of cell survival via a variety of mechanisms, particularly the phosphorylation and inactivation of proapoptotic proteins such as BAD (Koppers et al. 2011). If PI3K activity is suppressed with an inhibitor such as wortmannin, then AKT1 rapidly dephosphoryates, leading, in turn, to the dephosphorylation of BAD and the activation of sperm apoptosis (Koppers et al. 2011). Conversely, any pro-survival factor that stimulates PI3K activity leads to inhibition of apoptosis and prolongs the functional life span of the spermatozoa. There are, in all probability, many such pro-survival factors operating in vivo in order to achieve the prolonged survival of spermatozoa in both the male and female reproductive tracts. One of the first such pro-survival factors to be identified is prolactin (Pujianto et al. 2010). Spermatozoa possess several splice variants of the prolactin receptor, while addition of this hormone to human sperm suspensions has been found to prolong survival in association with an increase in AKT1 phosphorylation (Pujianto et al. 2010).

From the foregoing, it should be clear that oxidative stress is a feature of sperm cell biology responsible for both the induction of sperm capacitation and the eventual demise of these cells as a consequence of a senescence process that is associated with activation of the intrinsic apoptotic cascade. In vitro, sperm senescence and apoptosis are generally promoted by the lack of pro-survival factors in the culture medium. In addition, any factors that promote oxidative stress in spermatozoa have the potential to accelerate the apoptotic process. For example, just exposing spermatozoa to H_2O_2 will precipitate an apoptotic response (Lozano et al. 2009). Cryostorage of spermatozoa will also generate a high level of apoptosis associated with oxidative stress (Thomson et al. 2009) as will exposure to a number of free radical-generating xenobiotics (Aly 2013) including lifestyle factors such as cigarette smoke and radio-frequency electromagnetic radiation (Fraga et al. 1996; De Iuliis et al. 2009; Liu et al. 2013).

In addition it has been observed that a superabundance of polyunsaturated fatty acids will also stimulate mitochondrial ROS generation and create a state of oxidative stress in human spermatozoa; the greater the level of unsaturation, the greater the stimulatory effect (Koppers et al. 2010). Esterification of the fatty acid counters this prooxidant effect suggesting that it is the amphiphilic properties of these molecules that are central to their ROS-inducing activity, possibly by defining the orientation of the fatty acids in relation to the mitochondrial electron transport chain. In this context, it is significant that defective human spermatozoa generating excessive levels of ROS possess abnormally high cellular contents of free polyunsaturated fatty acids, the levels of which are positively correlated with mitochondrial superoxide generation (Aitken et al. 2006). Why some men should possess spermatozoa generating particularly high levels of ROS in association with high cellular contents of polyunsaturated fatty acids is not currently known; however, such associations may reflect the impact of diet or genetic factors on reproductive function.

2.2.2.2 The Truncated Nature of Apoptosis on Spermatozoa

The reliance of spermatozoa on the intrinsic apoptotic cascade is a distinguishing characteristic of these cells that reflects their status as terminally differentiated, disposable cells. Another unique attribute of apoptosis in spermatozoa is that this process is truncated (Koppers et al. 2011). The reason for such restriction lies in the unique architecture of these cells. Unique among all cell types, spermatozoa are distinguished by the fact that all of the mitochondria and most of the cytoplasm are physically separated from the sperm nucleus (Fig. 2.3). Thus, the conventional apoptosis paradigm, involving nucleases activated in the cytoplasm or released by the mitochondria that move into a centrally located sperm nucleus in order to destroy the DNA and create the cleaved DNA ladders that characterize this process, cannot apply to this cell type. Nucleases generated in the cytoplasm or mitochondria of the sperm midpiece are physically prevented from gaining access to the nuclear DNA located in the sperm head (Koppers et al. 2011). Since chemically, DNA fragmentation can only be induced by free radicals or nucleases, the above rationale explains why most DNA damage in spermatozoa is initially oxidative in nature.

2.2.3 *Oxidative Stress in Spermatozoa*

High levels of oxidative DNA damage have been repeatedly observed in the spermatozoa of subfertile males (Kodama et al. 1997; Irvine et al. 2000). Oxidative damage occurs primarily at the guanine bases and causes the formation of adducts, the most common of which are 8-hydroxy-2′-deoxyguanosine (8-OHdG) and 8–oxo-7,8-dihydro-2′-deoxyguanosine (8-oxodG), which are commonly used as biomarkers for oxidative stress. Using a flow cytometric assay for 8OHdG, Aitken et al. (2010) demonstrated that the population of males attending an infertility clinic possessed significantly elevated levels of 8OHdG in their spermatozoa compared with an unselected control donor population. The role of ROS in causing such

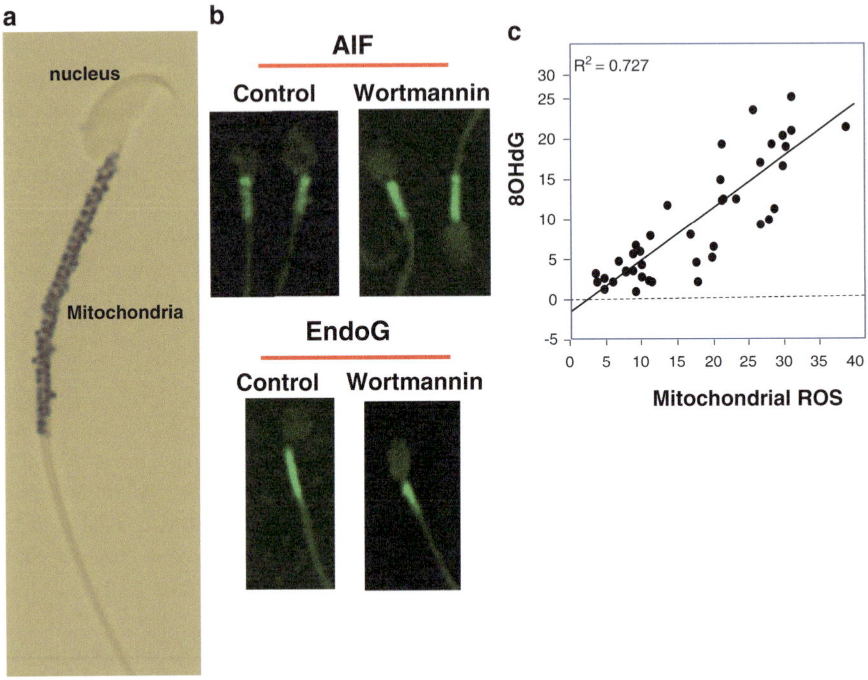

Fig. 2.3 The unique architecture of spermatozoa has an impact on the nature of apoptosis in these cells. (**a**) Unlike conventional somatic cells spermatozoa are designed in such a way that the sperm nucleus is in a separate physical compartment from all of the mitochondria and most of the cytoplasm. In this image of a mouse spermatozoon, the mitochondrial gyres have been highlighted black using a histochemical stain. (**b**) As a result of this unusual architecture, apoptosis can be induced with, for example, wortmannin, but nucleases that become activated during this process such as endonuclease G (EndoG) or apoptosis-inducing factor (AIF) remain resolutely locked in the midpiece of the cell and cannot penetrate into the sperm nucleus. This is why most DNA damage in spermatozoa is initiated by oxidative stress rather than intracellular nucleases; the only product of apoptosis that can cross from the midpiece to the sperm head and attack the nuclear DNA is membrane-permeant ROS such as H_2O_2. (**c**) As a result of such factors, the release of ROS from the sperm mitochondria is highly correlated with the induction of oxidative DNA damage in these cells

increases in oxidative DNA damage in spermatozoa is suggested by the beneficial impact of antioxidants such as melatonin on the levels of 8OHdG formation observed in human spermatozoa (Bejarano et al. 2014). The source of the ROS responsible for the induction of oxidative DNA damage may be a reflection of three factors: the activation of ROS-generating phagocytic leukocytes, the generation of ROS by the spermatozoa themselves, and a lack of ROS-scavenging enzymes and small molecular mass antioxidants.

2.2.3.1 Leukocytes as a Cause of Oxidative Stress

Infiltrating phagocytic leukocytes are potentially capable of inducing oxidative stress in the male reproductive tract; however, there does not appear to be a strong direct relationship between leukocytospermia and DNA damage in spermatozoa, with data being presented to both support and refute this proposal (Saleh et al. 2002 Moskovtsev et al. 2007). The situation is complex because the degree of oxidative stress experienced by the spermatozoa will depend on the number and type leukocytes present in the ejaculate, when they entered the seminal compartment, whether the leukocytes are activated and, if so, when and how they are activated (Aitken and Baker 1995). Every human ejaculate contains a small number of leukocytes (in the order of $2–5 \times 10^4$/ml) that appear to be in a free radical-generating, activated state (Aitken et al. 1995; Aitken and Baker 2013). However, the presence of such leukocytes does not seem to have a powerful effect on sperm function or create particularly high levels of oxidative stress. The reason for this may be that the leukocytes enter the seminal fluid at the level of the secondary sexual glands, particularly when infection is involved. In such instances, the first time the spermatozoa will come into contact with infiltrating leukocytes will be at the moment of ejaculation. At this juncture, the spermatozoa will be protected from extracellular ROS by the powerful antioxidant properties of seminal plasma (Jones et al. 1979; Rhemrev et al. 2000; van Overveld et al. 2000), and the impact of the leukocyte-derived ROS will be minimal. However, quite a different picture emerges when the seminal plasma is removed in the context of preparing spermatozoa for assisted conception therapy. Under these circumstances the protective impact of seminal plasma is lost, and any activated phagocytes that are present in the sperm suspension will have free rein to launch an oxidative attack on the spermatozoa, curtailing the motility of these cells and compromising their DNA integrity. As a consequence of these relationships, negative impacts have been observed between levels of leukocyte contamination in washed sperm preparation and fertilization rates in IVF programs (Krausz et al. 1994).

In order to circumvent such damage occurring during IVF treatment cycles, it is imperative that high-quality spermatozoa are isolated from the ejaculate while these cells are still protected by the antioxidants present in seminal plasma (Aitken and Clarkson 1988). For this reason, procedures such as swim-up from a washed pellet are known to be associated with high levels of oxidative stress and the iatrogenic induction of DNA damage in the spermatozoa (Twigg et al. 1998). One of the best ways to minimize oxidative DNA damage during sperm preparation for IVF is to use a simple, swim-up-from-raw-semen approach (Twigg et al. 1998). Alternative strategies for preparing spermatozoa for assisted conception therapy include the centrifugation of spermatozoa through discontinuous colloidal silicon gradients that isolate the highest quality spermatozoa on the basis of their physical density. Although such techniques are clearly successful in isolating subpopulations of spermatozoa with high levels of motility and good morphology (Aitken and Clarkson 1988), they have been found to actually increase the levels of oxidative DNA damage seen in the spermatozoa (Aitken et al. 2014). The reason for such DNA damage in otherwise high-quality sperm populations has been a mystery until recently, when it was revealed that commercial sperm preparation media are commonly contaminated with metals, including iron and copper, at concentrations that induce high levels of oxidative DNA damage in human spermatozoa (Aitken et al. 2014). Fortunately, the chemical chelation of such metal contaminants is sufficient to eliminate this threat to sperm DNA integrity and render such media suitable for use in an IVF context (Aitken et al. 2014).

2.2.3.2 ROS Generation by Spermatozoa

As indicated above, oxidative stress is commonly caused in spermatozoa by the excess generation of mitochondrial ROS as a consequence of their entry into the intrinsic apoptotic pathway. The induction of mitochondrial ROS generation with, for example, radio-frequency electromagnetic radiation leads to a marked increase in 8OHdG formation followed by DNA fragmentation (De Iuliis et al. 2009). There are many conditions that will trigger the entry of spermatozoa into this pathway including senescence, temperature, exposure to toxic chemicals, and aromatic amino acids (Tosic and Walton 1950; Aitken et al. 2012a, b).

2.2.3.3 Antioxidant Deficiency

Conditions that lead to a loss of antioxidant protection for the spermatozoa such as smoking, poor diet, or prolonged incubation in culture medium lacking antioxidant supplementation can also lead to oxidative stress and oxidative DNA damage in populations of spermatozoa (Fraga et al. 1996; Dalzell et al. 2003; Aitken et al. 2009).

2.2.4 Oxidative DNA Damage in Spermatozoa

In order to explain the etiology of DNA damage in the male germ line, we recently proposed a two-step hypothesis according to which two conditions have to be met in order for such damage to appear: (1) a source of ROS and (2) a state of heightened vulnerability to free radical attack on the part of the spermatozoa. In addition to the presence of oxidizable substrates such as polyunsaturated fatty acids, another key factor defining the vulnerability of spermatozoa to oxidative attack is the status of sperm chromatin. During spermiogenesis there is an extensive remodeling of nuclear chromatin in order to compact the entire haploid genome into the volume of a sperm head. This is achieved by the progressive replacement of nuclear histones with small positively charge proteins known as protamines. In eutherian mammals the proteins have evolved to contain cysteine residues that participate in the creation of intermolecular and intramolecular cross-links during epididymal transit that reinforce the compaction of the chromatin and render the DNA highly resistant to damage.

Interestingly, metatherian mammals do not possess cysteine-rich protamines, and as a result their spermatozoa do not undergo this post-testicular phase of nuclear remodeling. As a result, metatherian spermatozoa are much more vulnerable to oxidative DNA damage than their eutherian counterparts (Bennetts and Aitken 2005).

Sperm nuclear DNA is organized into doughnut-shaped toroids linked by nuclease-sensitive interlinker regions, bound to the nuclear matrix, which connect the toroids together (Ward 2010). These interlinker regions are thought to be associated with the retention of histones and to be particularly vulnerable to oxidative attack. Treating sperm with DNase 1 has been found to break apart the toroid linker regions but leaves the DNA within the toroid undamaged. Destruction of the linker regions in this manner has been shown to delay the replication of paternal DNA postfertilization and to impair embryogenesis, suggesting that the DNA contained within these particular regions of sperm chromatin encodes genes that are of critical importance in orchestrating the early stages of embryonic development (Gawecka et al. 2013). Laboratory experiments have demonstrated it is possible to remove protamines and histone-bound nucleosomes by treatment with high salt and reducing agent. This leaves only the sperm nuclear matrix with associated loop domains attached and a resulting nuclear structure that is called a "sperm nuclear halo" (Nadel et al. 1995; Kramer and Krawetz 1996). These halos are useful in giving us an idea as to which part of the chromatin is necessary for the first few cell divisions following fertilization. When sperm halos were injected into oocytes, pronucleus formation was normal and DNA replication was initiated (Shaman et al. 2007). Replication would still occur even if 50 % of the DNA that was not attached to the matrix was removed. As replication will proceed as long as these nuclear matrix-associated regions are present, the interlinker domains must be where DNA replication begins. If this is the case, then oxidative damage to these non-protaminated vulnerable regions of sperm nuclear DNA would be expected to have a major impact on development. In the future, determining which particular genes are housed in these domains would be extremely helpful in understanding how oxidative DNA damage in spermatozoa can influence the developmental potential of the embryo.

Outside of these linker regions, the extent to which sperm DNA becomes protaminated varies between species and between individuals within a species. Thus, in mouse spermatozoa around 95 % of histones are removed and replaced by protamines, whereas in human spermatozoa the equivalent figure is around 15 % (Ward, 2010). Additional vulnerability is created in patient samples because of the inefficient protamination of sperm chromatin. Using the fluorescent probe chromomycin A3 (CMA3) to monitor the degree of sperm chromatin protamination, a very close correlation has been observed between levels of oxidative DNA damage in human spermatozoa and the efficiency of protamine deposition during spermiogenesis; the more poorly protaminated the chromatin, the higher the CMA3 signal and the greater the risk of oxidative DNA damage (De Iuliis et al. 2009).

These observations are important because they add weight to the above mentioned two-step hypothesis of oxidative DNA damage in human spermatozoa. The first step occurs during spermiogenesis and leads to poor protamination of the sperm chromatin, resulting in defective compaction of the DNA and an accompanying increased risk of oxidative attack. The second step then involves the realization of this oxidative attack as a result of the mechanisms described above (Aitken et al. 2009).

Once an oxidative attack on sperm DNA occurs, there are two pathways open to the cell. First, if the oxidative attack is severe, it will initiate the intrinsic apoptotic cascade, culminating in a loss of motility, the appearance of apoptotic markers on the sperm surface, and ultimately cell death. When spermatozoa enter the perimortem and the internal structure of these cells starts to break down, it is possible that sperm nuclear DNA fragmentation becomes accentuated via the activation of intracellular nucleases or the entry of nucleases from the extracellular fluids bathing the spermatozoa in the epididymis or vas deferens (Sotolongo et al. 2005; Boaz et al. 2008; Smith et al. 2013). The purpose of this last-gasp perimortem attack on the sperm nucleus is to facilitate the complete destruction of the sperm genome prior to the phagocytosis of these cells by the immune system. Second, if the spermatozoon is not badly damaged, the oxidative DNA damage may simply be repaired in readiness for fertilization. Counterintuitively it is this process of DNA repair that poses the greatest threat to human health because it is errors in this process that are thought to underpin the connection between oxidative DNA damage in the spermatozoa and the mutational load subsequently carried by children, with important implications for their long-term health trajectory.

2.3 Impacts of Oxidative DNA Damage in Sperm on Human Reproduction and Health

Despite the propensity for human spermatozoa to suffer from oxidative DNA damage, there is relatively little data to confirm that such damage has a major impact on male fertility. The most compelling evidence that this is the case comes from studies in which men exhibiting DNA damage or other evidence of oxidative stress in their spermatozoa have had their sperm DNA integrity or fertility improved by treatment with antioxidants. A study by Suleiman et al. (1996), for example, was distinguished by the careful selection of patients according to evidence of oxidative stress in their spermatozoa, as measured by a lipid peroxidation assay. This trial encouragingly recorded a decrease in lipid peroxidation and a resulting increase in sperm motility and fertility following treatment with vitamin E. Another study selected patients on the basis of DNA damage in their spermatozoa (measured with the TUNEL assay) and recorded a significant improvement in this criterion following treatment with an antioxidant preparation containing vitamin E as well as an accompanying increase in fertility (Greco et al. 2005). These data are certainly promising, but the data sets are too small to draw definitive conclusions about antioxidant therapy and male fertility. There is an urgent need for definitive, randomized, double-blind cross-over trials on this topic to determine the true value of antioxidant therapy in the treatment of males exhibiting high levels of oxidative DNA damage in their spermatozoa (Aitken et al. 2010). The therapeutic efficacy of male antioxidant treatment on the maintenance of subsequent pregnancy has not yet been adequately examined, despite clear evidence linking DNA damage in spermatozoa with an increase in the incidence of miscarriage (Showell et al. 2011; Gharagozloo and Aitken 2011 Robinson et al. 2012).

In order to understand the importance of oxidative DNA damage in spermatozoa on the developmental competence of the embryo and the maintenance of pregnancy, we first need to understand how the 8OHdG lesions formed as a consequence of oxidative stress are repaired in spermatozoa and the zygote.

2.3.1 DNA Repair in Spermatozoa and Early Zygote

Regardless of how well defended the DNA is in sperm chromatin, some form of oxidative damage is inevitable. The base excision repair (BER) pathway plays a vital role in repairing oxidized, alkylated, and deaminated DNA bases and removing small non-helix-distorting base lesions from the genome. In most cells the BER pathway can be broken down into five major steps with each step being performed by a specific enzyme or class of enzymes. The process is highly regulated through individual protein to protein interactions and the formation of repair complexes. Incorporated into the subcellular structure of the sperm nucleus and mitochondria is the first enzyme in the BER pathway, an 8-oxoguanine glycosylase, known as OGG1. When spermatozoa experience an oxidative attack, OGG1 immediately clips the 8OHdG residues out of the DNA generating an abasic site, releasing the oxidized base into the extracellular space (Smith et al. 2013). The next enzyme in the base excision repair pathway, APE1, incises DNA at the phosphate groups, 3′ and 5′ to the baseless site, leaving 3′-OH and 5′-phosphate termini ready for the insertion of a new base. Spermatozoa do not possess this enzyme (Smith et al. 2013). As a result, they carry their abasic sites into the oocyte for continuation of the repair process. Fortunately, the oocyte possesses the remaining elements of the BER pathway including APE1 and XRCC1 (T. Lord and R.J. Aitken, unpublished observations) and is able to carry the repair process through to completion. The repair of oxidative DNA lesions in spermatozoa therefore involves a high level of collision between the male and female germ lines and will be impacted not just by the levels of oxidative stress and 8OHdG formation in the spermatozoa but also the competence of the oocyte to complete the repair process initiated by the spermatozoa. The capacity of the BER pathway in the latter is clearly limited because the spermatozoa of subfertile males are known to carry significantly elevated levels of 8OHdG, which have not been excised by OGG1 (Aitken et al. 2010). It is the responsibility of the oocyte to detect these oxidized base lesions and engage in a round of DNA repair immediately after fertilization and put S-phase on hold until this activity has been completed (Gawecka et al. 2013). Interestingly, although the oocyte possesses an abundance of APE1 and XRCC1, it appears to exhibit a limited supply of OGG1 (T. Lord and R.J. Aitken, unpublished observations). As a result, we can anticipate that 8OHdG adducts will be poorly repaired by the oocyte and may persist into the first cleavage division of the embryo.

The BER pathway is not the only mechanism for dealing with oxidative DNA damage; however, it is thought to be the most important. While the mature spermatozoon has little alternative to OGG1 of the BER for repairing 8OHdG, oocytes are also known to possess alternative DNA repair pathways including nucleotide excision repair (NER) and mismatch repair (MMR) (Menezo et al. 2007); however, the role of these pathways in orchestrating the oocyte's response to oxidative DNA damage has not yet been determined.

Whatever repair mechanisms are invoked, the inefficient or aberrant repair of these 8OHdG lesions by the oocyte is known to have a negative impact on embryo quality (Meseguer et al. 2008) and has the potential to create de novo mutations or epimutations in the offspring that could have a profound impact on their health and well-being.

2.3.2 Consequences of Oxidatively Damaged DNA for Development

2.3.2.1 DNA Repair in the Zygote

Following fertilization, the integrity of the decondensed maternal and paternal chromatin is assessed by the zygote. The latter does not have transcription-coupled translation and relies on mRNA and proteins stored in the spermatozoon and oocyte for repair. There appears to be some cross-talk between the male and female pronuclei in effecting this repair, although the mechanisms are not yet understood. When irradiated mouse spermatozoa were used to fertilize oocytes, *both* pronuclei exhibited p53 apoptotic responses and replicated only around half of their DNA. Some zygotes did manage to progress to more advanced stages of development including implantation, but none came to term (Shimura et al. 2002; Shaman et al. 2007).

Fertilization with DNA-damaged spermatozoa may alter the expression of DNA repair genes in preimplantation embryos as early as the one-cell stage (Harrouk et al. 2000). Spermatozoa from rats subjected to cyclophosphamide resulted in zygotes with significantly higher DNA damage and higher transcripts for proteins from the nucleotide excision repair family (XPC, XPE, and PCNA), mismatch repair family (PMS1), recombination repair family (RAD50), as well as BER family members (UNG1, UNG2) (Harrouk et al. 2000). Furthermore, other studies have reported increases in gene expression from the two-cell stage onward. A gene expression profile analysis by Zeng et al. (2004) revealed that a small class of genes involved in the regulation of the cell cycle are overrepresented in two-cell mouse blastomeres compared to the zygotes. It is not currently understood why some DNA repair genes are highly expressed at particular stages of embryo development and not others. It is possible that different sources and types of DNA damage will elicit different DNA repair responses in the zygote. Regardless of which DNA repair mechanisms are involved, the time between fertilization and the first cell division is thought to be linked to the amount of DNA damage that the zygote needs to repair.

A zygote that is slow to initiate cleavage is therefore more likely to possess higher levels of DNA damage and exhibit a poorer potential for normal development (Liu et al. 2014).

The newly formed zygote is thought to be reluctant to respond to DNA damage by undergoing apoptosis (Fear and Hansen 2011) and, instead, prefers indefinite developmental arrest. It is not entirely certain if replication is arrested by the zygote to give it time to repair the DNA or because the DNA is too damaged for replication to proceed; much probably depends on the amount of DNA damage involved. Although some aspects of apoptosis, such as cytoplasmic fragmentation, will occur in the first cell cycle in mice and the second in humans, in general, the zygote will resist proapoptotic signals until it is at the 8–16 cell stage. In keeping with this proposal, Fear and Hansen (2011) found that bovine blastomeres have higher concentrations of mRNA for the antiapoptotic genes BCL2 and HSPA1A in two-cell embryos compared with their 16-cell counterparts. Thus, as the embryo develops through further cell divisions, the apoptosis pathway gradually becomes available; however, in early development it is DNA repair and developmental arrest that predominate. In a recent study of human development, for example, Burruel et al. (2014) determined that the duration of the 2nd to 3rd mitoses was most sensitive to fertilization by oxidatively damaged spermatozoa. As a result, embryos that displayed either too long or too short cytokineses at this stage of development demonstrated an increased failure to reach blastocyst stage and commit to further development.

This notion that zygotes engage in a demanding round of DNA damage recognition and repair shortly after fertilization has been exploited to develop a system whereby an embryo's status can be noninvasively assessed by monitoring its metabolic status. With a particular focus on amino acid metabolism, Sturmey et al. (2009) found that embryos with greater viability exhibited a lower or "quieter" amino acid metabolism than those that went into arrest. It was hypothesized that this relationship exists because embryos with greater DNA damage consume more nutrients to facilitate the internal repair processes. Such differences in amino acid metabolism are significant since they could ultimately prove to be a useful marker of DNA damage when selecting embryos for transfer in an IVF context.

2.3.2.2 Environmental Factors Cause Oxidative Stress

Given the vulnerability of spermatozoa to oxidative attack and the limited capacity of the oocyte to repair oxidized DNA base adducts, there is concern that the carriage of such base lesions into the zygote may compromise the developmental competence of the embryo. Evidence in support of such concerns has largely come from an analysis of the health consequences of paternal cigarette smoking. Smoking has a long list of associated health problems including reproductive impacts. Despite the efforts made to control tobacco consumption across the world, smoking is still common, and the highest prevalence of smoking is seen among young men between the ages of 20–39, when they are likely to be fathering a child (Li et al. 2011).

Paternal, rather than maternal, smoking is associated with a significant increase in the incidence of childhood cancer in the offspring (Lee et al. 2009), implying that paternal DNA damage is the source of carcinogenic mutations. Given the heavy reliance on OGG1 to cleave out DNA base adducts prior to fertilization, any factor that impairs OGG1 has the potential to affect DNA repair in the germ line and, thence, reproductive function. The classic inhibitor of OGG1 is cadmium. Cadmium exposure has been shown to increase levels of DNA damage in spermatozoa (Oliveira et al. 2009), and a positive correlation has been found between 8OHdG in spermatozoa and cadmium concentration in seminal plasma (Xu et al. 2003). The impact of smoking on 8OHdG is also exacerbated by the presence of Ser326Cys polymorphism in the OGG1 gene (Ji et al. 2013). Individuals with variant Cys/Cys homozygosity for OGG1 had higher levels of sperm 8OHdG than wild-type homozygote carriers (Ser/Ser). In addition to the impact of genetic constitution and cadmium on OGG1, men who smoke heavily are also known to be deficient in antioxidants such as vitamin C (Fraga et al. 1996). The net result of all these factors is that the spermatozoa of men who smoke heavily possess high levels of oxidative DNA damage. These lesions do not have a dramatic impact on fertility; however, they are associated with the abovementioned increase in the incidence of cancer in the progeny (Lee et al. 2009). Although causation has not been formally established between 8OHdG lesions in spermatozoa and cancer in the children of men who smoke heavily, the circumstantial evidence is compelling.

Of course, smoking is not the only lifestyle or environmental factor that can influence levels of oxidative DNA damage in spermatozoa—another is paternal age. As males age, there is a downregulation of DNA repair genes (particularly BER), and levels of oxidative DNA damage in spermatozoa increase (Paul et al. 2011) such that a man aged over 35 will have three times the levels of DNA damage in his spermatozoa as a male below this age (Singh et al. 2003). One of the major consequences of this age-dependent increase in sperm DNA damage is a linear increase in the mutational load carried by the progeny over a paternal age range that stretches from 15 to 45 (Kong et al. 2012). As the mutational load carried by children increases, so does the incidence of diseases that are known to be associated with the age of the father at the moment of conception. Thus, paternal aging is known to be linked with many different kinds of adverse clinical conditions including elevated rates of miscarriage, increased incidences of dominant genetic disease such as achondroplasia and Apert syndrome, and an enhanced risk of neuropathology in the offspring including bipolar disease, autism, spontaneous schizophrenia, and epilepsy (Aitken et al. 2013).

Of the other factors that might cause oxidative DNA damage in the male germ line, it has become very apparent that obesity negatively influences many biological functions, and gamete health is no exception. Recent studies have demonstrated that obesity is associated with decreased semen parameters (Chavarro et al. 2010, Kort et al. 2006) and damage to germ cells in the testes. Several studies have shown a link between men with high BMI and decreased sperm DNA integrity due to oxidative stress (Tunc et al. 2011; Bakos et al. 2011). Men who are classified as overweight or obese are frequently relying on ART and ICSI to father children. Considering the plethora of health problems associated with obesity, it would be very beneficial to understand if using ICSI to circumvent infertility passes health burdens onto successive generations. More studies, particularly from the perspective of ICSI progeny of obese fathers, are clearly required.

Spermatozoa of males seeking ART may be cryopreserved for various reasons such as the preservation of fertility as a prelude to the initiation of aggressive cancer chemotherapy. Cryopreservation has been revealed to both generate and exacerbate the extent of DNA damage in spermatozoa. The exact mechanism of cryoinjury is not yet fully understood although there appears to be a strong element of oxidative damage, as ROS generation is increased in cryopreserved sperm (Wang et al. 1997) and levels of oxidized DNA damage are similarly elevated under these conditions (Thomson et al. 2009). Importantly, the addition of antioxidants such as genistein and quercetin has been shown to ameliorate the amount of DNA damage in these cells, whereas caspase inhibitors have no effect (Thomson et al. 2009; Zribi et al. 2012). So, although cryopreservation can induce apoptosis in spermatozoa and generate a concomitant increase in caspase activity (Paasch et al. 2004; Wündrich et al. 2006), such changes are probably a consequence of oxidative stress and DNA damage, not a cause.

2.4 Conclusions

The integrity of sperm DNA is vital for the subsequent health trajectory of the offspring. The most common cause of DNA damage in spermatozoa is oxidative stress induced by mitochondrial ROS. The spermatozoon has a limited range of antioxidant strategies, but the tight packaging of the nuclear DNA with protamines is able to protect most of the spermatozoon's nuclear genes from damage (Sawyer et al. 2003). However, certain regions of sperm chromatin, particularly the toroid interlinker domains, remain vulnerable (Ward 2010). After decondensation in the oocyte, these regions appear necessary for the initiation of DNA replication and successful cell division. The first cleavage division of the newly formed embryo is particularly important as any DNA changes induced at this time will continue to be replicated in all subsequent divisions. The early zygote possesses protection against the initiation of apoptosis and instead tends to put cell division on hold and during this delay attempts to repair the DNA using the BER, NER, MMR, HNEJ, and HR pathways. Thus, while most oxidative DNA damage is contributed to the zygote by the fertilizing spermatozoon, most of the responsibility for effecting adequate repair rests with the oocyte. Any mistakes made by the oocyte at this vital stage of development have the potential to result in mutations or epimutations that will influence the entire course of embryonic development.

Elucidating the mechanisms responsible for detecting and addressing DNA damage in the early zygote is important for our understanding of the developmental origins of human disease. In this context, it is clearly imperative that we identify the range of environmental and lifestyle factors responsible for inducing high levels of oxidative DNA damage in the male germ line. We already know that cigarette smoking, obesity, and advanced paternal age are associated with high levels of oxidative DNA damage in spermatozoa, and we strongly suspect that there are other factors capable of inducing such lesions, including radio-frequency electromagnetic radiation and exposure to a range of environmental toxicants and chemotherapeutic agents. Categorizing these causative factors, resolving their impact on the genetic and epigenetic profile of the progeny, and putting in place preventative measures to reduce risks to the genetic integrity of the progeny are significant tasks for gamete biologists—now and in the future. This responsibility is particularly significant given the current widespread use of ART to achieve conceptions in vitro that could not have occurred in vivo; until the genetic consequences of such trends are understood, we may be inadvertently creating a health burden for our species that future generations will have to solve.

Acknowledgments We gratefully acknowledge the support given by the University of Newcastle and the National Health and Medical Research Council that has enabled the studies summarized in this chapter to be undertaken.

References

Aitken RJ (2011) The capacitation-apoptosis highway: oxysterols and mammalian sperm function. Biol Reprod 85(1):9–12. doi:10.1095/biolreprod.111.092528

Aitken RJ, Baker HW (1995) Seminal leukocytes: passengers, terrorists or good samaritans? Hum Reprod 10(7):1736–1739

Aitken RJ, Baker MA (2013) Oxidative stress, spermatozoa and leukocytic infiltration: relationships forged by the opposing forces of microbial invasion and the search for perfection. J Reprod Immunol 100(1):11–19. doi:10.1016/j.jri.2013.06.005

Aitken RJ, Clarkson JS (1987) Cellular basis of defective sperm function and its association with the genesis of reactive oxygen species by human spermatozoa. J Reprod Fertil 81(2):459–469

Aitken RJ, Clarkson JS (1988) Significance of reactive oxygen species and antioxidants in defining the efficacy of sperm preparation techniques. J Androl 9(6):367–376

Aitken RJ, De Iuliis G (2010) On the possible origins of DNA damage in human spermatozoa. Mol Hum Reprod 16(1):3–13. doi:10.1093/molehr/gap059

Aitken RJ, Krausz C (2001) Oxidative stress, DNA damage and the Y chromosome. Reproduction 122(4):497–506

Aitken RJ, Buckingham DW, Brindle J et al (1995) Analysis of sperm movement in relation to the oxidative stress created by leukocytes in washed sperm preparations and seminal plasma. Hum Reprod 10(8):2061–2071

Aitken RJ, Buckingham DW, Carreras A et al (1996) Superoxide dismutase in human sperm suspensions: relationship with cellular composition, oxidative stress, and sperm function. Free Radic Biol Med 21(4):495–504

Aitken RJ, Harkiss D, Knox W et al (1998) A novel signal transduction cascade in capacitating human spermatozoa characterised by a redox-regulated, cAMP-mediated induction of tyrosine phosphorylation. J Cell Sci 111:645–656

Aitken RJ, Koopma P, Lewis SE (2004) Seeds of concern. Nature 432(7013):48–52
Aitken RJ, Wingate JK, De Iuliis GN et al (2006) Cis-unsaturated fatty acids stimulate reactive oxygen species generation and lipid peroxidation in human spermatozoa. J Clin Endocrinol Metab 91(10):4154–4163
Aitken RJ, De Iuliis GN, McLachlan RI (2009) Biological and clinical significance of DNA damage in the male germ line. Int J Androl 32(1):46–56. doi:10.1111/j.1365
Aitken RJ, De Iuliis GN, Finnie JM et al (2010) Analysis of the relationships between oxidative stress, DNA damage and sperm vitality in a patient population: development of diagnostic criteria. Hum Reprod 25(10):2415–2426. doi:10.1093/humrep/deq214
Aitken RJ, Gibb Z, Mitchell LA et al (2012a) Sperm motility is lost in vitro as a consequence of mitochondrial free radical production and the generation of electrophilic aldehydes but can be significantly rescued by the presence of nucleophilic thiols. Biol Reprod 87(5):110. doi:10.1095/biolreprod.112.102020
Aitken RJ, Whiting S, De Iuliis GN et al (2012b) Electrophilic aldehydes generated by sperm metabolism activate mitochondrial reactive oxygen species generation and apoptosis by targeting succinate dehydrogenase. J Biol Chem 287(39):33048–33060
Aitken RJ, Bronson R, Smith TB et al (2013) The source and significance of DNA damage in human spermatozoa; diagnostic strategies and straw man fallacies. Mol Hum Reprod 19(8):475–485. doi:10.1093/molehr/gat025
Aitken RJ, Finnie JM, Muscio L et al (2014) Potential importance of transition metals in the induction of DNA damage by sperm preparation media. Hum Reprod 29(10):2136–2147. doi:10.1093/humrep/deu204
Aly HA (2013) Aroclor 1254 induced oxidative stress and mitochondria mediated apoptosis in adult rat sperm in vitro. Environ Toxicol Pharmacol 36(2):274–283. doi:10.1016/j.etap.2013.04.006
Bakos HW, Mitchell M, Setchell BP et al (2011) The effect of paternal diet-induced obesity on sperm function and fertilization in a mouse model. Int J Androl 34(5):402–410
Bejarano I, Monllor F, Marchena AM et al (2014) Exogenous melatonin supplementation prevents oxidative stress-evoked DNA damage in human spermatozoa. J Pineal Res 57(3):333–339. doi:10.1111/jpi.12172
Bennetts LE, Aitken RJ (2005) A comparative study of oxidative DNA damage in mammalian spermatozoa. Mol Reprod Dev 71(1):77–87
Boaz SM, Dominguez K, Shaman JA et al (2008) Mouse spermatozoa contain a nuclease that is activated by pretreatment with EGTA and subsequent calcium incubation. J Cell Biochem 103(5):1636–1645
Brouwers JF, Boerke A, Silva PF et al (2011) Mass spectrometric detection of cholesterol oxidation in bovine sperm. Biol Reprod 85:128–36
Burruel V, Klooster K, Barker CM et al (2014) Abnormal early cleavage events predict early embryo demise: sperm oxidative stress and early abnormal cleavage. Sci Rep 4:6598. doi:10.1038/srep06598
Chavarro JE, Toth TL, Wright DL et al (2010) Body mass index in relation to semen quality, sperm DNA integrity, and serum reproductive hormone levels among men attending an infertility clinic. Fertil Steril 93(7):2222–2231. doi:10.1016/j.fertnstert.2009.01.100
Crow JF (2000) The origins, patterns and implications of human spontaneous mutation. Nat Rev Genet 1(1):40–47
Dalzell LH, Thompson-Cree ME, McClure N et al (2003) Effects of 24-hour incubation after freeze-thawing on DNA fragmentation of testicular sperm from infertile and fertile men. Fertil Steril 79(Suppl 3):1670–1672
De Iuliis GN, Newey RJ, King BV et al (2009) Mobile phone radiation induces reactive oxygen species production and DNA damage in human spermatozoa in vitro. PLoS One 4(7), e6446. doi:10.1371/journal.pone.0006446
de Lamirande E, Gagnon C (1993) A positive role for the superoxide anion in triggering hyperactivation and capacitation of human spermatozoa. Int J Androl 16(1):21–25

de Lamirande E, Lamothe G (2009) Reactive oxygen-induced reactive oxygen formation during human sperm capacitation. Free Radic Biol Med 46(4):502–510. doi:10.1016/j.freeradbiomed.2008.11.004
Evenson DP, Jost LK, Marshall D et al (1999) Utility of sperm chromatin structure assay as a diagnostic and prognostic tool in the human fertility clinic. Hum Reprod 14(4):1039–1049
Fear JM, Hansen PJ (2011) Developmental changes in expression of genes involved in regulation of apoptosis in the bovine preimplantation embryo. Biol Reprod 84(1):43–51. doi:10.1095/biolreprod.110.086249
Fraga CG, Motchnik PA, Wyrobek AJ et al (1996) Smoking and low antioxidant levels increase oxidative damage to DNA. Mutat Res 351:199–203
Gawecka JE, Marh J, Ortega M et al (2013) Mouse zygotes respond to severe sperm DNA damage by delaying paternal DNA replication and embryonic development. PLoS One 8(2), e56385. doi:10.1371/journal.pone.0056385
Gharagozloo P, Aitken RJ (2011) The role of sperm oxidative stress in male infertility and the significance of oral antioxidant therapy. Hum Reprod 26(7):1628–40. doi:10.1093/humrep/der132
Gomez E, Buckingham DW, Brindle J et al (1996) Development of an image analysis system to monitor the retention of residual cytoplasm by human spermatozoa, correlation with biochemical markers of the cytoplasmic space, oxidative stress and sperm function. J Androl 17:276–287
Goriely A, McVean GA, Röjmyr M et al (2003) Evidence for selective advantage of pathogenic FGFR2 mutations in the male germ line. Science 301(5633):643–646
Greco E, Romano S, Iacobelli M et al (2005) ICSI in cases of sperm DNA damage: beneficial effect of oral antioxidant treatment. Hum Reprod 20(9):2590–2594
Hakimi H, Geary I, Pacey A et al (2006) Spermicidal activity of bacterial lipopolysaccharide is only partly due to lipid A. J Androl 27(6):774–779
Harrouk W, Codrington A, Vinson R, Robaire B, Hales BF (2000) Paternal exposure to cyclophosphamide induces DNA damage and alters the expression of DNA repair genes in the rat preimplantation embryo. Mutant Res 461(3):229–241
Hochreiter-Hufford A, Ravichandran KS (2013) Clearing the dead: apoptotic cell sensing, recognition, engulfment, and digestion. Cold Spring Harb Perspect Biol 5(1):a008748. doi:10.1101/cshperspect.a008748
Hurst LD, Ellegren H (2002) Human genetics: mystery of the mutagenic male. Nature 420(6914):365–6
Irvine DS, Twigg JP, Gordon EL et al (2000) DNA integrity in human spermatozoa: relationships with semen quality. J Androl 21(1):33–44
Ji G, Yan L, Liu W et al (2013) OGG1 Ser326Cys polymorphism interacts with cigarette smoking to increase oxidative DNA damage in human sperm and the risk of male infertility. Toxicol Lett 218(2):144–9. doi:10.1016/j.toxlet.2013.01.017
Jones R, Mann T, Sherins RJ (1979) Peroxidative breakdown of phospholipids in human spermatozoa: spermicidal effects of fatty acid peroxides and protective action of seminal plasma. Fertil Steril 31(5):531–537
Kodama H, Yamaguchi R, Fukuda J et al (1997) Increased deoxyribonucleic acid damage in the spermatozoa of infertile male patients. Fertil Steril 68(3):519–524
Kong A, Frigge ML, Masson G et al (2012) Rate of de novo mutations and the importance of father's age to disease risk. Nature 488(7412):471–475. doi:10.1038/nature11396
Koppers AJ, De Iuliis GN, Finnie JM et al (2008) Significance of mitochondrial reactive oxygen species in the generation of oxidative stress in spermatozoa. J Clin Endocrinol Metab 93(8):3199–3207. doi:10.1210/jc.2007-2616
Koppers AJ, Garg ML, Aitken RJ (2010) Stimulation of mitochondrial reactive oxygen species production by unesterified, unsaturated fatty acids in defective human spermatozoa. Free Radic Biol Med 48(1):112–119. doi:10.1016/j.freeradbiomed.2009.10.033

Koppers AJ, Mitchell LA, Wang P et al (2011) Phosphoinositide 3-kinase signalling pathway involvement in a truncated apoptotic cascade associated with motility loss and oxidative DNA damage in human spermatozoa. Biochem J 436(3):687–698. doi:10.1042/BJ20110114

Kort HI, Massey JB, Elsner CW et al (2006) Impact of body mass index values on sperm quantity and quality. J Androl 27(3):450–452

Kramer JA, Krawetz SA (1996) Nuclear matrix interactions within the sperm genome. J Biol Chem 271(20):11619–11622

Krausz C, Mills C, Rogers S et al (1994) Stimulation of oxidant generation by human sperm suspensions using phorbol esters and formyl peptides: relationships with motility and fertilization in vitro. Fertil Steril 62(3):599–605

Larson KL, Dejonge CJ, Barnes CM et al (2000) Sperm chromatin structure assay parameters as predictors of failed pregnancy following assisted reproductive techniques. Hum Reprod 15(8):1717–1719

Lee KM, Ward MH, Han S et al (2009) Paternal smoking, genetic polymorphisms in CYP1A1 and childhood leukemia risk. Leuk Res 33(2):250–258

Li Y, Lin H, Li Y et al (2011) Association between socio-psycho-behavioral factors and male semen quality: systematic review and meta-analyses. Fertil Steril 95(1):116–123. doi:10.1016/j.fertnstert.2010.06.031

Liu C, Duan W, Xu S et al (2013) Exposure to 1800 MHz radiofrequency electromagnetic radiation induces oxidative DNA base damage in a mouse spermatocyte-derived cell line. Toxicol Lett 218(1):2–9. doi:10.1016/j.toxlet.2013.01.003

Liu Y, Chapple V, Roberts P et al (2014) Time-lapse videography of human oocytes following intracytoplasmic sperm injection: events up to the first cleavage division. Reprod Biol 14(4):249–256. doi:10.1016/j.repbio.2014.08.003

Lopes S, Jurisicova A, Sun JG et al (1998) Reactive oxygen species: potential cause for DNA fragmentation in human spermatozoa. Hum Reprod 13(4):896–900

Lozano GM, Bejarano I, Espino J et al (2009) Relationship between caspase activity and apoptotic markers in human sperm in response to hydrogen peroxide and progesterone. J Reprod Dev 55(6):615–621

Menezo Y Jr, Russo G, Tosti E et al (2007) Expression profile of genes coding for DNA repair in human oocytes using pangenomic microarrays, with a special focus on ROS linked decays. J Assist Reprod Genet 24(11):513–520

Meseguer M, Martínez-Conejero JA, O'Connor JE et al (2008) The significance of sperm DNA oxidation in embryo development and reproductive outcome in an oocyte donation program: a new model to study a male infertility prognostic factor. Fertil Steril 89(5):1191–1199

Moskovtsev SI, Willis J, White J et al (2007) Leukocytospermia: relationship to sperm deoxyribonucleic acid integrity in patients evaluated for male factor infertility. Fertil Steril 88(3):737–740

Nadel B, de Lara J, Finkernagel SW et al (1995) Cell-specific organization of the 5S ribosomal RNA gene cluster DNA loop domains in spermatozoa and somatic cells. Biol Reprod 53(5):1222–1228

Oliveira H, Spanò M, Santos C et al (2009) Adverse effects of cadmium exposure on mouse sperm. Reprod Toxicol 28(4):550–555. doi:10.1016/j.reprotox.2009.08.001

Paasch U, Grunewald S, Agarwal A et al (2004) Activation pattern of caspases in human spermatozoa. Fertil Steril 81(1):802–809

Paul C, Nagano M, Robaire B (2011) Aging results in differential regulation of DNA repair pathways in pachytene spermatocytes in the Brown Norway rat. Biol Reprod 85(6):1269–1278. doi:10.1095/biolreprod.111.094219

Pujianto DA, Curry BJ, Aitken RJ (2010) Prolactin exerts a prosurvival effect on human spermatozoa via mechanisms that involve the stimulation of Akt phosphorylation and suppression of caspase activation and capacitation. Endocrinology 151(3):1269–1279. doi:10.1210/en.2009-0964

Rhemrev JP, van Overveld FW, Haenen GR et al (2000) Quantification of the nonenzymatic fast and slow TRAP in a postaddition assay in human seminal plasma and the antioxidant contributions of various seminal compounds. J Androl 21(6):913–920
Robinson L, Gallos ID, Conner SJ et al (2012) The effect of sperm DNA fragmentation on miscarriage rates: a systematic review and meta-analysis. Hum Reprod 27(10):2908–2917. doi:10.1093/humrep/des261
Rodriguez PC, Valdez LB, Zaobornyj T et al (2011) Nitric oxide and superoxide anion production during heparin-induced capacitation in cryopreserved bovine spermatozoa. Reprod Domest Anim 46(1):74–81. doi:10.1111/j.1439-0531.2010.01583.x
Saleh RA, Agarwal A, Kandirali E et al (2002) Leukocytospermia is associated with increased reactive oxygen species production by human spermatozoa. Fertil Steril 78(6):1215–1224
Sanocka D, Miesel R, Jedrzejczak P et al (1997) Effect of reactive oxygen species and the activity of antioxidant systems on human semen; association with male infertility. Int J Androl 20(5): 255–264
Sawyer DE, Mercer BG, Wiklendt AM et al (2003) Quantitative analysis of gene-specific DNA damage in human spermatozoa. Mutat Res 529(1-2):21–34
Shaman JA, Yamauchi Y, Ward WS (2007) The sperm nuclear matrix is required for paternal DNA replication. J Cell Biochem 102(3):680–688
Shimura T, Inoue M, Taga M et al (2002) p53-dependent S-phase damage checkpoint and pronuclear cross talk in mouse zygotes with X-irradiated sperm. Mol Cell Biol 22(7):2220–2228
Shinde DN, Elmer DP, Calabrese P et al (2013) New evidence for positive selection helps explain the paternal age effect observed in achondroplasia. Hum Mol Genet 22(20):4117–4126. doi:10.1093/hmg/ddt260
Showell MG, Brown J, Yazdani A et al. (2011) Antioxidants for male subfertility. Cochrane Database Syst Rev (1): CD007411. doi: 10.1002/14651858.CD007411
Singh NP, Muller CH, Berger RE (2003) Effects of age on DNA double-strand breaks and apoptosis in human sperm. Fertil Steril 80(6):1420–1430
Smith TB, Dun MD, Smith ND et al (2013) The presence of a truncated base excision repair pathway in human spermatozoa that is mediated by OGG1. J Cell Sci 126(6):1488–1497. doi:10.1242/jcs.121657
Sotolongo B, Huang TT, Isenberger E et al (2005) An endogenous nuclease in hamster, mouse, and human spermatozoa cleaves DNA into loop-sized fragments. J Androl 26(2):272–80
Storey BT, Alvarez JG, Thompson KA (1998) Human sperm glutathione reductase activity in situ reveals limitation in the glutathione antioxidant defense system due to supply of NADPH. Mol Reprod Dev 49(4):400–407
Sturmey RG, Hawkhead JA, Barker EA et al (2009) DNA damage and metabolic activity in the preimplantation embryo. Hum Reprod 24(1):81–91
Suleiman SA, Ali ME, Zaki ZM et al (1996) Lipid peroxidation and human sperm motility: protective role of vitamin E. J Androl 17(5):530–537
Thomson LK, Fleming SD, Aitken RJ et al (2009) Cryopreservation-induced human sperm DNA damage is predominantly mediated by oxidative stress rather than apoptosis. Hum Reprod 24(9):2061–2070. doi:10.1093/humrep/dep214
Tosic J, Walton A (1950) Metabolism of spermatozoa. The formation and elimination of hydrogen peroxide by spermatozoa and effects on motility and survival. Biochem J 47(2):199–212
Tunc O, Bakos HW, Tremellen K (2011) Impact of body mass index on seminal oxidative stress. Andrologia 43(2):121–128
Twigg J, Irvine DS, Houston P et al (1998) Iatrogenic DNA damage induced in human spermatozoa during sperm preparation: protective significance of seminal plasma. Mol Hum Reprod 4(5):439–445
Uribe P, Boguen R, Treulen F et al (2015) Peroxynitrite-mediated nitrosative stress decreases motility and mitochondrial membrane potential in human spermatozoa. Mol Hum Reprod 21:237–43

van Overveld FW, Haenen GR, Rhemrev J et al (2000) Tyrosine as important contributor to the antioxidant capacity of seminal plasma. Chem Biol Interact 127:151–161

Wang Y, Sharma RK, Agarwal A (1997) Effect of cryopreservation and sperm concentration on lipid peroxidation in human semen. Urology 50(3):409–413

Ward WS (2010) Function of sperm chromatin structural elements in fertilization and development. Mol Hum Reprod 16(1):30–36. doi:10.1093/molehr/gap080

Wündrich K, Paasch U, Leicht M et al (2006) Activation of caspases in human spermatozoa during cryopreservation–an immunoblot study. Cell Tissue Bank 7(2):81–9

Xu DX, Shen HM, Zhu QX et al (2003) The associations among semen quality, oxidative DNA damage in human spermatozoa and concentrations of cadmium, lead and selenium in seminal plasma. Mutat Res 534(1-2):155–163

Zeng F, Baldwin D, Schultz RM (2004) Transcript profiling during preimplantation mouse development. Dev Biol 272(2):483–496

Zribi N, Chakroun NF, Ben Abdallah F et al (2012) Effect of freezing-thawing process and quercetin on human sperm survival and DNA integrity. Cryobiology 65(3):326–331. doi:10.1016/j.cryobiol.2012.09.003

Chapter 3
Small RNAs: Their Possible Roles in Reproductive Failure

Benjamin J. Hale, Aileen F. Keating, Cai-Xia Yang, and Jason W. Ross

Abstract Posttranscriptional gene regulation is a regulatory mechanism which occurs "above the genome" and confers different phenotypes and functions within a cell. Transcript and protein abundance above the level of transcription can be regulated via noncoding ribonucleic acid (ncRNA) molecules, which potentially play substantial roles in the regulation of reproductive function. MicroRNA (miRNA), endogenous small interfering RNA (endo-siRNA), and PIWI-interacting RNA (piRNA) are three primary classes of small ncRNA. Similarities and distinctions between their biogenesis and in the interacting protein machinery that facilitate their function distinguish these three classes. Characterization of the expression and importance of the critical components for the biogenesis of each class in different tissues contributes a clearer understanding of their contributions in specific reproductive tissues and their ability to influence fertility in both males and females. This chapter discusses the expression and potential roles of miRNA, endo-siRNA, and piRNA in the regulation of reproductive function. Additionally, this chapter elaborates on investigations aimed to address and characterize specific mechanisms through which miRNA may influence infertility and the use of miRNA as biomarkers associated with several reproductive calamities such as defective spermatogenesis in males, polycystic ovarian failure, endometriosis and obesity, and chemical-induced subfertility.

Keywords Small noncoding RNA • PIWI proteins • Transposons • Telomeres • Dicer • Posttranscriptional gene regulation • Spermatogenesis • Embryo implantation

---

B.J. Hale • A.F. Keating • C.-X. Yang • J.W. Ross (✉)
Department of Animal Science, Iowa State University,
2356 Kildee hall, Ames, IA 50011, USA
e-mail: jwross@iastate.edu

R. Bronson (ed.), *The Male Role in Pregnancy Loss and Embryo Implantation Failure*, Advances in Experimental Medicine and Biology 868,
DOI 10.1007/978-3-319-18881-2_3

3.1 Introduction

Reproductive organs and cells therein are unique in that they continually undergo substantial reorganization of both their transcriptome and proteome during development and eventually during female reproductive cycles. Understanding of the regulation of ribonucleic acid (RNA) and protein abundance has dramatically improved in recent years. Discovery of several classes of small noncoding RNA (ncRNA) and elucidation of their biological roles with respect to cellular function have substantially contributed to our biological understanding of reproductive function and fertility. RNA classes are broadly broken into coding RNA, those that result in the synthesis and production of a protein, and ncRNA. Traditionally, ncRNA such as transfer RNA, ribosomal RNA, and small nuclear and small nucleolar RNA is considered non-regulatory and primarily supports cell function without impacting cellular phenotype. However, identification and characterization of noncoding classes such as small interfering RNA (siRNA), PIWI-interacting RNA (piRNA), and microRNA (miRNA) have clearly demonstrated that ncRNA fulfills regulatory roles that substantially impact cellular phenotype and function. These ncRNAs differ in their origin, length, and Argonaute (AGO) protein partners through which they elicit their biological function (Babiarz and Blelloch 2008; Ishizu et al. 2012; Juliano et al. 2011; Kim et al. 2009; Thomson and Lin 2009). Knowledge of the biological roles of ncRNA, particularly miRNA, with respect to reproduction and fertility has dramatically increased over the past decade revealing numerous pathways and mechanisms through which fertility may be impacted.

3.2 Biogenesis of Small RNA

The presence or absence of critical components required for ncRNA biogenesis is one manner through which specific classes of ncRNA differ and impact specific reproductive cell types. MiRNA and endo-siRNA are processed from double-stranded precursors in a stepwise manner and result in 20–22 nucleotide mature ncRNAs. In contrast, piRNA biogenesis utilizes long single-stranded precursors to produce 26–31 nucleotide functional ncRNAs (Aravin et al. 2006; Grivna et al. 2006; Juliano et al. 2011). MiRNA can be further separated into two groups distinguished through their synthesis pathway: canonical or noncanonical. Both canonical (Fig. 3.1) and noncanonical miRNAs are initially transcribed as primary miRNA (pri-miRNA) by RNA polymerase II and depending on sequence complementation within the RNA molecule then form secondary RNA structures which produce 60–75 nucleotide (nt) hairpins which can also be found as clusters within a single pri-miRNA (Lee et al. 2002, 2004). In the canonical mechanism, DiGeorge syndrome critical region 8 (DGCR8) recognizes the miRNA hairpins and guides Drosha, an RNase III enzyme, to cleave the hairpin base resulting in a pre-miRNA (Denli et al. 2004; Gregory et al. 2004; Han et al. 2004). The pre-miRNA is composed of a

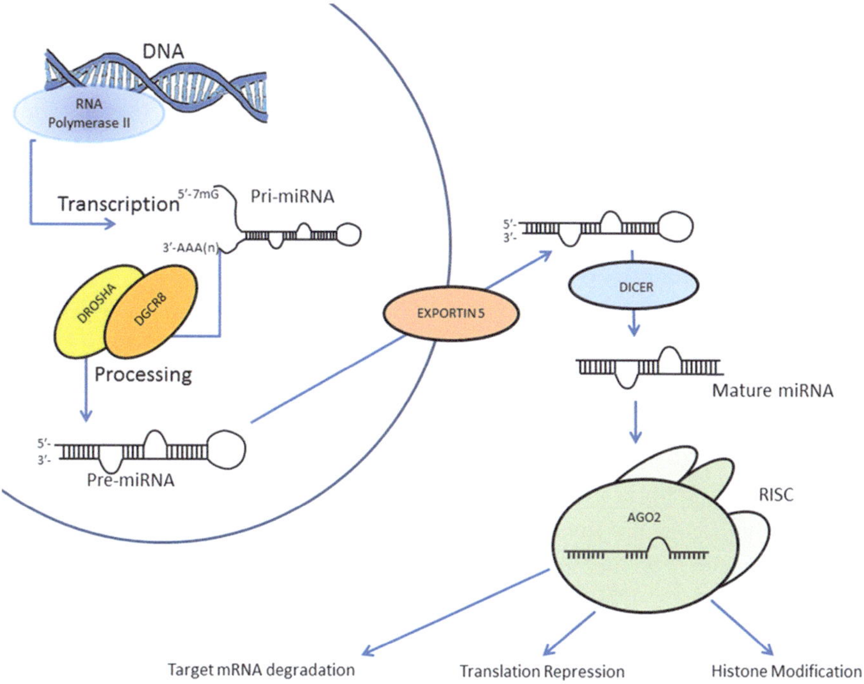

Fig. 3.1 MicroRNA biogenesis is initiated via the activity of RNA polymerase II resulting in synthesis of a primary miRNA transcript (Pri-miRNA) that is both capped and polyadenylated. The appropriate spatial complementation of specific nucleotides within the transcript results in the formation of hairpin secondary structures that are recognized by the RNA processing complex consisting of DROSHA and DGCR8. The enzymatic activity of this protein complex results in cleavage and removal of the hairpin structure (now considered a pre-miRNA) from the primary transcript. Exportin 5 facilitates the transport of pre-miRNA from the nucleus into the cytoplasm of the cell where it is recognized by Dicer, and the loop is cleaved leaving a short duplex mature miRNA molecule. Upon dissociation, either strand from the duplex miRNA structure can be utilized by the RNA-induced silencing complex to contribute to posttranscriptional gene regulation though impacting mRNA stability and/or translation efficiency in addition to contributing to chromatin modifications to control gene expression

hairpin with a 3′ overhang and is exported from the nucleus via Exportin-5 (Lund et al. 2004; Yi et al. 2003). In contrast, during noncanonical miRNA biogenesis, processing by the DGCR8/Drosha microprocessing complex is absent, and cleavage of hairpins occurs via other cellular endonucleases or through transcription directly as a short hairpin (Miyoshi et al. 2010; Schwab and Voinnet 2009). Regardless of whether their production is directed through the canonical or noncanonical pathway, all miRNAs reach the cytosol where they are cleaved by an RNase III enzyme, Dicer, resulting in a functionally mature miRNA (Castellano and Stebbing 2013; Lau et al. 2001; Lee et al. 2002). Mature miRNAs are capable of being loaded into an RNA-induced silencing complex (RISC) and contribute to posttranscriptional gene regulation (PTGR).

The endo-siRNA class is derived from long dsRNA, which can be either sense or antisense RNA pairs or long hairpins, and is also cleaved in the cytoplasm by Dicer (Ghildiyal et al. 2008; Okamura et al. 2008a, b; Tam et al. 2008). Synthesis of both endo-siRNA and miRNA occurs via Dicer activity and has a final mature product of an approximately 21 nucleotide single-stranded RNAs capable of association with members of the argonaute protein family (AGO 1–4) which enable the formation of the active RISC complex (Filipowicz 2005).

piRNA biogenesis is less well understood than miRNA and endo-siRNA, although they are produced from long single-stranded RNA precursors mediated through the action of Dicer (Houwing et al. 2007; Vagin et al. 2006). Mature piRNAs are approximately 26–31 nucleotides long and are expressed primarily in germ cells (Aravin et al. 2006; Houwing et al. 2007; Watanabe et al. 2006). Secondary processing of piRNA involves Miwi1, Miwi2, and Mili in mice (also known as Piwi1, Piwi2, and Pili) (Aravin et al. 2006, 2007; Klattenhoff and Theurkauf 2008). PiRNA action does not involve AGO protein association but is associated with transposon and repeat-associated siRNA inactivation (Brennecke et al. 2007; Gunawardane et al. 2007; Houwing et al. 2007; Vagin et al. 2006). The complex nature of piRNA sequences has been revealed via next-generation sequencing (Aravin et al. 2006; Girard et al. 2006). Class I piRNA are derived from clustered genomic loci in repeat sequences, indicating a role with respect to transposon defense (Houwing et al. 2008; Klattenhoff and Theurkauf 2008). Class II piRNAs are derived from transposon RNA cleavage, while the remainder of the piRNAs are thought to originate from diverse genomic regions perhaps including the 3′ untranslated region (UTR) of some mRNA as is the case in *Drosophila*, mouse, and *Xenopus* (Robine et al. 2009). The variation in piRNA biogenesis and sequence suggests their biological involvement in PTGR in addition to transposon repression.

The "ping-pong" model has been proposed as a regulatory model coupling piRNA generation and transposon repression (Brennecke et al. 2007). In Drosophila, Piwi and Aub proteins bind the genomic piRNA cluster loci-derived antisense primary piRNA. The proteins are then guided by the antisense primary piRNA, which binds and cleaves transposon mRNA to generate the 5′ ends of the sense secondary piRNA. AGO3 protein then binds the sense secondary piRNA resulting in transcript cleavage to produce the 5′ end of new primary piRNA (Brennecke et al. 2007). How 3′ ends of new piRNA are generated in the ping-pong model is not clear. The proportion of piRNAs generated via the ping-pong mechanism represent only a fraction of the piRNA population; therefore, how other piRNAs are generated from complex intergenic regions also remains uncharacterized (Cook and Blelloch 2013).

In mice, expression of piRNA is developmentally regulated (Aravin et al. 2007). During the prepachytene stage prior to meiosis in spermatogenesis, expression of one subtype of piRNA (26–28 nucleotides) is favored, whereas another piRNA subtype (29–31 nucleotides) is predominantly expressed during the pachytene meiotic stage. While piRNA expression is enriched in the mammalian and *Drosophila* germline, piRNAs are also expressed in somatic cells of *Drosophila* (Lin and Yin 2008), differentiated somatic cells of jellyfish (Seipel et al. 2004), and porcine cumulus cells (Yang et al. 2012a), suggesting additional piRNA roles in these cell types, which may be species dependent.

3.3 Small RNA Mechanism of Action

Mature miRNA and endo-siRNA interact with AGO proteins (AGO 1–4; also known as EIF2C1-4) to enable the association of RISC machinery (Filipowicz 2005). AGO proteins associated with either a miRNA or endo-siRNA interact with target mRNA via complementarity between the ncRNA and the 3′UTR of the mRNA target. If complete complementarity exists, cleavage of the target mRNA can occur. If there is incomplete complementarity, the primary mechanism of PTGR occurs by translation suppression (Fabian et al. 2010; Hutvagner and Zamore 2002; Martinez et al. 2002). There are, however, examples of weak complementation between the 3′ end of the ncRNA and the 3′UTR of the target gene which can still result in PTGR via mRNA degradation (Bagga et al. 2005).

PIWI proteins are located in nuclear and/or perinuclear nuage (Brennecke et al. 2007; Carmell et al. 2007; Wang et al. 2009) where they repress genetic elements (Lim and Kai 2007). PIWI proteins have an N-terminal PAZ domain followed by an MID (middle) domain, which together recognize and bind the 3′ end of piRNAs (Jinek and Doudna 2009). The C-terminal domain of PIWI proteins has RNase H activity capable of recognizing the 5′ end of piRNA and facilitates the cleavage of the target sequence (Frank et al. 2010; Wang et al. 2009). In the germline, the PIWI-piRNA complex silences transposons both transcriptionally and posttranscriptionally, through chromatin silencing of transposable elements via histone modification and by altering the DNA epigenetic status (Klenov et al. 2007; Kuramochi-Miyagawa et al. 2008). Posttranscriptionally, transposable elements are repressed through their active RNA being cleaved by PIWI- and AUB-primary piRNA complexes, thereby producing secondary piRNA through the aforementioned "ping-pong" model (Brennecke et al. 2007). This mechanism is supported by the observation that transposon RNA is expressed at a higher level in the presence of PIWI protein mutations (Li et al. 2009). In *Drosophila*, piRNAs are involved in telomere function, including telomere protection complex assembly, thereby maintaining chromosome integrity (Khurana et al. 2010). Expression of PIWI proteins in human somatic stem cells (Sharma et al. 2001) and neoplastic cells (Lee et al. 2006, 2010; Liu et al. 2006b, 2010; Taubert et al. 2007) suggests that piRNA may also be involved in stem cell function regulation and carcinogenesis. However, the mechanism, major functions, and pathways regulated by the Piwi-piRNA complex remain poorly understood.

3.4 Small RNA Expression and Function in Reproductive Tissues

3.4.1 Germ Cell Development and Maintenance

PIWI proteins are expressed in germ cells, stem cells, and oocytes of multiple species. In the female germline of *Xenopus laevis* (Xiwi and Xili), *Danio rerio* (Ziwi and Zili), and *Drosophila melanogaster* (Aubergine, Piwi, and Ago2) (Houwing et al. 2007; Li et al. 2009; Wilczynska et al. 2009), several PIWI regulatory members have been identified and characterized. In mice, *Miwi*, *Mili*, and *Miwi-2* exist (Deng and Lin 2002; Kuramochi-Miyagawa et al. 2008), and in pigs three *Piwi* genes (Piwil1, Piwil2, and Piwil4) are expressed in the testes, ovaries, and oocytes (Kowalczykiewicz et al. 2012). Utilizing bioinformatic analysis, human Piwi proteins (Hiwi, Hili, and Hiwi-2) have been identified (Gu et al. 2010), with *Hiwi* being expressed in hematopoietic stem cells and germ cells (Qiao et al. 2002; Sharma et al. 2001). During germ cell specification in mice (Deng and Lin 2002) and gonad development of pigs (Kowalczykiewicz et al. 2012), expression patterns of individual PIWI proteins differ. Partial or complete loss of germ cells and transposon derepression in the germline have been demonstrated in *Drosophila melanogaster* and mice (Carmell et al. 2007; Cox et al. 1998; Li et al. 2009; Vagin et al. 2006) due to the loss of PIWI proteins, suggesting a conserved role for PIWI proteins in the maintenance of germ cell viability.

3.4.2 Oocyte Development and Maturation

Mammals are born with a finite oocyte number that originates from the primordial germ cell pool following migration to the genital ridge during embryonic development (Tingen et al. 2009). Following recruitment from the primordial follicle pool and selection for maturation, oocytes undergo germinal vesicle breakdown (GVBD) facilitating their progression through meiosis until subsequent arrest at metaphase II. The oocyte is transcriptionally quiescent following GVBD until after fertilization and subsequent activation of the embryonic genome, which occurs around the two- to eight-cell stage of development depending upon the species. The inability of the developing embryo to elicit a transcriptional response prior to genome activation suggests a potentially important role of ncRNA during this period of development. MiRNA, endo-siRNA, and piRNA are expressed in oocytes of multiple species at various stages of development (Abd El Naby et al. 2013; Golden et al. 2008; Tesfaye et al. 2009; Watanabe et al. 2006, 2008; Xu et al. 2011; Yang et al. 2012a).

Conditional knockout mice have utility for deciphering functions and contributions of endo-siRNA and miRNA in both the maturing oocyte and developing embryo. Mice with a ZP3-driven conditional loss of Dgcr8 (responsible for canonical miRNA production) are fertile (Ma et al. 2010; Suh et al. 2010), but litter size was substantially reduced, suggesting that some miRNA may contribute to developmental competency of the subsequently produced embryo even if not required in the maturing oocyte (Suh et al. 2010). The observation that *Dgcr8* is not required for mouse oocyte maturation, since loss of *Dgcr8* has no noticeable effect on mRNA regulation, coupled with the observation that similarly deleting *Dicer* or *Ago2* does have a negative effect on maturing oocytes rendering them incompetent (Kaneda et al. 2009; Murchison et al. 2007; Suh et al. 2010; Tang et al. 2007), suggests that small ncRNAs are needed for the production of oocytes capable of producing viable embryos.

The small RNA population of both the oocyte and cumulus cells during in vitro maturation (IVM) have been sequenced and the portfolio of endo-siRNA, miRNA, and piRNA demonstrated in pigs (Yang et al. 2012a). During oocyte IVM and progression through meiosis, few alterations were evident, with the exception of miR-21 and miR-574-3p which were significantly upregulated and downregulated during IVM, respectively. In mice, granulosa cell miR-21 expression is luteinizing hormone dependent, and the in vivo use of oligonucleotide inhibitors against miR-21 suppressed ovulation rate but increased apoptosis, indicating a cell viability role for miR-21 (Carletti et al. 2010).

3.4.3 Fertilization and Early Embryo Development

Sperm possess a diverse portfolio of ncRNA (Amanai et al. 2006; Krawetz et al. 2011), and their abundance of miRNA may be related to the biogenesis and expression of small RNA contributing to spermatogenesis (Maatouk et al. 2008). Interestingly, miR-34c, a miRNA associated with the differentiation of male germ cells (Bouhallier et al. 2010), has been shown to influence embryonic development in mice following fertilization (Liu et al. 2012). Liu et al. (2012) identified six miRNAs, including miR-34c, as absent in mature oocytes but present in mouse sperm and zygotes. The impact on development occurs via the ability of miR-34c to interact with BCL2, thereby contributing to regulation of the first embryonic cleavage following oocyte activation (Liu et al. 2012).

Following fertilization and subsequent activation of the zygotic genome, the embryo begins to express RNA transcripts. Transcription of pri-miRNA has been observed in the mouse two-cell embryo, and mature transcripts are detectable at the four-cell stage (Tang et al. 2007; Zeng and Schultz 2005). Dicer and Dgcr8 activities are needed for epiblast formation (Kanellopoulou et al. 2005; Murchison et al. 2005; Wang et al. 2007, 2008), and in some organisms it is thought that miRNA contributes to maternal mRNA clearance before zygotic gene activation (ZGA) occurs (Giraldez et al. 2006; Hemberger et al. 2009; Sinkkonen et al. 2008; Svoboda and Flemr 2010).

Successful embryonic development requires broad transcriptional arrest and mRNA clearance to deplete maternally stored mRNA transcripts in coordination with both ZGA and subsequent mRNA and protein production. Some small RNAs, including miRNA, are abundantly expressed during oocyte maturation and early embryonic development in *Xenopus laevis* (Watanabe et al. 2005), *Drosophila* (Aboobaker et al. 2005; Biemar et al. 2005), zebra fish (Giraldez et al. 2006; Wienholds et al. 2005), mice (Tang et al. 2007), and pigs (Yang et al. 2012a). Maternal mRNA depletion is, in part, controlled via the 3′UTR of the expressed transcripts (Brevini et al. 2007; Tadros and Lipshitz 2005); thus, there is opportunity for small RNA to influence the posttranscriptional outcome of the resultant embryo.

3.4.4 Embryonic Stem Cells

Biogenesis of miRNA and the associated molecular machinery has been shown to be required for stem cell differentiation and the maintenance of pluripotency (Marson et al. 2008). The disruption of *Dicer* and *Dgcr8* leads to a defective capacity of ES cells to be cultured in both mice and humans (Bernstein et al. 2003; Han et al. 2004; Kanellopoulou et al. 2005; Murchison et al. 2005). Embryonic lethality is the consequence of *Dicer* deletion in mice (Bernstein et al. 2003) while *Dicer-null* embryonic stem (ES) cells are deficient in proper differentiation (Kanellopoulou et al. 2005). *Dgcr8* knockout mice ES cells show a phenotype similar to *Dicer-deficient* ES cells, with reduced proliferation and a defective capacity for differentiation (Wang et al. 2007). Taken together, these results support the requirement of canonical miRNA biogenesis for ES cell differentiation and proliferation.

There are a number of miRNA expression clusters that are likely to be involved in ES cell regulation including the let-7 family and the miR-290 and miR-17–92 cluster. Let-7 miRNA family members are highly expressed in differentiating cells (Gu et al. 2008; Lakshmipathy et al. 2007), and Let-7 g can be regulated by LIN28, which is highly expressed in pluripotent cells (Viswanathan et al. 2008), suggesting that let-7/*LIN28* regulatory mechanism contributes to the maintenance of pluripotency. Pluripotent factors such as *Oct4*, *Sox2*, *Nanog*, and *Tcf4* activate *LIN28* gene expression, which in turn inhibits differentiation, and since the 3′UTR of *LIN28* mRNA is targeted by Let-7, LIN28 inhibition of differentiation may be a result of let-7 g activity (Marson et al. 2008).

There are six miRNAs (miR-290 through miR-295) in the miR-290 cluster, all of which are transcribed in single polycistronic transcripts and regulated by a common promoter (Suh et al. 2004). All miR-290 members are expressed in undifferentiated mouse ES cells, but their abundance decreases after differentiation (Houbaviy et al. 2003). Embryonic lethality results in mice with a homozygous deletion of all six members of the miR-290 cluster (Ambros and Chen 2007), and exogenously delivered miR-290 cluster can partially rescue the self-renewal capacity of *Dicer*-null cells (Benetti et al. 2008; Sinkkonen et al. 2008).

The miR-17–92 cluster is strongly expressed in undifferentiated ES cells and forms a polycistronic transcript which generates miR-17, miR-18a, miR-19a, miR-20a, miR-19b-1, and miR-92a-1 (Gu et al. 2008; Houbaviy et al. 2003; Morin et al. 2008). The miR-17–92 cluster is activated by the oncogene, c-Myc (Chen and

Daley 2008), which in combination with the pluripotency factors Oct4, Sox2, and Klf4 can induce pluripotency (iPS cells). Thus, it is suggested that the miR-17–92 cluster plays roles in pluripotency and stem cell renewal (Melton et al. 2010).

3.4.5 Embryo Implantation and Interaction with Maternal Endometrium

During the peri-implantation period of pregnancy, uterine epithelial cells and the conceptus trophectoderm develop adhesion competence in unison to initiate an adhesion cascade within the window of receptivity. Upregulation of 32 miRNA has been demonstrated in mouse endometrium during the window of implantation. Of the identified miRNA, miR-101, miR-144, and miR-199a* are predicted to interact with cyclooxygenase-2 (Cox2) mRNA (Chakrabarty et al. 2007). Proinflammatory immune signaling is associated with uterine receptivity in multiple species (Cha et al. 2012), and miRNA with the potential to influence the expression of inflammatory and immune response mediators includes let7, miR-17-5p, miR-20a, miR-106a, miR-125b, miR-146, and miR-155 (Meng et al. 2007, 2008; O'Connell et al. 2007; Rodriguez et al. 2007; Tili et al. 2007). For example, both miR-125b and miR-155 are involved in the development of T, B, and dendritic cells, key cells of the immune system. Transcription of these miRNAs is proposed to be activated by the nuclear factor of kappa light polypeptide gene enhancer in B cells (NF-κB) (Pan and Chegini 2008). The presence of numerous small RNAs has been identified in the porcine uterine endometrium during the implantation window; however, the mechanism by which specific small RNAs contribute to uterine function and ultimately facilitate embryo implantation remains largely unknown. Conditional deletion of *Dicer* in mice driven by the progesterone receptor or anti-Müllerian hormone receptor type 2 promoters resulted in sterility, abnormal development, and altered signaling pathways in the uterus suggesting important biological contributions of small RNA to the function of the uterine endometrium (Hawkins et al. 2012; Nagaraja et al. 2008).

Postimplantation survival of a growing fetus is entirely dependent on the coordination of gas exchange, nutrient supply, and waste product removal via the placenta, while maintaining immunological protection of the embryo. Human placenta is abundant in miRNA, with distinctive expression profile patterns (Barad et al. 2004; Landgraf et al. 2007; Liang et al. 2007). In addition, miRNA biogenesis protein machinery also seems to be essential for placental development and function (Cheloufi et al. 2010).

Expression of chromosome 19 miRNA cluster (C19MC) and miR-371-3 cluster changes throughout pregnancy (Morales Prieto and Markert 2011) and differs in placental tissue from females who undergo preterm labor compared to normal term pregnancy (Mayor-Lynn et al. 2011). These clusters are located within imprinted genes that are known to be involved with embryonic development and cellular differentiation (Tsai et al. 2009). Interestingly, C19MC is one of the largest miRNA gene clusters in the human genome (Bentwich et al. 2005; Lin et al. 2010) and is expressed from the paternally inherited chromosome and controlled by upstream promoter region methylation (Noguer-Dance et al. 2010). Regulation of C19MC is not well characterized; however, the expression appears to be restricted to the reproductive system and the placenta (Liang et al. 2007; Lin et al. 2010) though expression of miR-498, a component of the C19MC cluster, has been shown to be expressed in the fetal brain (Flor and Bullerdiek 2012). Interestingly, no homologues of this cluster have been found in rat, mouse, or dog (Zhang et al. 2008). The miR-371-3 cluster consists mainly of three miRNAs (miR-371a-3p, miR-372, and miR-373-3p) sharing the same seed sequence, AAGUGC (Griffiths-Jones 2006), and is predominantly placentally expressed (Bentwich et al. 2005).

3.5 miRNA Regulation During Spermatogenesis

Spermatogenesis is the process by which diploid spermatogonium differentiates into motile spermatozoa. The process begins after birth when spermatogonial stem cells enter the differentiation pathway. The stages of spermatogenesis vary both in location within the seminiferous tubule and by morphological changes to the germ cell as stem cell-like spermatogonia near the basal lamina are recruited and undergo several rounds of mitotic divisions, followed by progression through meiosis, morphogenesis, and eventual release of mature spermatids into the lumen of the seminiferous tubule (Kotaja et al. 2004). MiRNA expression and function throughout numerous stages of spermatogenesis has been observed by numerous investigators (Fig. 3.2).

3.5.1 miRNA Biogenesis During Spermatogenesis

The miRNA biogenesis and RISC-associated proteins Dicer, Drosha, AGO1, AGO2, AGO3, and AGO4 have all been detected in pachytene stage spermatocytes, round and elongated spermatids, and Sertoli cells (Gonzalez-Gonzalez et al. 2008). The importance of small RNA regulation within the maturing sperm cell has been exhibited by extensive rodent knockout models. Primordial germ cells as well as spermatogonia from Dicer-deficient mice exhibit poor proliferation (Hayashi et al. 2008). Furthermore selective knockout of Dicer driven by promoters for DEAD box polypeptide 4 (*Ddx4*) or tissue-nonspecific alkaline phosphatase (*Tnap*) at the early onset of spermatogenesis also leads to infertility implying the importance of small RNA production and function to enable spermatogenesis. There is an observable loss of sperm differentiation, abnormal morphology, and loss of motility with the absence of Dicer (Maatouk et al. 2008; Romero et al. 2011). Neurogenin 3 (*Ngn3*) promoter-driven conditional knockout of Dicer results in reduced testis size and disruption of

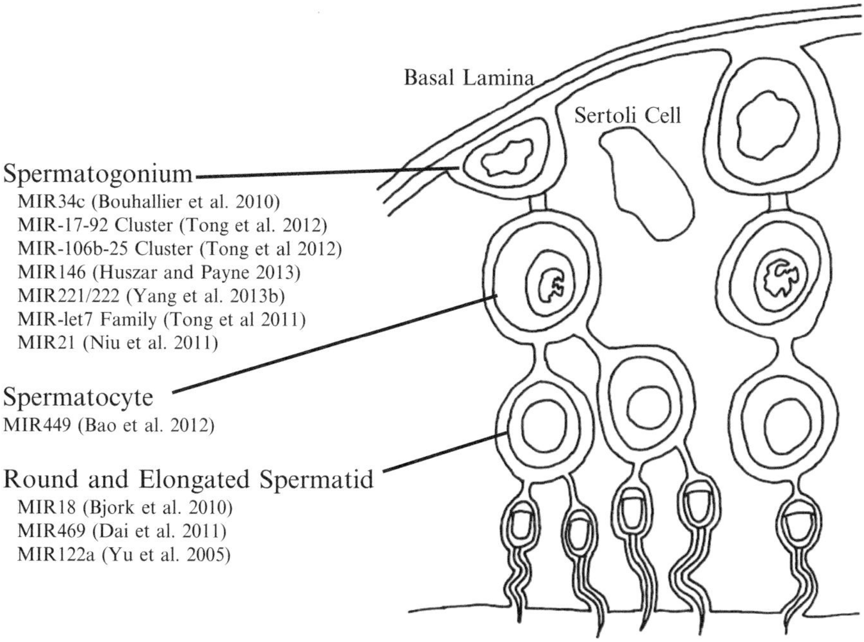

Fig. 3.2 Schematic of microRNA associated with different stages of spermatogenesis (Niu et al. 2011; Bao et al. 2012; Dai et al. 2000; Yu et al. 2005; Bjork et al. 2010)

spermatogenesis within the seminiferous tubules as well as the epididymis (Korhonen et al. 2011). Furthermore, selective knockout of Dicer in mice Sertoli cells driven by the *Mis* (anti- Müllerian hormone) promoter also leads to infertility due to complete absence of spermatozoa and progressive testicular degradation (Papaioannou et al. 2009). Other evidence that small RNA regulation is important in the maturing male gamete is the presence of chromatoid bodies. Chromatoid bodies are a unique cloud-like structure found within spermatids that are thought to contain miRNA, RISC-associated proteins, and potentially the mRNA targets of miRNA (Kotaja et al. 2006; Kotaja and Sassone-Corsi 2007; Meikar et al. 2011).

3.5.2 miRNA Expression Patterns During Spermatogenesis

MiRNAs are differentially expressed during testicular development (Yu et al. 2005). High-throughput sequencing experiments have detected 141 miRNAs expressing in the mouse testis and 29 novel miRNAs in the human testis (Ro et al. 2007), and another study detected 770 known and five novel miRNAs in the human testis (Yang et al. 2013a). Immunohistochemistry detected miRNAs in the Sertoli cell nucleus and in the dense body of the pachytene spermatocytes (Marcon et al. 2008).

Microarray experiments have shown dynamic changes in miRNA profiles when comparing immature and mature testis in mice, rhesus monkeys, and pigs (Luo et al. 2010; Yan et al. 2007, 2009).

When investigating specific miRNA or miRNA clusters, investigators have found changes in abundance of miRNA comparing stage of spermatogenesis (Fig. 3.2). There is high abundance of miR-34c in the spermatocyte and spermatid. There is also high abundance of miR-34c in the testis compared to the spleen, brain, or liver tissue (Bouhallier et al. 2010). When mouse embryonic stem cells (mESCs) are transfected with pri-miR-34c-GFP and combined with retinoic acid induction, miR-34c promotes mESCs to differentiate into spermatogenic-like cells (Zhang et al. 2012). However, in caprine germline stem cells, overexpression of miR-34c leads to apoptosis and suppressed proliferation (Li et al. 2013). The potential mechanisms of miR-34c during spermatogonial differentiation in mice are thought to be through targeting *Atf1* and inducing apoptosis (Liang et al. 2012) or by targeting *Nanos2* mRNA (Yu et al. 2014).

Both the miR-17–92 and miR-106b-25 clusters are downregulated during retinoic acid-induced spermatogonial differentiation both in vivo and in vitro (Tong et al. 2012). A mouse loss-of-function model for miR-17–92 cluster in male germ cells causes smaller testis, decreased number of epididymal sperm, and mild defects in spermatogenesis (Tong et al. 2012). The abundance of miR-146 is highly increased in mice undifferentiated spermatogonia (Huszar and Payne 2013), suggesting that miR-146 plays a role in differentiation. Impaired function of the X chromosome-clustered miR-221 and miR-222 in mouse undifferentiated spermatogonia induces loss of stem cell capacity to regenerate spermatogenesis and induces the transition from a KIT^- to a KIT^+ state (or expressing the CD117 receptor), a hallmark of the transition into differentiation. In accordance with this, growth factors that promote maintenance of undifferentiated spermatogonia increase miR-221/miR-222 abundance (Yang et al. 2013b). When comparing abundance of miR-135a in rat descended testis or undescended testis, abundance is greatest in descended testis. A specific example to PTGR occurring in the testis involves miR-135a and forkhead box protein O1 (FOXO1), both of which have been localized in spermatogonial stem cells. Transfection of miR-135a into spermatogonia in vitro resulted in decreased FOXO1 abundance (Moritoki et al. 2014). Another example of miRNA regulation during spermatogenesis is the significant increase in the expression of the miR-let7 family of miRNAs during mouse spermatogonial differentiation through suppression of *Lin28* (Tong et al. 2011).

3.5.3 miRNA Associated with Male Fertility

Infertility is estimated to affect 15 % of the couples worldwide, and male infertility is expected to be responsible for 50 % of this (Dohle et al. 2005; Hellani et al. 2006). Idiopathic, or spontaneously occurring, male infertility is accompanied by qualitative and quantitative abnormalities (Ferlin et al. 2007; Hargreave 2000). This has led research groups to be able to diagnose the molecular components and morphology of infertile men or species and compare miRNA abundance between the two. Abu-Halima et al. (Abu-Halima et al. 2013) compared the miRNA profiles in 27 diagnosed male patients: nine normozoospermia, nine asthenozoospermia, and nine oligoasthenozoospermia. When comparing the miRNA profiles of asthenozoospermia to normozoospermia patients through miRNA microarray, 50 miRNAs were upregulated and 27 were downregulated. In comparing oligoasthenozoospermia to normozoospermia, 42 miRNAs were upregulated and 44 were downregulated. Interestingly, miR-34c was downregulated in oligoasthenozoospermia compared to normozoospermia.

More recently Abu-Halima et al. (Abu-Halima et al. 2014) have suggested the use of five miRNAs, miR-34b*, miR-34b, miR-34c, miR-429, and miR-122, as potential biomarkers for the diagnosis and assessment of male infertility. In this study 80 semen samples from patients showing abnormal semen parameters were compared with 90 semen samples showing normal parameters. The abundance of the five miRNAs was compared using qRT-PCR, and miR-449 was increased and the other four were decreased in the abnormal semen groups compared to normal control subjects. Using support vector machine classification combined with these five miRNAs, the study was able to discriminate individuals with subfertility from control subjects with accuracy of 98.65 %, specificity of 98.83 %, and sensitivity of 98.44 % (Abu-Halima et al. 2014).

3.6 Small RNA and Female Reproductive Disorders

3.6.1 Relationship Between Obesity and Infertility Mediated by miRNA

Obesity has detrimental effects on female reproductive function, including increasing the likelihood for polycystic ovarian syndrome (PCOS), ovulation defects, reduced fecundity, and poor quality oocytes (Brewer and Balen 2010; Maheshwari et al. 2007; Rachon and Teede 2010). There is association between obesity and increased risk of birth defects, prematurity and stillbirths, and gestational diabetes (Bellver et al. 2007; Maheshwari et al. 2007). Obesity contributes to the development of type 2 diabetes, characterized by hyperglycemia and impaired insulin signaling (Akamine et al. 2010). In contrast to many peripheral tissues that display insulin resistance (Akamine et al. 2010; Kalra et al. 2006; Kashyap and Defronzo 2007), the ovary remains insulin sensitive during obesity-induced type 2 diabetes. Ovaries from female mice fed with a diet containing 60 % kcal of fat for 12 weeks maintained insulin sensitivity, despite that other classical tissues like the muscle and liver became insulin resistant (Wu et al. 2012).

The exact mechanism(s) by which obesity affects ovarian function remains poorly understood; however, systemic low-grade inflammation has been implicated in the development of infertility and other obesity-associated adverse reproductive health outcomes (Blencowe et al. 2012; Carmichael et al. 2010; Nestler 2000; Pasquali and Gambineri 2006; Pasquali et al. 2003; Pettigrew and Hamilton-Fairley 1997; Rubens et al. 2010). Chronic inflammation alters miRNA levels in immune cells (Fernandez-Valverde et al. 2011; Williams and Mitchell 2012; Xie et al. 2009), and miRNAs have been shown to regulate the activity of key cellular processes, including insulin release in pancreatic β cells, adipocyte differentiation (Williams and Mitchell 2012; Xie et al. 2009), and insulin sensitivity (Trajkovski et al. 2011), and are therefore potentially involved in obesity-induced infertility in females. We have demonstrated elevated tumor necrosis factor alpha (Tnfa) mRNA expression in the ovaries from obese mice (Nteeba et al. 2013a). miR-125b has been shown to negatively regulate *Tnfa* mRNA (Huang et al. 2012), while increased Tnfa itself downregulates miR-143 (Xie et al. 2009). Both miR-125b and miR-143 were decreased in ovarian tissue during obesity supporting that miRNAs may mediate physiological alterations involving an inflammatory mechanism (Nteeba et al. 2013a). Further, decreased circulating miR-125b has also been associated with increased fat mass in morbidly obese men (Ortega et al. 2013), while miR-143 is reported to be critical for the formation of the primordial follicle pool in utero (Zhang et al. 2013), raising concerns about the impact of obesity on the neonatal ovary.

Another mechanism by which miRNA can impact fertility may be through the phosphatidylinositol-3 kinases (PI3K) pathway. PI3K are lipid kinases that phosphorylate the 3′-OH group on the inositol ring of inositol phospholipids. Activation of PI3K results in conversion of the plasma membrane lipid phosphatidylinositol-4,5-bisphosphate (PIP_2) to phosphatidylinositol-3,4,5-triphosphate (PIP_3). PIP_3 can recruit proteins containing lipid-binding domains from the cytoplasm (Pawson and Nash 2000) such as the serine/threonine kinases 3′-phosphoinositide-dependent kinase-1 (PDK1) and AKT (Cantley 2002) to the plasma membrane where their proximity results in their phosphorylation. Once phosphorylated, AKT can translocate to the nucleus, where it regulates a number of cellular responses such as growth, survival, and cell cycle entry (Datta et al. 1999). PI3K signaling has many ovarian roles, being involved in steroidogenesis, oocyte viability (Brown et al. 2010), and primordial follicle activation (Jagarlamudi et al. 2009; Liu et al. 2006a), and can be regulated through PTGR via miRNA activity (Näär 2011; Trajkovski et al. 2011; Xu and Mo 2012; Yu et al. 2008). In order to evaluate if obesity has any effect on miRNA that could at least partially explain alterations observed with respect to PI3K signaling, levels of miR-103, miR-21, miR-184, and miR-205 in ovaries from obese female mice were investigated. Obese mice had decreased ovarian miR-21 and miR-103. MiR-21 expression is important for the regulation of apoptosis (Donadeu et al. 2012; McBride et al. 2012), and decreased miR-21 abundance has been reported to increase cell apoptosis in a variety of cell culture systems including the granulosa cells from mouse preovulatory follicles both in vivo and in vitro (Carletti et al. 2010). In addition, miR-21 has been identified as promoting follicular cell survival during ovulation, and miR-21 inhibition also has been reported to reduce ovulatory rates (Carletti et al. 2010). Although many different cell types undergo apoptosis in response to inhibition of miR-21 action, the miR-21 targets implicated vary widely for different cells, and the mechanism by which miR-21 suppresses apoptosis in granulosa cells remains to be identified (Kim et al. 2012; Sirotkin et al. 2010; Yang et al. 2008).

MiR-184 has been shown to act as a physiological suppressor of general secretory activity of progesterone and estradiol (Sirotkin et al. 2009, 2010). Obese females have increased levels of miR-184 (Nteeba et al. 2013b), and miR-184 is believed to play a critical role in development in addition to being a mediator of apoptosis. Upregulation of miR-184 has been reported to interfere with the ability of miR-205 to repress PI3K signaling (Yu et al. 2008). Several studies from human and rodent models have reported obesity-induced miR-103 upregulation (Perri et al. 2012; Rottiers and Näär 2012; Trajkovski et al. 2011). We observed a trend for decreased ovarian miR-103 due to obesity (Nteeba et al. 2013b). Importantly, silencing miR-103 has been proposed to improve insulin sensitivity in adipocytes mainly through increased caveolin-1 expression, which in turn leads to stabilization of the insulin receptor thus enhancing insulin signaling (Trajkovski et al. 2011). In contrast, miR-103 expression has been found to be downregulated in the mouse model of genetic insulin resistance and obesity (ob/ob mice) (Xie et al. 2009), consistent with our observations using a diet-induced obesity model.

3.6.2 Endometriosis

Endometriosis is a condition affecting approximately 10–15 % of females whereby cells of the endometrial lining of the uterus relocate via retrograde menstruation and localize to other regions outside of the endometrium. Recent studies investigating the miRNA profiles of endometriosis and associated tissues have shown that some miRNAs are associated with the induction and persistence of the disease. Transcriptional profiling has previously been used to identify numerous candidate genes that may contribute to the etiology of the disease (Kao et al. 2003). The large number of endometrial mRNA being altered in endometriosis may be due to miRNA function influencing the transcriptome.

Among the multitude of miRNAs associated with endometriosis, several appear to be promising markers in both diagnosis and progression of the disease. Zhao et al. demonstrated that miR-20a is overexpressed in patients with ovarian endometriosis, being greatest in patients in advanced stages of the disease (Zhao et al. 2014). Using small RNA sequencing of both endometriotic lesions and adjacent tissues, Saare et al. (2014) demonstrated that miR-499a, miR-200a, miR-200b, miR-141, and miR-34c were in significantly greater abundance in endometriotic lesions compared to neighboring healthy tissues (Saare et al. 2014). Using microarray analysis, Ohlsson Teague et al. (2009) were able to identify differential miRNA expression differences between ectopic and eutopic endometrial samples of women with endometriosis (Ohlsson Teague et al. 2009). The authors identified 14 miRNAs with elevated abundance and eight with reduced abundance in ectopic endometrium in comparison to eutopic endometrial samples. The miRNAs observed to be dysregulated in ectopic endometrium were predicted to interact with mRNA known to impact critical biological processes such as cell migration, invasion, and cell proliferation which are all putative mechanisms contributing to the endometriotic tissue formation (Ohlsson Teague et al. 2009).

3.6.3 Polycystic Ovarian Syndrome

Polycystic ovarian syndrome is an anovulatory disorder that is characterized by the pathological development of ovarian cysts. The underlying mechanisms resulting in PCOS have to this point been elusive; thus, multiple groups have worked towards the identification of miRNA associated with PCOS (Sorensen et al. 2014). Using microarray and quantitative PCR, five (let-7i-3 pm, miR-5706, miR-4463, miR-3665, miR-638) miRNAs were found in greater abundance in circulation of women with PCOS compared to healthy controls, while four (miR-124-3p, miR-128, miR-29a-3p, let-7c) were less abundant in women with PCOS (Ding et al. 2014). Others have taken similar approaches in the comparison of global miRNA expression profiling in women with PCOS compared to healthy controls. Sang et al. (2013) utilized deep sequencing to characterize miRNA content in microvesicles isolated from the follicular fluid of women with PCOS and identified miR-132 and miR-320 as being suppressed in the follicular fluid compared to control patients (Sang et al. 2013). In addition, the authors demonstrate the ability of miR-132 and miR-320 stimulation to increase estradiol production in a human granulosa-like tumor cell line, while inhibition of these miRNA had a suppressive effect on estradiol production.

Not only do aberrations in serum and follicular miRNA populations appear to be associated with women afflicted with PCOS, in some cases, differences are also observed in the developing embryo. Blastocysts derived from women afflicted with PCOS appear to have suppressed abundance of let-7a, miR-19a, miR-19b, miR-24, miR-92, and miR-93 (McCallie et al. 2010), suggesting that compromised embryos resulting from abnormalities during oocyte development may be a mechanism by which PCOS compromises fertility.

3.6.4 Chemically Induced Infertility Mediated Through miRNA

Exposure to environmental or occupational chemicals can disrupt female reproductive function (Mattison and Schulman 1980). Additionally, the reproductive age and status of an individual alter the susceptibility and the outcome following exposure to a reproductive toxicant (Fig. 3.3). A number of studies have shown that exposure to ovarian toxicants can lead to oocyte depletion [reviewed in Bhattacharya and

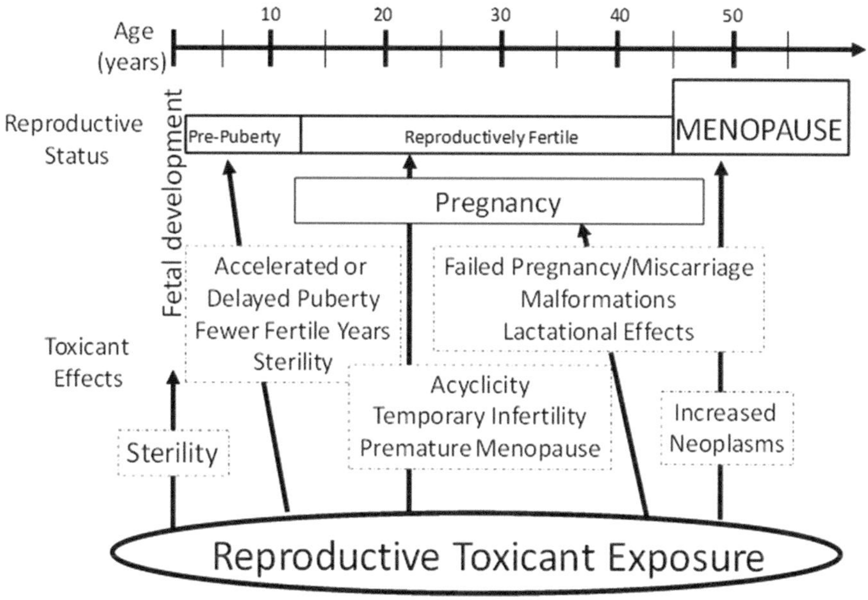

Fig. 3.3 Potential age-related effects of reproductive toxicants in females. The impact(s) of reproductive toxicants is partially dependent upon the reproductive status of the exposed individual. In most cases, direct ovarian toxicity can lead to premature ovarian failure and infertility (menopause). From Keating and Hoyer, 2009 with copyright permission

Keating (2012a) and Hoyer and Keating (2014)]. The stage of development at which the follicle is lost determines the reproductive impact; if large or antral follicles are depleted, temporary interruptions to reproductive function are observed since these follicles can be replaced by recruitment from the pool of primordial follicles (Hoyer and Keating 2014). Due to the irreplaceable nature of the ovarian reserve, chemicals that destroy oocytes contained in primordial follicles can lead to permanent infertility and premature ovarian failure (POF). Also, the level and duration of exposure to an environmental toxicant can influence the reproductive impact. Chronic, low-dose exposures, likely to be environmental in nature, are difficult to identify because their ovarian impact may go unrecognized for years. Ongoing selective damage of small preantral follicles may not initially raise concern until the onset of POF that will eventually result. Further, the age at which exposure occurs can impact the outcome. Prepubertal exposure may not cause the same extent of follicle loss as that postpubertal, due to the higher number of follicles present during childhood. However, damage to oocytes by chemical exposures in utero and/or during childhood presents a concern, which would not be detected until the reproductive years.

Ovotoxic chemicals can accelerate activation of primordial follicles from the ovarian reserve (Fernandez et al. 2008; Keating et al. 2009), leading to POF. Since this process is, at least partially, regulated by PI3K signaling, the involvement of miRNAs is plausible. It is known that miR-21 inhibits phosphatase and tensin homologue (PTEN), an antagonist of PI3K (Carletti et al. 2010; Ling et al. 2012). Also, increased abundance of miR-184 can interfere with AKT action, repressing PI3K action (Baley and Li 2012; Yu et al. 2008). Furthermore, miR-451 (Tian et al. 2011) and miR-7 (Fang et al. 2012) inhibit PI3K, while the miR-17–92 cluster (Ji et al. 2011) and miR-21 (Darido et al. 2011) activate PI3K.

It is important to note that the ovary has the capability to biotransform chemicals to more or less toxic metabolites, and these metabolic processes are highly active in ovarian tissues (Bhattacharya and Keating 2011, 2012b; Bhattacharya et al. 2012, 2013; Igawa et al. 2009; Keating et al. 2008a, b, 2010; Madden and Keating 2014) and contribute to the extent of ovotoxicity observed. We have demonstrated that many genes encoding ovarian chemical biotransformation enzymes are regulated by PI3K signaling (Bhattacharya and Keating 2012b; Bhattacharya et al. 2012), including aryl hydrocarbon receptor (*AhR*) and nuclear erythroid-related factor (*Nrf2*) (Bhattacharya and Keating 2012b), transcription factors that regulate xenobiotic metabolism. MiRNA-mediated regulation of Nrf2 (Eades et al. 2011; Yang et al. 2011) and Ahr (Huang et al. 2011) has recently been demonstrated. Thus, taken together, the potential for miRNA to have major roles in mediating chemical-induced ovotoxicity is being realized and is likely to be deciphered in greater detail in the coming years.

3.7 RNA-Binding Proteins Impact miRNA Function and Fertility

Despite their presence and activity in cells contributing to reproduction and development, the ability of ncRNAs to impact the cellular phenotype can be tempered by a variety of molecular regulators, and RNA-binding proteins, such as dead end homologue 1 (DND1), may be involved. The 3′ UTR of mRNA in close proximity to potential miRNA-binding sites is frequently the location of AU-rich elements (ARE) which contribute to mRNA stability (Chen and Shyu 1995; Fan et al. 1997). Since DND1 is capable of binding to regions where miRNA-mediated PTGR is conferred, it is thought that DND1 interaction with specific mRNA near miRNA-binding sites could potentially alter miRNA-induced PTGR (Kedde and Agami 2008; Kedde et al. 2007).

Germ cell viability for both male and female and early embryonic development requires DND1 function (Bhattacharya et al. 2007; Saga 2008). We have also demonstrated that DND1 abundance in the maturing pig oocyte and early embryo is greatest during the period of transcriptional quiescence following GVBD and prior ZGA (Yang et al. 2012b). The ability of DND1 to bind specific mRNA transcripts associated with pluripotency (POU5F1 (aka OCT4), SOX2, and LIN28) was demonstrated in the germinal vesicle stage oocyte (Yang et al. 2012b). Considering the numbers of miRNA present in the developing oocyte and the period of oocyte transcriptional quiescence following GVBD, it is possible that DND1 influences embryonic developmental competency by contributing to the mRNA and protein repertoire maintenance in the oocyte and early embryo. The biological roles of DND1 in addition to other RNA-binding proteins, such as fragile X mental retardation syndrome-related protein 1 (FXR1) which is also capable of mediating miRNA PTGR (Vasudevan and Steitz 2007; Vasudevan et al. 2007), will likely be continued areas of investigation in determining the biological ability of small RNA to influence reproductive function and fertility.

3.8 Conclusion

Computational and biochemical approaches have certainly improved the understanding of ncRNA regulation, and it is clear that ncRNA, in particular small RNA, impacts reproductive function. Advances in high-throughput sequencing technologies have rapidly amplified the discovery and characterization of multiple RNA classes in relation to specific biological functions of specific reproductive tissues and in relation to reproductive dysfunction and infertility. Together with knockout models deciphering the specific contributions of RNA classes in reproductive tissues, a more comprehensive understanding of small RNAs and their contribution to cellular function in reproductive tissues and fertility is being developed. Small RNAs, in particular endogenously produced miRNA, have distinct and critical biological roles contributing to fertility and reproductive success for both males and females. Additionally, RNA-binding proteins add biological complexity through their potential to also influence small RNA function in both a tissue-specific and target transcript-specific manner. As a result of the intricacies associated with small RNA biogenesis, the complexities of the interactions with specific transcripts, and to what degree those interactions result in PTGR based on what other contributing factors are present, bioinformatic predictions alone to characterize the role of small RNA in reproduction function are insufficient. Thus, there remains a dearth of information on how small RNA directly and indirectly influences fertility, and future studies are required to focus on determining the specific mechanistic function of particular small RNA in reproductive tissues.

References

Abd El Naby WS, Hagos TH, Hossain MM, Salilew-Wondim D, Gad AY, Rings F, Cinar MU, Tholen E, Looft C, Schellander K, Hoelker M, Tesfaye D (2013) Expression analysis of regulatory microRNAs in bovine cumulus oocyte complex and preimplantation embryos. Zygote 21:31–51

Aboobaker AA, Tomancak P, Patel N, Rubin GM, Lai EC (2005) Drosophila microRNAs exhibit diverse spatial expression patterns during embryonic development. Proc Natl Acad Sci USA 102(50):18017–18022

Abu-Halima M, Hammadeh M, Schmitt J, Leidinger P, Keller A, Meese E, Backes C (2013) Altered microRNA expression profiles of human spermatozoa in patients with different spermatogenic impairments. Fertil Steril 99(5):1249–1255

Abu-Halima M, Hammadeh M, Backes C, Fischer U, Leidinger P, Lubbad AM, Keller A, Meese E (2014) Panel of five microRNAs as potential biomarkers for the diagnosis and assessment of male infertility. Fertil Steril 102(4):989–997

Akamine EH, Marcal AC, Camporez JP, Hoshida MS, Caperuto LC, Bevilacqua E, Carvalho CR (2010) Obesity induced by high-fat diet promotes insulin resistance in the ovary. J Endocrinol 206(1):65–74

Amanai M, Brahmajosyula M, Perry AC (2006) A restricted role for sperm-borne microRNAs in mammalian fertilization. Biol Reprod 75(6):877–884

Ambros V, Chen X (2007) The regulation of genes and genomes by small RNAs. Development 134:1635–1641

Aravin A, Gaidatzis D, Pfeffer S, Lagos-Quintana M, Landgraf P, Iovino N, Morris P, Brownstein MJ, Kuramochi-Miyagawa S, Nakano T, Chien M, Russo JJ, Ju J, Sheridan R, Sander C, Zavolan M, Tuschl T (2006) A novel class of small RNAs bind to MILI protein in mouse testes. Nature 442:203–207

Aravin AA, Sachidanandam R, Girard A, Fejes-Toth K, Hannon GJ (2007) Developmentally regulated piRNA clusters implicate MILI in transposon control. Science 316:744–747

Babiarz JE, Blelloch R (2008) Small RNAs—their biogenesis, regulation and function in embryonic stem cells. StemBook. Harvard Stem Cell Institute, Cambridge, MA

Bagga S, Bracht J, Hunter S, Massirer K, Holtz J, Eachus R, Pasquinelli AE (2005) Regulation by let-7 and lin-4 miRNAs results in target mRNA degradation. Cell 122(4):553–563

Baley J, Li J (2012) MicroRNAs and ovarian function. J Ovarian Res 5:8

Bao J, Li D, Wang L, Wu J, Hu Y, Wang Z, Chen Y, Cao X, Jiang C, Yan W, Xu C (2012) MicroRNA-449 and microRNA-34b/c function redundantly in murine testes by targeting E2F transcription factor-retinoblastoma protein (E2F-pRb) pathway. J Biol Chem 287(26): 21686–21698

Barad O, Meiri E, Avniel A, Aharonov R, Barzilai A, Bentwich I, Einav U, Gilad S, Hurban P, Karov Y, Lobenhofer EK, Sharon E, Shiboleth YM, Shtutman M, Bentwich Z, Einat P (2004) MicroRNA expression detected by oligonucleotide microarrays: system establishment and expression profiling in human tissues. Genome Res 14(12):2486–2494

Bellver J, Melo MA, Bosch E, Serra V, Remohí J, Pellicer A (2007) Obesity and poor reproductive outcome: the potential role of the endometrium. Fertil Steril 88(2):446–451

Benetti R, Gonzalo S, Jaco I, Munoz P, Gonzalez S, Schoeftner S, Murchison E, Andl T, Chen T, Klatt P, Li E, Serrano M, Millar S, Hannon G, Blasco MA (2008) A mammalian microRNA cluster controls DNA methylation and telomere recombination via Rbl2-dependent regulation of DNA methyltransferases. Nat Struct Mol Biol 15(3):268–279

Bentwich I, Avniel A, Karov Y, Aharonov R, Gilad S, Barad O, Barzilai A, Einat P, Einav U, Meiri E, Sharon E, Spector Y, Bentwich Z (2005) Identification of hundreds of conserved and non-conserved human microRNAs. Nat Genet 37(7):766–770

Bernstein E, Kim SY, Carmell MA, Murchison EP, Alcorn H, Li MZ, Mills AA, Elledge SJ, Anderson KV, Hannon GJ (2003) Dicer is essential for mouse development. Nat Genet 35(3):215–217

Bhattacharya P, Keating AF (2011) Ovarian metabolism of xenobiotics. Exp Biol Med 236(7): 765–771

Bhattacharya P, Keating AF (2012a) Impact of environmental exposures on ovarian function and role of xenobiotic metabolism during ovotoxicity. Toxicol Appl Pharmacol 261:227–35

Bhattacharya P, Keating AF (2012b) Protective role for ovarian glutathione S-transferase isoform pi during 7,12-dimethylbenz[a]anthracene-induced ovotoxicity. Toxicol Appl Pharmacol 260(2):201–208

Bhattacharya C, Aggarwal S, Zhu R, Kumar M, Zhao M, Meistrich ML, Matin A (2007) The mouse dead-end gene isoform alpha is necessary for germ cell and embryonic viability. Biochem Biophys Res Commun 355(1):194–199

Bhattacharya P, Sen N, Hoyer PB, Keating AF (2012) Ovarian expressed microsomal epoxide hydrolase: Role in detoxification of 4-vinylcyclohexene diepoxide and regulation by phosphatidylinositol-3 kinase signaling. Toxicol Appl Pharmacol 258(1):118–123

Bhattacharya P, Madden JA, Sen N, Hoyer PB, Keating AF (2013) Glutathione S-transferase class mu regulation of apoptosis signal-regulating kinase 1 protein during VCD-induced ovotoxicity in neonatal rat ovaries. Toxicol Appl Pharmacol 267(1):49–56

Biemar F, Zinzen R, Ronshaugen M, Sementchenko V, Manak JR, Levine MS (2005) Spatial regulation of microRNA gene expression in the Drosophila embryo. Proc Natl Acad Sci USA 102(44):15907–15911

Bjork JK, Sandqvist A, Elsing AN, Kotaja N, Sistonen L (2010) miR-18, a member of Oncomir-1, targets heat shock transcription factor 2 in spermatogenesis. Development 137(19): 3177–3184

Blencowe H, Cousens S, Oestergaard MZ, Chou D, Moller AB, Narwal R, Adler A, Vera Garcia C, Rohde S, Say L, Lawn JE (2012) National, regional, and worldwide estimates of preterm birth rates in the year 2010 with time trends since 1990 for selected countries: a systematic analysis and implications. Lancet 379(9832):2162–2172

Bouhallier F, Allioli N, Lavial F, Chalmel F, Perrard MH, Durand P, Samarut J, Pain B, Rouault JP (2010) Role of miR-34c microRNA in the late steps of spermatogenesis. RNA 16(4):720–731

Brennecke J, Aravin AA, Stark A, Dus M, Kellis M, Sachidanandam R, Hannon GJ (2007) Discrete small RNA-generating loci as master regulators of transposon activity in Drosophila. Cell 128:1089–1103

Brevini TA, Cillo F, Antonini S, Tosetti V, Gandolfi F (2007) Temporal and spatial control of gene expression in early embryos of farm animals. Reprod Fertil Dev 19(1):35–42

Brewer CJ, Balen AH (2010) The adverse effects of obesity on conception and implantation. Reproduction 140(3):347–364

Brown C, LaRocca J, Pietruska J, Ota M, Anderson L, Smith SD, Weston P, Rasoulpour T, Hixon ML (2010) Subfertility caused by altered follicular development and oocyte growth in female mice lacking PKB alpha/Akt1. Biol Reprod 82(2):246–256

Cantley LC (2002) The phosphoinositide 3-kinase pathway. Science 296(5573):1655–1657

Carletti MZ, Fiedler SD, Christenson LK (2010) MicroRNA 21 blocks apoptosis in mouse periovulatory granulosa cells. Biol Reprod 83(2):286–295

Carmell MA, Girard A, van de Kant HJ, Bourc'his D, Bestor TH, de Rooij DG, Hannon GJ (2007) MIWI2 is essential for spermatogenesis and repression of transposons in the mouse male germline. Dev Cell 12:503–514

Carmichael SL, Rasmussen SA, Shaw GM (2010) Prepregnancy obesity: a complex risk factor for selected birth defects. Birth Defects Res A Clin Mol Teratol 88(10):804–810

Castellano L, Stebbing J (2013) Deep sequencing of small RNAs identifies canonical and non-canonical miRNA and endogenous siRNAs in mammalian somatic tissues. Nucleic Acids Res 41:3339–51

Cha J, Sun X, Dey SK (2012) Mechanisms of implantation: strategies for successful pregnancy. Nat Med 18(12):1754–1767

Chakrabarty A, Tranguch S, Daikoku T, Jensen K, Furneaux H, Dey SK (2007) MicroRNA regulation of cyclooxygenase-2 during embryo implantation. Proc Natl Acad Sci USA 104(38): 15144–15149

Cheloufi S, Dos Santos CO, Chong MM, Hannon GJ (2010) A dicer-independent miRNA biogenesis pathway that requires Ago catalysis. Nature 465(7298):584–589

Chen L, Daley GQ (2008) Molecular basis of pluripotency. Hum Mol Genet 17(R1):R23–27

Chen CY, Shyu AB (1995) AU-rich elements: characterization and importance in mRNA degradation. Trends Biochem Sci 20(11):465–470

Cook MS, Blelloch R (2013) Small RNAs in germline development. Curr Top Dev Biol 102: 159–205

Cox DN, Chao A, Baker J, Chang L, Qiao D, Lin H (1998) A novel class of evolutionarily conserved genes defined by piwi are essential for stem cell self-renewal. Genes Dev 12(23): 3715–3727

Dai Y, Lee C, Hutchings A, Sun Y, Moor R (2000) Selective requirement for Cdc25C protein synthesis during meiotic progression in porcine oocytes. Biol Reprod 62(3):519–532

Darido C, Georgy SR, Wilanowski T, Dworkin S, Auden A, Zhao Q, Rank G, Srivastava S, Finlay MJ, Papenfuss AT, Pandolfi PP, Pearson RB, Jane SM (2011) Targeting of the tumor suppressor GRHL3 by a miR-21-dependent proto-oncogenic network results in PTEN loss and tumorigenesis. Cancer Cell 20(5):635–648

Datta SR, Brunet A, Greenberg ME (1999) Cellular survival: a play in three Akts. Genes Dev 13(22):2905–2927

Deng W, Lin H (2002) miwi, a murine homolog of piwi, encodes a cytoplasmic protein essential for spermatogenesis. Dev Cell 2:819–830

Denli AM, Tops BB, Plasterk RH, Ketting RF, Hannon GJ (2004) Processing of primary microRNAs by the Microprocessor complex. Nature 432:231–235

Ding CF, Chen WQ, Zhu YT, Bo YL, Hu HM, Zheng RH (2014) Circulating microRNAs in patients with polycystic ovary syndrome. Hum Fertil 18:22–9

Dohle GR, Colpi GM, Hargreave TB, Papp GK, Jungwirth A, Weidner W (2005) EAU guidelines on male infertility. Eur Urol 48(5):703–711

Donadeu FX, Schauer S, Sontakke S (2012) Involvement of miRNAs in ovarian follicular and luteal development. J Endocrinol 215:323–34

Eades G, Yang M, Yao Y, Zhang Y, Zhou Q (2011) miR-200a regulates Nrf2 activation by targeting Keap1 mRNA in breast cancer cells. J Biol Chem 286(47):40725–40733

Fabian MR, Sonenberg N, Filipowicz W (2010) Regulation of mRNA translation and stability by microRNAs. Annu Rev Biochem 79:351–379

Fan XC, Myer VE, Steitz JA (1997) AU-rich elements target small nuclear RNAs as well as mRNAs for rapid degradation. Genes Dev 11(19):2557–2568

Fang YX, Xue JL, Shen Q, Chen J, Tian L (2012) miR-7 inhibits tumor growth and metastasis by targeting the PI3K/AKT pathway in hepatocellular carcinoma. Hepatology 55:1852–62

Ferlin A, Arredi B, Speltra E, Cazzadore C, Selice R, Garolla A, Lenzi A, Foresta C (2007) Molecular and clinical characterization of Y chromosome microdeletions in infertile men: a 10-year experience in Italy. J Clin Endocrinol Metab 92(3):762–770

Fernandez SM, Keating AF, Christian PJ, Sen N, Hoying JB, Brooks HL, Hoyer PB (2008) Involvement of the KIT/KITL signaling pathway in 4-vinylcyclohexene diepoxide-induced ovarian follicle loss in rats. Biol Reprod 79(2):318–327

Fernandez-Valverde SL, Taft RJ, Mattick JS (2011) MicroRNAs in beta-cell biology, insulin resistance, diabetes and its complications. Diabetes 60(7):1825–1831

Filipowicz W (2005) RNAi: the nuts and bolts of the RISC machine. Cell 122:17–20

Flor I, Bullerdiek J (2012) The dark side of a success story: microRNAs of the C19MC cluster in human. J Pathol 227(3):270–274

Frank F, Sonenberg N, Nagar B (2010) Structural basis for 5′-nucleotide base-specific recognition of guide RNA by human AGO2. Nature 465(7299):818–822

Ghildiyal M, Seitz H, Horwich MD, Li C, Du T, Lee S, Xu J, Kittler EL, Zapp ML, Weng Z, Zamore PD (2008) Endogenous siRNAs derived from transposons and mRNAs in Drosophila somatic cells. Science 320:1077–1081

Giraldez AJ, Mishima Y, Rihel J, Grocock RJ, Van Dongen S, Inoue K, Enright AJ, Schier AF (2006) Zebrafish MiR-430 promotes deadenylation and clearance of maternal mRNAs. Science 312(5770):75–79

Girard A, Sachidanandam R, Hannon GJ, Carmell MA (2006) A germline-specific class of small RNAs binds mammalian Piwi proteins. Nature 442(7099):199–202

Golden DE, Gerbasi VR, Sontheimer EJ (2008) An inside job for siRNAs. Mol Cell 31:309–312

Gonzalez-Gonzalez E, Lopez-Casas PP, del Mazo J (2008) The expression patterns of genes involved in the RNAi pathways are tissue-dependent and differ in the germ and somatic cells of mouse testis. Biochim Biophys Acta 1779(5):306–311

Gregory RI, Yan KP, Amuthan G, Chendrimada T, Doratotaj B, Cooch N, Shiekhattar R (2004) The microprocessor complex mediates the genesis of microRNAs. Nature 432:235–240

Griffiths-Jones S (2006) miRBase: the microRNA sequence database. Methods Mol Biol 342: 129–138

Grivna ST, Beyret E, Wang Z, Lin H (2006) A novel class of small RNAs in mouse spermatogenic cells. Genes Dev 20(13):1709–1714

Gu P, Reid JG, Gao X, Shaw CA, Creighton C, Tran PL, Zhou X, Drabek RB, Steffen DL, Hoang DM, Weiss MK, Naghavi AO, El-daye J, Khan MF, Legge GB, Wheeler DA, Gibbs RA, Miller JN, Cooney AJ, Gunaratne PH (2008) Novel microRNA candidates and miRNA-mRNA pairs in embryonic stem (ES) cells. PLoS One 3(7), e2548

Gu A, Ji G, Shi X, Long Y, Xia Y, Song L, Wang S, Wang X (2010) Genetic variants in Piwi-interacting RNA pathway genes confer susceptibility to spermatogenic failure in a Chinese population. Hum Reprod 25(12):2955–2961

Gunawardane LS, Saito K, Nishida KM, Miyoshi K, Kawamura Y, Nagami T, Siomi H, Siomi MC (2007) A slicer-mediated mechanism for repeat-associated siRNA 5′ end formation in Drosophila. Science 315:1587–1590

Han J, Lee Y, Yeom KH, Kim YK, Jin H, Kim VN (2004) The Drosha-DGCR8 complex in primary microRNA processing. Genes Dev 18:3016–3027

Hargreave TB (2000) Genetic basis of male fertility. Br Med Bull 56(3):650–671

Hawkins SM, Andreu-Vieyra CV, Kim TH, Jeong JW, Hodgson MC, Chen R, Creighton CJ, Lydon JP, Gunaratne PH, DeMayo FJ, Matzuk MM (2012) Dysregulation of uterine signaling pathways in progesterone receptor-Cre knockout of dicer. Mol Endocrinol 26(9):1552–1566

Hayashi K, de Sousa C, Lopes SM, Kaneda M, Tang F, Hajkova P, Lao K, O'Carroll D, Das PP, Tarakhovsky A, Miska EA, Surani MA (2008) MicroRNA biogenesis is required for mouse primordial germ cell development and spermatogenesis. PLoS One 3(3), e1738

Hellani A, Al-Hassan S, Iqbal MA, Coskun S (2006) Y chromosome microdeletions in infertile men with idiopathic oligo- or azoospermia. J Exp Clin Assist Reprod 3:1

Hemberger M, Dean W, Reik W (2009) Epigenetic dynamics of stem cells and cell lineage commitment: digging Waddington's canal. Nat Rev Mol Cell Biol 10:526–537

Houbaviy HB, Murray MF, Sharp PA (2003) Embryonic stem cell-specific MicroRNAs. Dev Cell 5(2):351–358

Houwing S, Kamminga LM, Berezikov E, Cronembold D, Girard A, van den Elst H, Filippov DV, Blaser H, Raz E, Moens CB, Plasterk RH, Hannon GJ, Draper BW, Ketting RF (2007) A role for Piwi and piRNAs in germ cell maintenance and transposon silencing in Zebrafish. Cell 129:69–82

Houwing S, Berezikov E, Ketting RF (2008) Zili is required for germ cell differentiation and meiosis in zebrafish. EMBO J 27(20):2702–2711

Hoyer PB, Keating AF (2014) Xenobiotic effects in the ovary: temporary versus permanent infertility. Expert Opin Drug Metab Toxicol 10(4):511–523

Huang TC, Chang HY, Chen CY, Wu PY, Lee H, Liao YF, Hsu WM, Huang HC, Juan HF (2011) Silencing of miR-124 induces neuroblastoma SK-N-SH cell differentiation, cell cycle arrest and apoptosis through promoting AHR. FEBS Lett 585(22):3582–3586

Huang HC, Yu HR, Huang LT, Huang HC, Chen RF, Lin IC, Ou CY, Hsu TY, Yang KD (2012) miRNA-125b regulates TNF-alpha production in CD14+ neonatal monocytes via post-transcriptional regulation. J Leukoc Biol 92(1):171–182

Huszar JM, Payne CJ (2013) MicroRNA 146 (Mir146) modulates spermatogonial differentiation by retinoic acid in mice. Biol Reprod 88(1):15

Hutvagner G, Zamore PD (2002) A microRNA in a multiple-turnover RNAi enzyme complex. Science 297:2056–2060

Igawa Y, Keating AF, Rajapaksa KS, Sipes IG, Hoyer PB (2009) Evaluation of ovotoxicity induced by 7, 12-dimethylbenz[a]anthracene and its 3,4-diol metabolite utilizing a rat in vitro ovarian culture system. Toxicol Appl Pharmacol 234(3):361–369

Ishizu H, Siomi H, Siomi MC (2012) Biology of PIWI-interacting RNAs: new insights into biogenesis and function inside and outside of germlines. Genes Dev 26(21):2361–2373

Jagarlamudi K, Liu L, Adhikari D, Reddy P, Idahl A, Ottander U, Lundin E, Liu K (2009) Oocyte-specific deletion of Pten in mice reveals a stage-specific function of PTEN/PI3K signaling in oocytes in controlling follicular activation. PLoS One 4(7), e6186

Ji M, Rao E, Ramachandrareddy H, Shen Y, Jiang C, Chen J, Hu Y, Rizzino A, Chan WC, Fu K, McKeithan TW (2011) The miR-17-92 microRNA cluster is regulated by multiple mechanisms in B-cell malignancies. Am J Pathol 179(4):1645–1656

Jinek M, Doudna JA (2009) A three-dimensional view of the molecular machinery of RNA interference. Nature 457(7228):405–412

Juliano C, Wang J, Lin H (2011) Uniting germline and stem cells: the function of Piwi proteins and the piRNA pathway in diverse organisms. Annu Rev Genet 45:447–469

Kalra A, Nair S, Rai L (2006) Association of obesity and insulin resistance with dyslipidemia in Indian women with polycystic ovarian syndrome. Indian J Med Sci 60(11):447–453

Kaneda M, Tang F, O'Carroll D, Lao K, Surani MA (2009) Essential role for Argonaute2 protein in mouse oogenesis. Epigenetics Chromatin 2:9

Kanellopoulou C, Muljo SA, Kung AL, Ganesan S, Drapkin R, Jenuwein T, Livingston DM, Rajewsky K (2005) Dicer-deficient mouse embryonic stem cells are defective in differentiation and centromeric silencing. Genes Dev 19:489–501

Kao LC, Germeyer A, Tulac S, Lobo S, Yang JP, Taylor RN, Osteen K, Lessey BA, Giudice LC (2003) Expression profiling of endometrium from women with endometriosis reveals candidate genes for disease-based implantation failure and infertility. Endocrinology 144(7):2870–2881

Kashyap SR, Defronzo RA (2007) The insulin resistance syndrome: physiological considerations. Diab Vasc Dis Res 4(1):13–19

Keating AF, Rajapaksa KS, Sipes IG, Hoyer PB (2008a) Effect of CYP2E1 gene deletion in mice on expression of microsomal epoxide hydrolase in response to VCD exposure. Toxicol Sci 105(2):351–359

Keating AF, Sipes IG, Hoyer PB (2008b) Expression of ovarian microsomal epoxide hydrolase and glutathione S-transferase during onset of VCD-induced ovotoxicity in B6C3F(1) mice. Toxicol Appl Pharmacol 230(1):109–116

Keating AF, Mark CJ, Sen N, Sipes IG, Hoyer PB (2009) Effect of phosphatidylinositol-3 kinase inhibition on ovotoxicity caused by 4-vinylcyclohexene diepoxide and 7, 12-dimethylbenz[a] anthracene in neonatal rat ovaries. Toxicol Appl Pharmacol 241(2):127–134

Keating AF, Sen N, Sipes IG, Hoyer PB (2010) Dual protective role for glutathione S-transferase class pi against VCD-induced ovotoxicity in the rat ovary. Toxicol Appl Pharmacol 247(2): 71–75

Kedde M, Agami R (2008) Interplay between microRNAs and RNA-binding proteins determines developmental processes. Cell Cycle 7(7):899–903

Kedde M, Strasser MJ, Boldajipour B, Oude Vrielink JA, Slanchev K, le Sage C, Nagel R, Voorhoeve PM, van Duijse J, Orom UA, Lund AH, Perrakis A, Raz E, Agami R (2007) RNA-binding protein Dnd1 inhibits microRNA access to target mRNA. Cell 131(7):1273–1286

Khurana JS, Xu J, Weng Z, Theurkauf WE (2010) Distinct functions for the Drosophila piRNA pathway in genome maintenance and telomere protection. PLoS Genet 6(12), e1001246

Kim VN, Han J, Siomi MC (2009) Biogenesis of small RNAs in animals. Nat Rev Mol Cell Biol 10:126–139

Kim YJ, Hwang SH, Cho HH, Shin KK, Bae YC, Jung JS (2012) MicroRNA 21 regulates the proliferation of human adipose tissue-derived mesenchymal stem cells and high-fat diet-induced obesity alters microRNA 21 expression in white adipose tissues. J Cell Physiol 227(1):183–193

Klattenhoff C, Theurkauf W (2008) Biogenesis and germline functions of piRNAs. Development 135:3–9

Klenov MS, Lavrov SA, Stolyarenko AD, Ryazansky SS, Aravin AA, Tuschl T, Gvozdev VA (2007) Repeat-associated siRNAs cause chromatin silencing of retrotransposons in the Drosophila melanogaster germline. Nucleic Acids Res 35(16):5430–5438

Korhonen HM, Meikar O, Yadav RP, Papaioannou MD, Romero Y, Da Ros M, Herrera PL, Toppari J, Nef S, Kotaja N (2011) Dicer is required for haploid male germ cell differentiation in mice. PLoS One 6(9), e24821

Kotaja N, Sassone-Corsi P (2007) The chromatoid body: a germ-cell-specific RNA-processing centre. Nat Rev Mol Cell Biol 8(1):85–90

Kotaja N, Kimmins S, Brancorsini S, Hentsch D, Vonesch JL, Davidson I, Parvinen M, Sassone-Corsi P (2004) Preparation, isolation and characterization of stage-specific spermatogenic cells for cellular and molecular analysis. Nat Methods 1(3):249–254

Kotaja N, Bhattacharyya SN, Jaskiewicz L, Kimmins S, Parvinen M, Filipowicz W, Sassone-Corsi P (2006) The chromatoid body of male germ cells: similarity with processing bodies and presence of Dicer and microRNA pathway components. Proc Natl Acad Sci USA 103(8):2647–2652
Kowalczykiewicz D, Pawlak P, Lechniak D, Wrzesinski J (2012) Altered expression of porcine Piwi genes and piRNA during development. PLoS One 7(8), e43816
Krawetz SA, Kruger A, Lalancette C, Tagett R, Anton E, Draghici S, Diamond MP (2011) A survey of small RNAs in human sperm. Hum Reprod 26(12):3401–3412
Kuramochi-Miyagawa S, Watanabe T, Gotoh K, Totoki Y, Toyoda A, Ikawa M, Asada N, Kojima K, Yamaguchi Y, Ijiri TW, Hata K, Li E, Matsuda Y, Kimura T, Okabe M, Sakaki Y, Sasaki H, Nakano T (2008) DNA methylation of retrotransposon genes is regulated by Piwi family members MILI and MIWI2 in murine fetal testes. Genes Dev 22(7):908–917
Lakshmipathy U, Love B, Goff LA, Jornsten R, Graichen R, Hart RP, Chesnut JD (2007) MicroRNA expression pattern of undifferentiated and differentiated human embryonic stem cells. Stem Cells Dev 16(6):1003–1016
Landgraf P, Rusu M, Sheridan R, Sewer A, Iovino N, Aravin A, Pfeffer S, Rice A, Kamphorst AO, Landthaler M, Lin C, Socci ND, Hermida L, Fulci V, Chiaretti S, Foa R, Schliwka J, Fuchs U, Novosel A, Muller RU, Schermer B, Bissels U, Inman J, Phan Q, Chien M, Weir DB, Choksi R, De Vita G, Frezzetti D, Trompeter HI, Hornung V, Teng G, Hartmann G, Palkovits M, Di Lauro R, Wernet P, Macino G, Rogler CE, Nagle JW, Ju J, Papavasiliou FN, Benzing T, Lichter P, Tam W, Brownstein MJ, Bosio A, Borkhardt A, Russo JJ, Sander C, Zavolan M, Tuschl T (2007) A mammalian microRNA expression atlas based on small RNA library sequencing. Cell 129(7):1401–1414
Lau NC, Lim LP, Weinstein EG, Bartel DP (2001) An abundant class of tiny RNAs with probable regulatory roles in Caenorhabditis elegans. Science 294:858–862
Lee Y, Jeon K, Lee JT, Kim S, Kim VN (2002) MicroRNA maturation: stepwise processing and subcellular localization. EMBO J 21(17):4663–4670
Lee Y, Kim M, Han J, Yeom KH, Lee S, Baek SH, Kim VN (2004) MicroRNA genes are transcribed by RNA polymerase II. EMBO J 23:4051–4060
Lee JH, Schutte D, Wulf G, Fuzesi L, Radzun HJ, Schweyer S, Engel W, Nayernia K (2006) Stem-cell protein Piwil2 is widely expressed in tumors and inhibits apoptosis through activation of Stat3/Bcl-XL pathway. Hum Mol Genet 15(2):201–211
Lee JH, Jung C, Javadian-Elyaderani P, Schweyer S, Schutte D, Shoukier M, Karimi-Busheri F, Weinfeld M, Rasouli-Nia A, Hengstler JG, Mantilla A, Soleimanpour-Lichaei HR, Engel W, Robson CN, Nayernia K (2010) Pathways of proliferation and antiapoptosis driven in breast cancer stem cells by stem cell protein piwil2. Cancer Res 70(11):4569–4579
Li C, Vagin VV, Lee S, Xu J, Ma S, Xi H, Seitz H, Horwich MD, Syrzycka M, Honda BM, Kittler EL, Zapp ML, Klattenhoff C, Schulz N, Theurkauf WE, Weng Z, Zamore PD (2009) Collapse of germline piRNAs in the absence of Argonaute3 reveals somatic piRNAs in flies. Cell 137(3):509–521
Li M, Yu M, Liu C, Zhu H, He X, Peng S, Hua J (2013) miR-34c works downstream of p53 leading to dairy goat male germline stem-cell (mGSCs) apoptosis. Cell Prolif 46(2):223–231
Liang Y, Ridzon D, Wong L, Chen C (2007) Characterization of microRNA expression profiles in normal human tissues. BMC Genomics 8:166
Liang X, Zhou D, Wei C, Luo H, Liu J, Fu R, Cui S (2012) MicroRNA-34c enhances murine male germ cell apoptosis through targeting ATF1. PLoS One 7(3), e33861
Lim AK, Kai T (2007) Unique germ-line organelle, nuage, functions to repress selfish genetic elements in Drosophila melanogaster. Proc Natl Acad Sci USA 104(16):6714–6719
Lin H, Yin H (2008) A novel epigenetic mechanism in Drosophila somatic cells mediated by Piwi and piRNAs. Cold Spring Harb Symp Quant Biol 73:273–281
Lin S, Cheung WK, Chen S, Lu G, Wang Z, Xie D, Li K, Lin MC, Kung HF (2010) Computational identification and characterization of primate-specific microRNAs. Comput Biol Chem 34(4):232–241
Ling HY, Hu B, Hu XB, Zhong J, Feng SD, Qin L, Liu G, Wen GB, Liao DF (2012) MiRNA-21 reverses high glucose and high insulin induced insulin resistance in 3 T3-L1 adipocytes through targeting phosphatase and tensin homologue. Exp Clin Endocrinol Diabetes 120(9):553–559

Liu K, Rajareddy S, Liu L, Jagarlamudi K, Boman K, Selstam G, Reddy P (2006a) Control of mammalian oocyte growth and early follicular development by the oocyte PI3 kinase pathway: new roles for an old timer. Dev Biol 299(1):1–11
Liu X, Sun Y, Guo J, Ma H, Li J, Dong B, Jin G, Zhang J, Wu J, Meng L, Shou C (2006b) Expression of hiwi gene in human gastric cancer was associated with proliferation of cancer cells. Int J Cancer 118(8):1922–1929
Liu JJ, Shen R, Chen L, Ye Y, He G, Hua K, Jarjoura D, Nakano T, Ramesh GK, Shapiro CL, Barsky SH, Gao JX (2010) Piwil2 is expressed in various stages of breast cancers and has the potential to be used as a novel biomarker. Int J Clin Exp Pathol 3(4):328–337
Liu WM, Pang RT, Chiu PC, Wong BP, Lao K, Lee KF, Yeung WS (2012) Sperm-borne microRNA-34c is required for the first cleavage division in mouse. Proc Natl Acad Sci USA 109(2): 490–494
Lund E, Guttinger S, Calado A, Dahlberg JE, Kutay U (2004) Nuclear export of microRNA precursors. Science 303:95–98
Luo L, Ye L, Liu G, Shao G, Zheng R, Ren Z, Zuo B, Xu D, Lei M, Jiang S, Deng C, Xiong Y, Li F (2010) Microarray-based approach identifies differentially expressed microRNAs in porcine sexually immature and mature testes. PLoS One 5(8), e11744
Ma J, Flemr M, Stein P, Berninger P, Malik R, Zavolan M, Svoboda P, Schultz RM (2010) MicroRNA activity is suppressed in mouse oocytes. Curr Biol 20(3):265–270
Maatouk DM, Loveland KL, McManus MT, Moore K, Harfe BD (2008) Dicer1 is required for differentiation of the mouse male germline. Biol Reprod 79(4):696–703
Madden JA, Keating AF (2014) Ovarian xenobiotic biotransformation enzymes are altered during phosphoramide mustard-induced ovotoxicity. Toxicol Sci 41:441–52
Maheshwari A, Stofberg L, Bhattacharya S (2007) Effect of overweight and obesity on assisted reproductive technology–a systematic review. Hum Reprod Update 13(5):433–444
Marcon E, Babak T, Chua G, Hughes T, Moens PB (2008) miRNA and piRNA localization in the male mammalian meiotic nucleus. Chromosome Res 16(2):243–260
Marson A, Levine SS, Cole MF, Frampton GM, Brambrink T, Johnstone S, Guenther MG, Johnston WK, Wernig M, Newman J, Calabrese JM, Dennis LM, Volkert TL, Gupta S, Love J, Hannett N, Sharp PA, Bartel DP, Jaenisch R, Young RA (2008) Connecting microRNA genes to the core transcriptional regulatory circuitry of embryonic stem cells. Cell 134(3):521–533
Martinez J, Patkaniowska A, Urlaub H, Luhrmann R, Tuschl T (2002) Single-stranded antisense siRNAs guide target RNA cleavage in RNAi. Cell 110:563–574
Mattison DR, Schulman JD (1980) How xenobiotic chemicals can destroy oocytes. Am J Ind Med 4:157–169
Mayor-Lynn K, Toloubeydokhti T, Cruz AC, Chegini N (2011) Expression profile of microRNAs and mRNAs in human placentas from pregnancies complicated by preeclampsia and preterm labor. Reprod Sci 18(1):46–56
McBride D, Carré W, Sontakke SD, Hogg CO, Law A, Donadeu FX, Clinton M (2012) Identification of miRNAs associated with the follicular-luteal transition in the ruminant ovary. Reproduction 144(2):221–233
McCallie B, Schoolcraft WB, Katz-Jaffe MG (2010) Aberration of blastocyst microRNA expression is associated with human infertility. Fertil Steril 93(7):2374–2382
Meikar O, Da Ros M, Korhonen H, Kotaja N (2011) Chromatoid body and small RNAs in male germ cells. Reproduction 142(2):195–209
Melton C, Judson RL, Blelloch R (2010) Opposing microRNA families regulate self-renewal in mouse embryonic stem cells. Nature 463(7281):621–626
Meng F, Henson R, Wehbe-Janek H, Smith H, Ueno Y, Patel T (2007) The MicroRNA let-7a modulates interleukin-6-dependent STAT-3 survival signaling. J Biol Chem 282(11): 8256–8264
Meng F, Wehbe-Janek H, Henson R, Smith H, Patel T (2008) Epigenetic regulation of microRNA-370 by interleukin-6 in malignant human. Oncogene 27(3):378–386
Miyoshi K, Miyoshi T, Siomi H (2010) Many ways to generate microRNA-like small RNAs: non-canonical pathways for microRNA production. Mol Genet Genomics 284(2):95–103

Morales Prieto DM, Markert UR (2011) MicroRNAs in pregnancy. J Reprod Immunol 88(2): 106–111
Morin RD, O'Connor MD, Griffith M, Kuchenbauer F, Delaney A, Prabhu AL, Zhao Y, McDonald H, Zeng T, Hirst M, Eaves CJ, Marra MA (2008) Application of massively parallel sequencing to microRNA profiling and discovery in human embryonic stem cells. Genome Res 18(4): 610–621
Moritoki Y, Hayashi Y, Mizuno K, Kamisawa H, Nishio H, Kurokawa S, Ugawa S, Kojima Y, Kohri K (2014) Expression profiling of microRNA in cryptorchid testes: miR-135a contributes to the maintenance of spermatogonial stem cells by regulating FoxO1. J Urol 191(4):1174–1180
Murchison EP, Partridge JF, Tam OH, Cheloufi S, Hannon GJ (2005) Characterization of Dicer-deficient murine embryonic stem cells. Proc Natl Acad Sci USA 102:12135–12140
Murchison EP, Stein P, Xuan Z, Pan H, Zhang MQ, Schultz RM, Hannon GJ (2007) Critical roles for dicer in the female germline. Genes Dev 21:682–693
Näär AM (2011) MiRs with a sweet tooth. Cell Metab 14(2):149–150
Nagaraja AK, Andreu-Vieyra C, Franco HL, Ma L, Chen R, Han DY, Zhu H, Agno JE, Gunaratne PH, DeMayo FJ, Matzuk MM (2008) Deletion of dicer in somatic cells of the female reproductive tract causes sterility. Mol Endocrinol 22(10):2336–2352
Nestler JE (2000) Obesity, insulin, sex steroids and ovulation. Int J Obes Relat Metab Disord 24(Suppl 2):S71–73
Niu Z, Goodyear SM, Rao S, Wu X, Tobias JW, Avarbock MR, Brinster RL (2011) MicroRNA-21 regulates the self-renewal of mouse spermatogonial stem cells. Proc Natl Acad Sci USA 108(31):12740–12745
Noguer-Dance M, Abu-Amero S, Al-Khtib M, Lefevre A, Coullin P, Moore GE, Cavaille J (2010) The primate-specific microRNA gene cluster (C19MC) is imprinted in the placenta. Hum Mol Genet 19(18):3566–3582
Nteeba J, Ortinau LC, Perfield JW 2nd, Keating AF (2013a) Diet-induced obesity alters immune cell infiltration and expression of inflammatory cytokine genes in mouse ovarian and peri-ovarian adipose depot tissues. Mol Reprod Dev 80(11):948–958
Nteeba J, Ross JW, Perfield JW 2nd, Keating AF (2013b) High fat diet induced obesity alters ovarian phosphatidylinositol-3 kinase signaling gene expression. Reprod Toxicol 42:68–77
O'Connell RM, Taganov KD, Boldin MP, Cheng G, Baltimore D (2007) MicroRNA-155 is induced during the macrophage inflammatory response. Proc Natl Acad Sci USA 104(5): 1604–1609
Ohlsson Teague EM, Van der Hoek KH, Van der Hoek MB, Perry N, Wagaarachchi P, Robertson SA, Print CG, Hull LM (2009) MicroRNA-regulated pathways associated with endometriosis. Mol Endocrinol 23(2):265–275
Okamura K, Balla S, Martin R, Liu N, Lai EC (2008a) Two distinct mechanisms generate endogenous siRNAs from bidirectional transcription in Drosophila melanogaster. Nat Struct Mol Biol 15:581–590
Okamura K, Chung WJ, Ruby JG, Guo H, Bartel DP, Lai EC (2008b) The Drosophila hairpin RNA pathway generates endogenous short interfering RNAs. Nature 453:803–806
Ortega FJ, Mercader JM, Catalan V, Moreno-Navarrete JM, Pueyo N, Sabater M, Gomez-Ambrosi J, Anglada R, Fernandez-Formoso JA, Ricart W, Fruhbeck G, Fernandez-Real JM (2013) Targeting the circulating microRNA signature of obesity. Clin Chem 59(5):781–792
Pan Q, Chegini N (2008) MicroRNA signature and regulatory functions in the endometrium during normal and disease states. Semin Reprod Med 26(6):479–493
Papaioannou MD, Pitetti JL, Ro S, Park C, Aubry F, Schaad O, Vejnar CE, Kuhne F, Descombes P, Zdobnov EM, McManus MT, Guillou F, Harfe BD, Yan W, Jegou B, Nef S (2009) Sertoli cell dicer is essential for spermatogenesis in mice. Dev Biol 326(1):250–259
Pasquali R, Gambineri A (2006) Metabolic effects of obesity on reproduction. Reprod Biomed Online 12(5):542–551
Pasquali R, Pelusi C, Genghini S, Cacciari M, Gambineri A (2003) Obesity and reproductive disorders in women. Hum Reprod Update 9(4):359–372

Pawson T, Nash P (2000) Protein–protein interactions define specificity in signal transduction. Genes Dev 14(9):1027–1047
Perri R, Nares S, Zhang S, Barros SP, Offenbacher S (2012) MicroRNA modulation in obesity and periodontitis. J Dent Res 91(1):33–38
Pettigrew R, Hamilton-Fairley D (1997) Obesity and female reproductive function. Br Med Bull 53(2):341–358
Qiao D, Zeeman AM, Deng W, Looijenga LH, Lin H (2002) Molecular characterization of hiwi, a human member of the piwi gene family whose overexpression is correlated to seminomas. Oncogene 21(25):3988–3999
Rachon D, Teede H (2010) Ovarian function and obesity–interrelationship, impact on women's reproductive lifespan and treatment options. Mol Cell Endocrinol 316(2):172–179
Ro S, Park C, Sanders KM, McCarrey JR, Yan W (2007) Cloning and expression profiling of testis-expressed microRNAs. Dev Biol 311(2):592–602
Robine N, Lau NC, Balla S, Jin Z, Okamura K, Kuramochi-Miyagawa S, Blower MD, Lai EC (2009) A broadly conserved pathway generates 3′UTR-directed primary piRNAs. Curr Biol 19(24):2066–2076
Rodriguez A, Vigorito E, Clare S, Warren MV, Couttet P, Soond DR, van Dongen S, Grocock RJ, Das PP, Miska EA, Vetrie D, Okkenhaug K, Enright AJ, Dougan G, Turner M, Bradley A (2007) Requirement of bic/microRNA-155 for normal immune function. Science 316(5824): 608–611
Romero Y, Meikar O, Papaioannou MD, Conne B, Grey C, Weier M, Pralong F, De Massy B, Kaessmann H, Vassalli JD, Kotaja N, Nef S (2011) Dicer1 depletion in male germ cells leads to infertility due to cumulative meiotic and spermiogenic defects. PLoS One 6(10), e25241
Rottiers V, Näär AM (2012) MicroRNAs in metabolism and metabolic disorders. Nat Rev Mol Cell Biol 13(4):239–250
Rubens CE, Gravett MG, Victora CG, Nunes TM, Group GR (2010) Global report on preterm birth and stillbirth (7 of 7): mobilizing resources to accelerate innovative solutions (Global Action Agenda). BMC Pregnancy Childbirth 10(Suppl 1):S7
Saare M, Rekker K, Laisk-Podar T, Soritsa D, Roost AM, Simm J, Velthut-Meikas A, Samuel K, Metsalu T, Karro H, Soritsa A, Salumets A, Peters M (2014) High-throughput sequencing approach uncovers the miRNome of peritoneal endometriotic lesions and adjacent healthy tissues. PLoS One 9(11), e112630
Saga Y (2008) Mouse germ cell development during embryogenesis. Curr Opin Genet Dev 18(4): 337–341
Sang Q, Yao Z, Wang H, Feng R, Zhao X, Xing Q, Jin L, He L, Wu L, Wang L (2013) Identification of microRNAs in human follicular fluid: characterization of microRNAs that govern steroidogenesis in vitro and are associated with polycystic ovary syndrome in vivo. J Clin Endocrinol Metab 98(7):3068–3079
Schwab R, Voinnet O (2009) MiRNA processing turned upside down. EMBO J 28:3633–3634
Seipel K, Yanze N, Schmid V (2004) The germ line and somatic stem cell gene Cniwi in the jellyfish Podocoryne carnea. Int J Dev Biol 48(1):1–7
Sharma AK, Nelson MC, Brandt JE, Wessman M, Mahmud N, Weller KP, Hoffman R (2001) Human CD34(+) stem cells express the hiwi gene, a human homologue of the Drosophila gene piwi. Blood 97(2):426–434
Sinkkonen L, Hugenschmidt T, Berninger P, Gaidatzis D, Mohn F, Artus-Revel CG, Zavolan M, Svoboda P, Filipowicz W (2008) MicroRNAs control de novo DNA methylation through regulation of transcriptional repressors in mouse embryonic stem cells. Nat Struct Mol Biol 15:259–267
Sirotkin AV, Ovcharenko D, Grossmann R, Lauková M, Mlyncek M (2009) Identification of microRNAs controlling human ovarian cell steroidogenesis via a genome-scale screen. J Cell Physiol 219(2):415–420
Sirotkin AV, Lauková M, Ovcharenko D, Brenaut P, Mlyncek M (2010) Identification of microRNAs controlling human ovarian cell proliferation and apoptosis. J Cell Physiol 223(1):49–56

Sorensen AE, Wissing ML, Salo S, Englund AL, Dalgaard LT (2014) MicroRNAs related to polycystic ovary syndrome (PCOS). Genes (Basel) 5(3):684–708
Suh MR, Lee Y, Kim JY, Kim SK, Moon SH, Lee JY, Cha KY, Chung HM, Yoon HS, Moon SY, Kim VN, Kim KS (2004) Human embryonic stem cells express a unique set of microRNAs. Dev Biol 270(2):488–498
Suh N, Baehner L, Moltzahn F, Melton C, Shenoy A, Chen J, Blelloch R (2010) MicroRNA function is globally suppressed in mouse oocytes and early embryos. Curr Biol 20(3):271–277
Svoboda P, Flemr M (2010) The role of miRNAs and endogenous siRNAs in maternal-to-zygotic reprogramming and the establishment of pluripotency. EMBO Rep 11:590–597
Tadros W, Lipshitz HD (2005) Setting the stage for development: mRNA translation and stability during oocyte maturation and egg activation in Drosophila. Dev Dyn 232(3):593–608
Tam OH, Aravin AA, Stein P, Girard A, Murchison EP, Cheloufi S, Hodges E, Anger M, Sachidanandam R, Schultz RM, Hannon GJ (2008) Pseudogene-derived small interfering RNAs regulate gene expression in mouse oocytes. Nature 453:534–538
Tang F, Kaneda M, O'Carroll D, Hajkova P, Barton SC, Sun YA, Lee C, Tarakhovsky A, Lao K, Surani MA (2007) Maternal microRNAs are essential for mouse zygotic development. Genes Dev 21:644–648
Taubert H, Greither T, Kaushal D, Wurl P, Bache M, Bartel F, Kehlen A, Lautenschlager C, Harris L, Kraemer K, Meye A, Kappler M, Schmidt H, Holzhausen HJ, Hauptmann S (2007) Expression of the stem cell self-renewal gene Hiwi and risk of tumour-related death in patients with soft-tissue sarcoma. Oncogene 26(7):1098–1100
Tesfaye D, Worku D, Rings F, Phatsara C, Tholen E, Schellander K, Hoelker M (2009) Identification and expression profiling of microRNAs during bovine oocyte maturation using heterologous approach. Mol Reprod Dev 76(7):665–677
Thomson T, Lin H (2009) The biogenesis and function of PIWI proteins and piRNAs: progress and prospect. Annu Rev Cell Dev Biol 25:355–376
Tian Y, Nan Y, Han L, Zhang A, Wang G, Jia Z, Hao J, Pu P, Zhong Y, Kang C (2011) MicroRNA miR-451 downregulates the PI3K/AKT pathway through CAB39. Int J Oncol 40:1105–12
Tili E, Michaille JJ, Cimino A, Costinean S, Dumitru CD, Adair B, Fabbri M, Alder H, Liu CG, Calin GA, Croce CM (2007) Modulation of miR-155 and miR-125b levels following lipopolysaccharide/TNF-alpha. J Immunol 179(8):5082–5089
Tingen C, Kim A, Woodruff TK (2009) The primordial pool of follicles and nest breakdown in mammalian ovaries. Mol Hum Reprod 15:795–803
Tong MH, Mitchell D, Evanoff R, Griswold MD (2011) Expression of Mirlet7 family microRNAs in response to retinoic acid-induced spermatogonial differentiation in mice. Biol Reprod 85(1):189–197
Tong MH, Mitchell DA, McGowan SD, Evanoff R, Griswold MD (2012) Two miRNA clusters, Mir-17-92 (Mirc1) and Mir-106b-25 (Mirc3), are involved in the regulation of spermatogonial differentiation in mice. Biol Reprod 86(3):72
Trajkovski M, Hausser J, Soutschek J, Bhat B, Akin A, Zavolan M, Heim MH, Stoffel M (2011) MicroRNAs 103 and 107 regulate insulin sensitivity. Nature 474(7353):649–653
Tsai KW, Kao HW, Chen HC, Chen SJ, Lin WC (2009) Epigenetic control of the expression of a primate-specific microRNA cluster in. Epigenetics 4(8):587–592
Vagin VV, Sigova A, Li C, Seitz H, Gvozdev V, Zamore PD (2006) A distinct small RNA pathway silences selfish genetic elements in the germline. Science 313:320–324
Vasudevan S, Steitz JA (2007) AU-rich-element-mediated upregulation of translation by FXR1 and Argonaute 2. Cell 128(6):1105–1118
Vasudevan S, Tong Y, Steitz JA (2007) Switching from repression to activation: microRNAs can up-regulate translation. Science 318(5858):1931–1934
Viswanathan SR, Daley GQ, Gregory RI (2008) Selective blockade of microRNA processing by Lin28. Science 320(5872):97–100
Wang Y, Medvid R, Melton C, Jaenisch R, Blelloch R (2007) DGCR8 is essential for microRNA biogenesis and silencing of embryonic stem cell self-renewal. Nat Genet 39:380–385

Wang Y, Baskerville S, Shenoy A, Babiarz JE, Baehner L, Blelloch R (2008) Embryonic stem cell-specific microRNAs regulate the G1-S transition and promote rapid proliferation. Nat Genet 40:1478–1483

Wang J, Saxe JP, Tanaka T, Chuma S, Lin H (2009) Mili interacts with tudor domain-containing protein 1 in regulating spermatogenesis. Curr Biol 19(8):640–644

Watanabe T, Takeda A, Mise K, Okuno T, Suzuki T, Minami N, Imai H (2005) Stage-specific expression of microRNAs during Xenopus development. FEBS Lett 579(2):318–324

Watanabe T, Takeda A, Tsukiyama T, Mise K, Okuno T, Sasaki H, Minami N, Imai H (2006) Identification and characterization of two novel classes of small RNAs in the mouse germline: retrotransposon-derived siRNAs in oocytes and germline small RNAs in testes. Genes Dev 20(13):1732–1743

Watanabe T, Totoki Y, Toyoda A, Kaneda M, Kuramochi-Miyagawa S, Obata Y, Chiba H, Kohara Y, Kono T, Nakano T, Surani MA, Sakaki Y, Sasaki H (2008) Endogenous siRNAs from naturally formed dsRNAs regulate transcripts in mouse oocytes. Nature 453:539–543

Wienholds E, Kloosterman WP, Miska E, Alvarez-Saavedra E, Berezikov E, de Bruijn E, Horvitz HR, Kauppinen S, Plasterk RH (2005) MicroRNA expression in zebrafish embryonic development. Science 309(5732):310–311

Wilczynska A, Minshall N, Armisen J, Miska EA, Standart N (2009) Two Piwi proteins, Xiwi and Xili, are expressed in the Xenopus female germline. RNA 15(2):337–345

Williams MD, Mitchell GM (2012) MicroRNAs in insulin resistance and obesity. Exp Diabetes Res 2012:484696

Wu S, Divall S, Wondisford F, Wolfe A (2012) Reproductive tissues maintain insulin sensitivity in diet-induced obesity. Diabetes 61(1):114–123

Xie H, Lim B, Lodish HF (2009) MicroRNAs induced during adipogenesis that accelerate fat cell development are downregulated in obesity. Diabetes 58(5):1050–1057

Xu M, Mo YY (2012) The Akt-associated microRNAs. Cell Mol Life Sci 69:3601–12

Xu YW, Wang B, Ding CH, Li T, Gu F, Zhou C (2011) Differentially expressed micoRNAs in human oocytes. J Assist Reprod Genet 28(6):559–566

Yan N, Lu Y, Sun H, Tao D, Zhang S, Liu W, Ma Y (2007) A microarray for microRNA profiling in mouse testis tissues. Reproduction 134(1):73–79

Yan N, Lu Y, Sun H, Qiu W, Tao D, Liu Y, Chen H, Yang Y, Zhang S, Li X, Ma Y (2009) Microarray profiling of microRNAs expressed in testis tissues of developing primates. J Assist Reprod Genet 26(4):179–186

Yang H, Kong W, He L, Zhao JJ, O'Donnell JD, Wang J, Wenham RM, Coppola D, Kruk PA, Nicosia SV, Cheng JQ (2008) MicroRNA expression profiling in human ovarian cancer: miR-214 induces cell survival and cisplatin resistance by targeting PTEN. Cancer Res 68(2): 425–433

Yang M, Yao Y, Eades G, Zhang Y, Zhou Q (2011) MiR-28 regulates Nrf2 expression through a Keap1-independent mechanism. Breast Cancer Res Treat 129(3):983–991

Yang CX, Du ZQ, Wright EC, Rothschild MF, Prather RS, Ross JW (2012a) Small RNA profile of the cumulus oocyte complex and early embryos in the pig. Biol Reprod 87:117

Yang CX, Wright EC, Ross JW (2012b) Expression of RNA-binding proteins DND1 and FXR1 in the porcine ovary, and during oocyte maturation and early embryo development. Mol Reprod Dev 79(8):541–552

Yang Q, Hua J, Wang L, Xu B, Zhang H, Ye N, Zhang Z, Yu D, Cooke HJ, Zhang Y, Shi Q (2013a) MicroRNA and piRNA profiles in normal human testis detected by next generation sequencing. PLoS One 8(6), e66809

Yang QE, Racicot KE, Kaucher AV, Oatley MJ, Oatley JM (2013b) MicroRNAs 221 and 222 regulate the undifferentiated state in mammalian male germ cells. Development 140(2):280–290

Yi R, Qin Y, Macara IG, Cullen BR (2003) Exportin-5 mediates the nuclear export of pre-microRNAs and short hairpin RNAs. Genes Dev 17:3011–3016

Yu Z, Raabe T, Hecht NB (2005) MicroRNA Mirn122a reduces expression of the posttranscriptionally regulated germ cell transition protein 2 (Tnp2) messenger RNA (mRNA) by mRNA cleavage. Biol Reprod 73(3):427–433

Yu J, Ryan DG, Getsios S, Oliveira-Fernandes M, Fatima A, Lavker RM (2008) MicroRNA-184 antagonizes microRNA-205 to maintain SHIP2 levels in epithelia. Proc Natl Acad Sci USA 105(49):19300–19305

Yu M, Mu H, Niu Z, Chu Z, Zhu H, Hua J (2014) miR-34c enhances mouse spermatogonial stem cells differentiation by targeting Nanos2. J Cell Biochem 115(2):232–242

Zeng F, Schultz RM (2005) RNA transcript profiling during zygotic gene activation in the preimplantation mouse embryo. Dev Biol 283:40–57

Zhang R, Wang YQ, Su B (2008) Molecular evolution of a primate-specific microRNA family. Mol Biol Evol 25(7):1493–1502

Zhang S, Yu M, Liu C, Wang L, Hu Y, Bai Y, Hua J (2012) MIR-34c regulates mouse embryonic stem cells differentiation into male germ-like cells through RARg. Cell Biochem Funct 30(8):623–632

Zhang J, Ji X, Zhou D, Li Y, Lin J, Liu J, Luo H, Cui S (2013) miR-143 is critical for the formation of primordial follicles in mice. Front Biosci 18:588–597

Zhao M, Tang Q, Wu W, Xia Y, Chen D, Wang X (2014) miR-20a contributes to endometriosis by regulating NTN4 expression. Mol Biol Rep 41(9):5793–5797

Chapter 4
The Sperm Epigenome, Male Aging, and Potential Effects on the Embryo

Timothy G. Jenkins, Kenneth I. Aston, Tyson Meyer, and Douglas T. Carrell

Abstract The effect of paternal aging on fertility, embryo quality, and offspring health is an important area of study that has received far less attention than the age effect in women. This is, in part, due to the fact that in females there are dramatic alterations to fertility and pregnancy outcomes that abruptly occur as a female ages. Such abrupt alterations to pregnancy success and/or embryonic and offspring health are not seen in males. Instead, there are subtle alterations to pregnancy success and offspring phenotypes that occur as a man ages. It is believed that, at least in part, these alterations can be explained by perturbations to the sperm epigenome that occur over time. This chapter will explore the effect of aging on the sperm epigenome and the potential impacts these perturbations may have on embryonic development and ultimately offspring health.

Keywords Sperm epigenome • DNA methylation • Embryogenesis • Paternal germline

T.G. Jenkins, Ph.D. • K.I. Aston, Ph.D. • T. Meyer
Andrology and IVF Laboratories, Department of Surgery, University of Utah School of Medicine, 675 Arapeen Dr. Suite 201, Salt Lake City, UT 84108, USA

D.T. Carrell, Ph.D. (✉)
Andrology and IVF Laboratories, Department of Surgery, University of Utah School of Medicine, 675 Arapeen Dr. Suite 201, Salt Lake City, UT 84108, USA

Department of Obstetrics and Gynecology, University of Utah School of Medicine, 675 Arapeen Dr. Suite 201, Salt Lake City, UT 84108, USA

Department of Genetics, University of Utah School of Medicine, 675 Arapeen Dr. Suite 201, Salt Lake City, UT 84108, USA
e-mail: douglas.carrell@hsc.utah.edu

R. Bronson (ed.), *The Male Role in Pregnancy Loss and Embryo Implantation Failure*, Advances in Experimental Medicine and Biology 868,
DOI 10.1007/978-3-319-18881-2_4

4.1 Introduction

The human sperm is a highly specialized cell, elegantly equipped with the minimum necessary to deliver a haploid genome through the female reproductive tract to the oocyte. Upon fusion with the oolemma, the sperm deposits not only half of the genetic material into the oocyte but also initiates signal transduction cascades responsible for completion of meiosis in the egg and the initiation of embryogenesis. The role of the sperm in delivering DNA and activating the oocyte has long been appreciated. In addition, a growing body of data indicates that the epigenetic, as well as the genetic, landscape of the sperm has direct effects on embryogenesis and offspring phenotypes and that paternal epigenetic contributions can, in some cases, confer transgenerational effects (Milekic et al. 2014; Govorko et al. 2012; Carone et al. 2010; Hammoud et al. 2009).

A variety of natural and extraneous influences can impact the sperm epigenome with potential downstream consequences (Guerrero-Bosagna et al. 2012; Hare and Moran 1979; Hemminki et al. 1999; Marczylo et al. 2012). This chapter will focus on the effects of male age on sperm epigenetics. Age has been shown to consistently and predictably affect the epigenetic profiles of numerous cell types (Richardson 2003; Christensen et al. 2009; Day et al. 2013). Remarkably, the age-induced epigenetic changes observed in sperm appear to be greater in magnitude and often more consistent than changes reported in other cell types (Jenkins et al. 2013). While much remains to be learned about the epigenetic contributions of the sperm to the early embryo, a growing body of evidence suggests that some alterations in the sperm epigenome escape the early waves of epigenetic reprogramming. These changes may explain some of the increased risks of certain diseases that are observed more frequently in the offspring of older fathers (Hemminki et al. 1999; Frans et al. 2008, 2013).

4.2 The Sperm Epigenome

The sperm is morphologically and functionally distinct from any other cell type. Perhaps the greatest distinction between the sperm cell and other cell types is the nuclear structure. While the DNA of somatic cells is packaged around histones, the majority of sperm histones are displaced in a two-step process during spermiogenesis, first by transition proteins, which are subsequently replaced by protamines 1 and 2 (P1 and P2) to form a tight toroidal structure that compresses the nucleus 6–20 times tighter than the somatic cell nucleus (Fig. 4.1) (Balhorn 2007; Ward and Coffey 1991). In normal fertile men, the ratio of P1:P2 is approximately 1:1. Importantly, infertile men often display an altered P1:P2 ratio, and deviations from the normal ratio are associated with abnormal semen parameters, increased DNA damage, and reduced fertilization and implantation rates (Aoki et al. 2005, 2006).

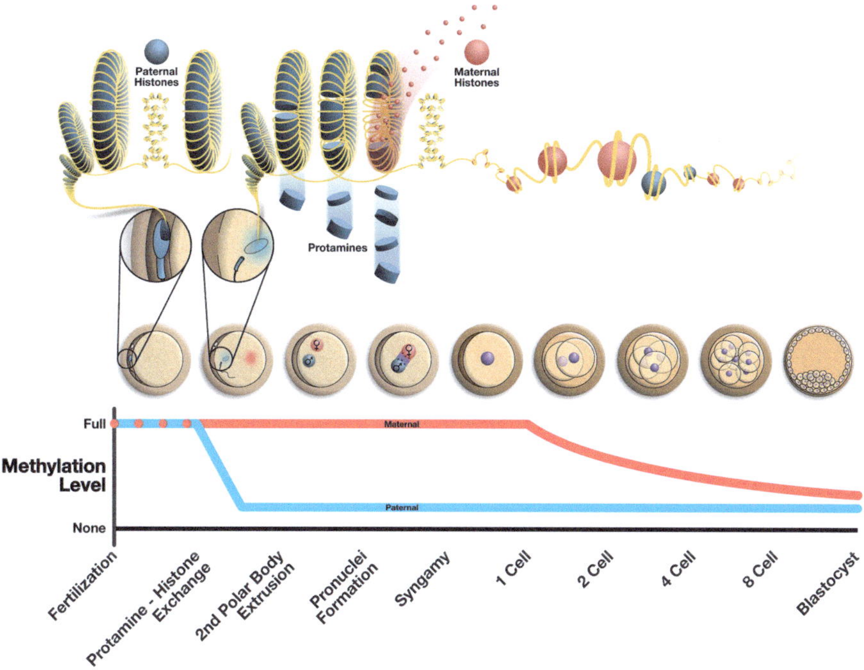

Fig. 4.1 Illustration of epigenetic structure in the mature sperm and the dramatic organization that occurs in the early embryo immediately following fertilization. The *top panel* shows the protamine-bound mature sperm, undergoing chromatin decondensation marked by the removal of protamine proteins. The *bottom panel* shows the active demethylation that occurs in the paternal pronucleus, as well as the passive, replication-dependent demethylation that occurs in the maternal pronucleus

These observations were the first to suggest that the epigenetic status of sperm might be important for early development. While the mature sperm nucleus is comprised primarily of protamine-bound DNA, about 5 % of the DNA remains bound to histones (Hammoud et al. 2009). Until recently, it was unclear whether the persistent histones were the result of incomplete histone replacement or whether they served a functional purpose. Several years ago, our lab demonstrated that histones are consistently retained at specific loci including developmental gene promoters, genes encoding microRNAs, and imprinted loci. In addition, it was found that the retained histones often display bivalency, the presence of both activating and silencing modifications within the same region, which is reminiscent of stem cell signatures (Hammoud et al. 2009). These findings suggest that the epigenetic status of sperm is tightly regulated and likely mechanistically important for embryogenesis and early development. Following fertilization, the sperm nucleus undergoes decondensation and pronuclear development, and the protamines are replaced by oocyte-derived histones. During this process, the majority of DNA methylation marks are removed to restore totipotency to the sperm and oocyte genomes (Fig. 4.1), which clearly raises questions regarding the importance of pre-fertilization epigenetic marks; however, two important considerations are warranted. First, the identity of unmodified loci remains uncertain, raising the possibility that key sperm loci remain unchanged and functionally important during embryogenesis. Second, data suggest that epigenetic abnormalities in male gametes may affect embryo development, and there is evidence to suggest that these abnormalities can affect offspring phenotype. Even less data are available on the impacts of age-associated sperm epigenetic alterations and their impacts on the embryo. Despite this, there are many indications that age-associated epigenetic alterations may play a role in both embryogenesis and offspring health.

4.3 Delayed Parenthood

Advanced paternal age has recently become a heavily investigated topic as a result of multiple studies demonstrating ties between advanced paternal age and various offspring abnormalities. Additional trends contributing to the increasing interest in the role of advanced paternal age in reproduction is the trend in delayed parentage (Mills et al. 2011). Though this trend is justified by increasing life expectancies in both sexes, advanced paternal age may affect general semen parameters and sperm quality ultimately altering fecundity and offspring health. While many couples consider the risks associated with advanced maternal age in family planning decisions, very little thought has been given to the age of male partners. In recent history, paternal age has steadily increased, particularly in developed countries. This trend is believed to be associated with increased life expectancy, socioeconomic pressures, and divorce rates with subsequent remarriage at older ages (Kuhnert and Nieschlag 2004). During a 10-year span (1993–2003) in Great Britain, the percent of fathers within the age range of 35–54 increased from 25 % of total births to 40 %. Associated with this trend was a decrease in the number of births to fathers less than 35 years of age from 74 % of total births to only 60 % (Bray et al. 2006). Over two decades in Australia (1988–2008), the average age of fathers has increased by approximately 3 years (Australian Bureau of Statistics 2009). Similarly, the average age of fathers in Germany increased by 2 years over a 10-year period (Kuhnert and Nieschlag 2004). Congruent trends can be found in the United States and many other developed countries. As average paternal age continues to increase, it is becoming increasingly important to characterize the potential consequences of advanced paternal age on embryonic development and offspring health.

4.4 Heritability of Epigenetic Alteration Through the Paternal Germline

Though poorly understood, there is clear evidence that demonstrates a unique mechanism of heritability through the paternal germline. This idea initially became of great interest to many different scientific fields as a result of findings from growing catalogs of epidemiologic data coupled with landmark studies in mouse models. Specifically, data collected during and following massive crop failures in Sweden in the late 1800s and early 1900s was used to perform large retrospective studies in human populations. From these studies, it was found that paternal diet, independent of other factors, contributes to offspring disease susceptibility and general health in ways never before identified (Kaati et al. 2007; Pembrey et al. 2006). Though the nature of the data set made it impossible to understand any biological mechanisms that underlie these alterations, many believe it plausible that alterations of heritable epigenetic marks in gametes play some role in the process. In support of this idea is the work on the agouti viable yellow gene in mouse models, which demonstrated that nutrition can affect offspring phenotype through heritable altered epigenetic marks (Waterland and Jirtle 2003). This and other work have stimulated the study of transgenerational inheritance as we see it today.

Many intriguing studies have come as a result of the increased emphasis on transgenerational inheritance in the literature. One recent study found that male mice fed a low-protein diet, when mated with a normal female, sire offspring with altered expression of many genes important in metabolism and cholesterol synthesis (Carone et al. 2010). Similarly, metabolic alterations, specifically changes in insulin sensitivity, were also seen in the female offspring of male rats fed high-fat diets (Ng et al. 2010).

Although no concrete mechanism for inheritance of altered metabolic activity has been identified or any other nongenetic inheritance from the male germline, there are intriguing candidates including epigenetic inheritance through altered sperm DNA methylation. A recent study strongly supports the idea of transgenerational inheritance. Govorko et al. demonstrated that the offspring of male mice exposed in utero to alcohol had altered hypothalamic proopiomelanocortin (POMC) gene activity as a result of hypermethylation at the POMC promoter and that these deficits were passed down through the F3 generation (Govorko et al. 2012). Interestingly, although the methylation marks at the POMC promoter were similar in the F1 female and male (both exposed to prenatal alcohol), the alterations were not inherited via the maternal germline, suggesting a unique mechanism of epigenetic inheritance through the paternal germline (Govorko et al. 2012). Taken together, these data demonstrate the likelihood that the sperm epigenome plays an essential role in embryogenesis and is capable of contributing to offspring health.

4.5 Age-Associated Sperm Epigenetic Alterations

An important consideration in the role of paternal aging on embryo quality and offspring health is the degree to which the sperm is susceptible to genomic or epigenomic perturbation that could lead to embryonic or offspring dysfunction and disease. Because of the plastic nature of epigenetic marks in the sperm and the potential heritability of any perturbations, sperm epigenetics, in particular DNA methylation, has become one of the main candidates on which studies have focused.

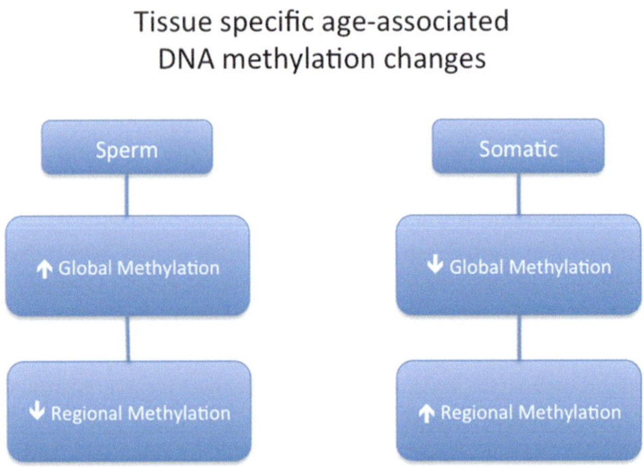

Fig. 4.2 General tissue-specific age-associated alterations that occur in sperm and other somatic tissues. Sperm tend to have slight increases in global methylation with age, while regionally there is a bias toward methylation loss. In somatic cells the opposite is true, as global methylation decreases and regional methylation increases with age

Only recently has data become available to describe the epigenetic landscape of the aged sperm, and these have focused primarily on DNA methylation in both human and mouse models. It is informative to describe this in context of somatic cell alteration associated with age where it is known that DNA methylation is altered in many somatic cell types with age in relatively consistent patterns (Wilson and Jones 1983; Oakes et al. 2003). From the few studies that have been performed, it appears that sperm methylation patterns resultant from aging are far different and of greater magnitude than what is seen in somatic cells (Jenkins et al. 2013, 2014). In fact, these cells display a virtually opposite profile of epigenetic change with age (Fig. 4.2). Although this may appear counterintuitive, it is important to note that other genomic alterations, such as telomere length, follow similar trends between these two tissue types (Eisenberg 2011). Furthermore, the idea that the magnitude of methylation alteration is greater in sperm as compared to somatic cells with aging is not without precedence. Work in support of this idea demonstrates that frequently dividing cells have more striking methylation changes associated with age than do cells that divide less frequently. As sperm undergo large amounts of division over the lifespan of an individual, it is not surprising that the magnitude of epigenetic change is greater in sperm over time than in other human tissues.

Two recent studies on human sperm from anonymous donors have revealed distinct patterns of methylation alteration with age (Jenkins et al. 2013, 2014). These studies utilized sperm donors who collected two samples many years apart (between 10 and 20 years approximately). This allowed the authors to analyze paired data to determine the intraindividual impact of aging on the sperm methylome. It was discovered that there is an increase in the global level of methylation in human sperm,

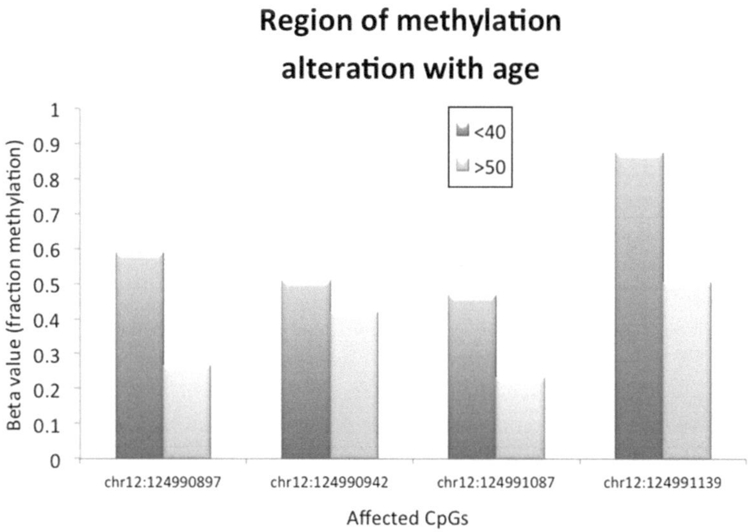

Fig. 4.3 An example of sperm-specific regional methylation. At this relatively small genomic window (approximate 250 bps), there is a significant decrease in fraction methylation (*y* axis) at each CpG (*x* axis) that occurs within men over 50 ($n=9$) when compared to men under 40 ($n=12$). These data represent an example of one of many loci significantly affected by age in (Jenkins et al. 2014)

a surprising finding based on the baseline hypermethylation in the mature sperm and the contrasting global hypomethylation that occurs with age in somatic cells (Jenkins et al. 2013, 2014). A number of regional alterations (approximately 1,000 bps in length) were also significantly altered with age and displayed a strong bias toward demethylation. This finding is, again, in opposition to what has been described in somatic cells where there is a bias toward regional hypermethylation (Jenkins et al. 2014). Alterations at these sites were confirmed with the use of targeted bisulfite sequencing in an independent cohort of unpaired general population sperm samples. These findings were remarkably consistent at the identified regions of alteration (Fig. 4.3). Intriguingly, it appeared that the age-associated regional alterations identified were enriched at genes known to be associated with neuropsychiatric disease. Similar results were identified in mice where regional hypomethylation was common in the sperm of aged mice though no global changes were identified (Milekic et al. 2014). Interestingly, all offspring of older males had similar alterations to methylation patterns in brain tissue coupled with alterations in social behaviors. Taken together, age-associated methylation perturbations represent a plausible mechanism by which the increased incidence of disease in the offspring of older fathers may be transmitted.

4.6 Embryo Quality, Pregnancy Outcomes, and Offspring Health

The effects of paternal age on pregnancy outcome and embryo quality are controversial. This controversy is mainly a result of the scant data available on the subject. Some reports suggest that there is a significant decline in fertility (as measured by time to pregnancy) with age, while others report no such associations (Hassan and Killick 2003; Begueria et al. 2014). Additional data does suggest that paternal age is a significant factor when compounded with maternal age (de la Rochebrochard and Thonneau 2002). Other studies support these data by suggesting an increased frequency of fetal loss, increased time to pregnancy, and decreased probability of conception in older men (Selvin and Garfinkel 1976; Ford et al. 2000; Dunson et al. 2002). However, there are conflicting data which suggest little to no effect of paternal age on fertility in natural conception or with the use of assisted reproductive technologies (ART) (Begueria et al. 2014; Olsen 1990; Bellver et al. 2008). Similar controversy exists in the effect of paternal age on embryo quality with the use of ART with some studies showing no effect (Bellver et al. 2008; Ferreira et al. 2010) and some suggesting decreased quality of embryos sired by older fathers on day 3, 4, and 5 (Luna et al. 2009; Frattarelli et al. 2008). The most compelling indication that paternal age may affect embryo quality is data on miscarriage. In general, the consensus from the available data is that advanced paternal age is a risk factor for miscarriage though no real mechanisms for this finding have been elucidated (Kleinhaus et al. 2006; Slama et al. 2005). Other studies evaluating ART with donor eggs (to completely remove the influence of maternal factors) found no associations between paternal age and risk of miscarriage (Begueria et al. 2014). Taken together, much work remains to determine what, if any, effect advanced paternal age has on male fertility and embryo quality.

The subtlety of the effect of age on male fertility, and particularly pregnancy outcomes, is in striking contrast to the dramatic decline seen in female fertility. In fact, even men of advanced age are able to sire offspring with little difficulty, though possibly with slightly reduced efficiency which is why paternal age has largely been ignored and has received far less attention in the clinical setting than the age of the female partner. The fact that males are still fertile at advanced ages may present, and potentially complicate, another issue that, while subtle, is far more consistent, namely, the effect of paternal age on offspring health and disease susceptibility. It has been shown that the offspring of older fathers have increased incidence of various forms of cancer, including hematological and central nervous system tumors (Hemminki et al. 1999; Oksuzyan et al. 2012; Murray et al. 2002; Yip et al. 2006), though the data remains somewhat controversial. Furthermore, it has long been suggested that advanced paternal age is a risk factor for schizophrenia (Hare and Moran 1979; Miller et al. 2011; Matheson et al. 2011; Wohl and Gorwood 2007). More recently, it has been suggested that advanced paternal age is significantly associated with many forms of neuropsychiatric or neurocognitive diseases including autism spectrum disorders (ASD) (Gardener et al. 2009; Hultman et al. 2011), bipolar disorder (Frans et al. 2008; Menezes et al. 2010), and general increases in behavioral issues (Kuja-Halkola et al. 2012; Saha et al. 2009a) in children of older fathers though some controversy exists. In addition, some studies indicate that children of older fathers display slightly reduced IQ compared with children of younger fathers (Malaspina et al. 2005; Saha et al. 2009b), although the differences are small, and conflicting reports exist (Svensson et al. 2011).

Taken together, it is clear that advanced paternal age does not have a dramatic affect on pregnancy outcomes, embryo quality, or fertility in general, but it may impact offspring health and disease susceptibility. While the lack of striking age-associated fertility declines in males has garnered it little attention in the study of fertility, it is this same maintenance of fertility that might require more study in the field of transgenerational inheritance. Age-associated alterations to sperm, which appear to affect offspring health, do not seem to be catastrophic to spermatogenesis or cause declines in fertility. This, in turn, means that aged sperm are entirely competent to fertilize an oocyte and produce viable offspring, while harboring alterations that may potentially affect offspring health.

4.7 Future Directions

To gain a more complete understanding of the epigenetic alterations in sperm that are capable of embryonic or offspring phenotype alterations, much work is still needed. A number of genomic regions have been identified that have both methylation alterations with age and are important in various cell processes and diseases known to have increased occurrence in the offspring of older males. To determine if these marks can contribute to disease susceptibility in the offspring or affect events in the embryo, a number of unanswered questions must be addressed.

What is the impact of altered methylation profiles at our regions of interest? To completely understand the alterations which have been identified and their impact on offspring phenotype or embryo development, it must be determined if these alterations are associated with transcriptional changes. Future work can target genomic sites that are known to be altered with age in mouse models to determine if (1) there are changes to transcription in the embryo and (2) determine if there are altered transcript levels in various tissues in the offspring (should the sperm be competent to generate viable offspring).

Do the altered methylation marks seen in sperm escape, or impact in any way, embryonic epigenetic reprogramming? This is an essential question to fully understand the impact of an altered epigenetic profile. It is feasible that an alteration could affect embryogenesis in one of two ways. First, it could directly affect transcription of an important developmental factor. Second, the epigenetic abnormality may result in not a targeted perturbation but in the global alteration of epigenetic reprogramming, effectively altering an important aspect of embryogenesis, likely to the point of embryonic arrest.

Do methylation perturbations contribute to neuropsychiatric disorders in the offspring or perturbations to embryogenesis? To date, there are many intriguing studies that have provided some small degree of insight into the effect of aging on the sperm epigenome. However, much of the potential impacts are simply extrapolation of the available data without any real targeted studies to prove causative relationships. While the data is exciting, future targeted work is still required to enable us to reach these further conclusions.

4.8 Conclusions

The role of the paternal epigenome in embryogenesis should not be downplayed. It appears from a growing body of evidence that the sperm epigenetic landscape is essential in facilitating gene poising and general transcription regulation at genes important in embryonic development (Hammoud et al. 2009). However, with our current understanding, we are unable to definitively determine that sperm epigenetic alterations associated with age are causative of any poor pregnancy outcomes or decreased embryo quality declines. In fact, the aged male remains remarkably fertile with, at most, only modest declines in fecundity. When we consider this fact coupled with the data regarding known and consistent age-associated alterations to the paternal epigenome, it is easy to contemplate the implications of these alterations beyond embryogenesis. Specifically, a great deal of focus has now been given to the increased incidence of diseases seen in the offspring of older fathers and the transgenerational impacts that they impose. This is of particular concern in developed countries where the age of paternity is steadily increasing. While the questions regarding paternal age and epigenetic alterations that may affect embryogenesis are essential and must be addressed further, the impact of these alterations on the offspring appears to be a more relevant question due to the fact that the alterations identified with aging do not appear to affect (at least in a great degree) the competency of sperm to yield viable offspring.

Many important questions must still be addressed in regard to the epigenetic findings associated with advanced paternal age. While we know that there are real alterations that occur with remarkable consistency, the impact of these alterations is unknown. Mouse data suggesting similar methylation patterns in the brain of offspring sired by older fathers is intriguing (particularly when coupled with the identified behavioral abnormalities), but this also requires a great degree of further study. We currently have a great deal of genomic targets that are known to be altered in the sperm of men with advanced age, and these can be used to analyze potential implications in the embryo and the offspring. Taken together, while having learned much about the impacts of advanced age in the recent past, there is still a great deal of work that needs to be performed to truly elucidate the impact of age-associated sperm epigenetic alterations on the embryo and beyond.

References

Aoki VW et al (2005) DNA integrity is compromised in protamine-deficient human sperm. J Androl 26:741–748. doi:10.2164/jandrol.05063

Aoki VW, Emery BR, Liu L, Carrell DT (2006) Protamine levels vary between individual sperm cells of infertile human males and correlate with viability and DNA integrity. J Androl 27: 890–898. doi:10.2164/jandrol.106.000703

Balhorn R (2007) The protamine family of sperm nuclear proteins. Genome Biol 8:227. doi:10.1186/gb-2007-8-9-227, gb-2007-8-9-227 [pii]

Begueria R et al (2014) Paternal age and assisted reproductive outcomes in ICSI donor oocytes: is there an effect of older fathers? Hum Reprod 29:2114–2122. doi:10.1093/humrep/deu189

Bellver J, Garrido N, Remohi J, Pellicer A, Meseguer M (2008) Influence of paternal age on assisted reproduction outcome. Reprod Biomed Online 17:595–604

Australian Bureau of Statistics (2009) Births, Australia (2008) Summary of findings. Births

Bray I, Gunnell D, Davey Smith G (2006) Advanced paternal age: how old is too old? J Epidemiol Community Health 60:851–853. doi:10.1136/jech.2005.045179, 60/10/851 [pii]

Carone BR et al (2010) Paternally induced transgenerational environmental reprogramming of metabolic gene expression in mammals. Cell 143:1084–1096. doi:10.1016/j.cell.2010.12.008

Christensen BC et al (2009) Aging and environmental exposures alter tissue-specific DNA methylation dependent upon CpG island context. PLoS Genet 5, e1000602. doi:10.1371/journal.pgen.1000602

Day K et al (2013) Differential DNA methylation with age displays both common and dynamic features across human tissues that are influenced by CpG landscape. Genome Biol 14:R102. doi:10.1186/gb-2013-14-9-r102

de la Rochebrochard E, Thonneau P (2002) Paternal age and maternal age are risk factors for miscarriage; results of a multicentre European study. Hum Reprod 17:1649–1656

Dunson DB, Colombo B, Baird DD (2002) Changes with age in the level and duration of fertility in the menstrual cycle. Hum Reprod 17:1399–1403

Eisenberg DT (2011) An evolutionary review of human telomere biology: the thrifty telomere hypothesis and notes on potential adaptive paternal effects. Am J Hum Biol 23:149–167. doi:10.1002/ajhb.21127

Ferreira RC et al (2010) Negative influence of paternal age on clinical intracytoplasmic sperm injection cycle outcomes in oligozoospermic patients. Fertil Steril 93:1870–1874. doi:10.1016/j.fertnstert.2008.12.043

Ford WC et al (2000) Increasing paternal age is associated with delayed conception in a large population of fertile couples: evidence for declining fecundity in older men. The ALSPAC study team (Avon longitudinal study of pregnancy and childhood). Hum Reprod 15:1703–1708

Frans EM et al (2008) Advancing paternal age and bipolar disorder. Arch Gen Psychiatry 65:1034–1040. doi:10.1001/archpsyc.65.9.1034

Frans EM et al (2013) Autism risk across generations: a population-based study of advancing grandpaternal and paternal age. JAMA Psychiatry 70:516–521. doi:10.1001/jamapsychiatry.2013.1180

Frattarelli JL, Miller KA, Miller BT, Elkind-Hirsch K, Scott RT Jr (2008) Male age negatively impacts embryo development and reproductive outcome in donor oocyte assisted reproductive technology cycles. Fertil Steril 90:97–103. doi:10.1016/j.fertnstert.2007.06.009

Gardener H, Spiegelman D, Buka SL (2009) Prenatal risk factors for autism: comprehensive meta-analysis. Br J Psychiatry 195:7–14. doi:10.1192/bjp.bp.108.051672

Govorko D, Bekdash RA, Zhang C, Sarkar DK (2012) Male germline transmits fetal alcohol adverse effect on hypothalamic proopiomelanocortin gene across generations. Biol Psychiatry 72:378–388. doi:10.1016/j.biopsych.2012.04.006

Guerrero-Bosagna C et al (2012) Epigenetic transgenerational inheritance of vinclozolin induced mouse adult onset disease and associated sperm epigenome biomarkers. Reprod Toxicol 34:694–707. doi:10.1016/j.reprotox.2012.09.005

Hammoud SS et al (2009) Distinctive chromatin in human sperm packages genes for embryo development. Nature 460:473–478. doi:10.1038/nature08162, nature08162 [pii]
Hare EH, Moran PA (1979) Raised parental age in psychiatric patients: evidence for the constitutional hypothesis. Br J Psychiatry 134:169–177
Hassan MA, Killick SR (2003) Effect of male age on fertility: evidence for the decline in male fertility with increasing age. Fertil Steril 79(Suppl 3):1520–1527, S0015028203003662 [pii]
Hemminki K, Kyyronen P, Vaittinen P (1999) Parental age as a risk factor of childhood leukemia and brain cancer in offspring. Epidemiology 10:271–275
Hultman CM, Sandin S, Levine SZ, Lichtenstein P, Reichenberg A (2011) Advancing paternal age and risk of autism: new evidence from a population-based study and a meta-analysis of epidemiological studies. Mol Psychiatry 16:1203–1212. doi:10.1038/mp.2010.121
Jenkins TG, Aston KI, Cairns BR, Carrell DT (2013) Paternal aging and associated intraindividual alterations of global sperm 5-methylcytosine and 5-hydroxymethylcytosine levels. Fertil Steril 100:945–951. doi:10.1016/j.fertnstert.2013.05.039
Jenkins TG, Aston KI, Pflueger C, Cairns BR, Carrell DT (2014) Age-associated sperm DNA methylation alterations: possible implications in offspring disease susceptibility. PLoS Genet 10, e1004458. doi:10.1371/journal.pgen.1004458
Kaati G, Bygren LO, Pembrey M, Sjostrom M (2007) Transgenerational response to nutrition, early life circumstances and longevity. Eur J Hum Genet 15:784–790. doi:10.1038/sj.ejhg.5201832
Kleinhaus K et al (2006) Paternal age and spontaneous abortion. Obstet Gynecol 108:369–377. doi:10.1097/01.AOG.0000224606.26514.3a
Kuhnert B, Nieschlag E (2004) Reproductive functions of the ageing male. Hum Reprod Update 10:327–339. doi:10.1093/humupd/dmh030, dmh030 [pii]
Kuja-Halkola R, Pawitan Y, D'Onofrio BM, Langstrom N, Lichtenstein P (2012) Advancing paternal age and offspring violent offending: a sibling-comparison study. Dev Psychopathol 24:739–753. doi:10.1017/S095457941200034X
Luna M et al (2009) Paternal age and assisted reproductive technology outcome in ovum recipients. Fertil Steril 92:1772–1775. doi:10.1016/j.fertnstert.2009.05.036
Malaspina D et al (2005) Paternal age and intelligence: implications for age-related genomic changes in male germ cells. Psychiatr Genet 15:117–125
Marczylo EL, Amoako AA, Konje JC, Gant TW, Marczylo TH (2012) Smoking induces differential miRNA expression in human spermatozoa: a potential transgenerational epigenetic concern? Epigenetics 7:432–439. doi:10.4161/epi.19794
Matheson SL, Shepherd AM, Laurens KR, Carr VJ (2011) A systematic meta-review grading the evidence for non-genetic risk factors and putative antecedents of schizophrenia. Schizophr Res 133:133–142. doi:10.1016/j.schres.2011.09.020
Menezes PR et al (2010) Paternal and maternal ages at conception and risk of bipolar affective disorder in their offspring. Psychol Med 40:477–485. doi:10.1017/S003329170999064X
Milekic MH et al (2014) Age-related sperm DNA methylation changes are transmitted to offspring and associated with abnormal behavior and dysregulated gene expression. Mol Psychiatry. doi:10.1038/mp.2014.84
Miller B et al (2011) Meta-analysis of paternal age and schizophrenia risk in male versus female offspring. Schizophr Bull 37:1039–1047. doi:10.1093/schbul/sbq011
Mills M, Rindfuss RR, McDonald P, te Velde E (2011) Why do people postpone parenthood? Reasons and social policy incentives. Hum Reprod Update 17:848–860. doi:10.1093/humupd/dmr026, dmr026 [pii]
Murray L et al (2002) Association of early life factors and acute lymphoblastic leukaemia in childhood: historical cohort study. Br J Cancer 86:356–361. doi:10.1038/sj.bjc.6600012
Ng SF et al (2010) Chronic high-fat diet in fathers programs beta-cell dysfunction in female rat offspring. Nature 467:963–966. doi:10.1038/nature09491, nature09491 [pii]
Oakes CC, Smiraglia DJ, Plass C, Trasler JM, Robaire B (2003) Aging results in hypermethylation of ribosomal DNA in sperm and liver of male rats. Proc Natl Acad Sci USA 100:1775–1780. doi:10.1073/pnas.0437971100, 0437971100 [pii]

Oksuzyan S, Crespi CM, Cockburn M, Mezei G, Kheifets L (2012) Birth weight and other perinatal characteristics and childhood leukemia in California. Cancer Epidemiol 36:e359–e365. doi:10.1016/j.canep.2012.08.002

Olsen J (1990) Subfecundity according to the age of the mother and the father. Dan Med Bull 37:281–282

Pembrey ME et al (2006) Sex-specific, male-line transgenerational responses in humans. Eur J Hum Genet 14:159–166. doi:10.1038/sj.ejhg.5201538

Richardson B (2003) Impact of aging on DNA methylation. Ageing Res Rev 2:245–261

Saha S, Barnett AG, Buka SL, McGrath JJ (2009a) Maternal age and paternal age are associated with distinct childhood behavioural outcomes in a general population birth cohort. Schizophr Res 115:130–135. doi:10.1016/j.schres.2009.09.012

Saha S et al (2009b) Advanced paternal age is associated with impaired neurocognitive outcomes during infancy and childhood. PLoS Med 6:e40. doi:10.1371/journal.pmed.1000040

Selvin S, Garfinkel J (1976) Paternal age, maternal age and birth order and the risk of a fetal loss. Hum Biol 48:223–230

Slama R et al (2005) Influence of paternal age on the risk of spontaneous abortion. Am J Epidemiol 161:816–823. doi:10.1093/aje/kwi097

Svensson AC, Abel K, Dalman C, Magnusson C (2011) Implications of advancing paternal age: does it affect offspring school performance? PLoS One 6:e24771. doi:10.1371/journal.pone.0024771

Ward WS, Coffey DS (1991) DNA packaging and organization in mammalian spermatozoa: comparison with somatic cells. Biol Reprod 44:569–574

Waterland RA, Jirtle RL (2003) Transposable elements: targets for early nutritional effects on epigenetic gene regulation. Mol Cell Biol 23:5293–5300

Wilson VL, Jones PA (1983) DNA methylation decreases in aging but not in immortal cells. Science 220:1055–1057

Wohl M, Gorwood P (2007) Paternal ages below or above 35 years old are associated with a different risk of schizophrenia in the offspring. Eur Psychiatry 22:22–26. doi:10.1016/j.eurpsy.2006.08.007

Yip BH, Pawitan Y, Czene K (2006) Parental age and risk of childhood cancers: a population-based cohort study from Sweden. Int J Epidemiol 35:1495–1503. doi:10.1093/ije/dyl177

Chapter 5
The Role of Uterine NK Cells in Normal Reproduction and Reproductive Disorders

Judith N. Bulmer and Gendie E. Lash

Abstract The human endometrium contains a substantial population of leucocytes which vary in distribution during the menstrual cycle and pregnancy. An unusual population of natural killer (NK) cells, termed uterine NK (uNK) cells, are the most abundant of these cells in early pregnancy. The increase in number of uNK cells in the mid-secretory phase of the cycle with further increases in early pregnancy has focused attention on the role of uNK cells in early pregnancy. Despite many studies, the in vivo role of these cells is uncertain. This chapter reviews current information regarding the role of uNK cells in healthy human pregnancy and evidence indicating their importance in various reproductive and pregnancy problems. Studies in humans are limited by the availability of suitable tissues and the limitations of extrapolation from animal models.

Keywords Endometrial leucocyte • Uterine NK cells • T lymphocytes • CD56 • Cytokines • Chemokines • Extravillous trophoblast • Angiogenic growth factors • Miscarriage • Preeclampsia

5.1 Introduction

The human uterus fulfils a unique role, allowing implantation of the semi-allogeneic fetoplacental unit and, in normal human pregnancy, survival in safety for 40 weeks of gestation. Outside of pregnancy, the endometrium lining the uterine cavity is exposed to spermatozoa and seminal fluid on a regular basis and has to be able to resist infection potentially ascending through the cervix. The endometrium undergoes profound morphological changes during each menstrual cycle under the control of ovarian steroid hormones. After epithelial and stromal proliferation in the oestrogen-dependent follicular phase, after ovulation, influenced by progesterone, the glands become secretory and the stroma shows predecidual change, initially

J.N. Bulmer (✉) • G.E. Lash
Reproductive and Vascular Biology Group, Institute of Cellular Medicine, Newcastle University, Newcastle upon Tyne, UK
e-mail: j.n.bulmer@ncl.ac.uk; Gendie.lash@ncl.ac.uk

R. Bronson (ed.), *The Male Role in Pregnancy Loss and Embryo Implantation Failure*, Advances in Experimental Medicine and Biology 868,
DOI 10.1007/978-3-319-18881-2_5

around spiral arterioles. In addition to epithelial and stromal cells, the endometrium contains a substantial leucocyte population (Bulmer et al. 1991), mainly within the stroma but with a small population of intraepithelial leucocytes (Pace et al. 1991). The endometrial stromal leucocyte population varies dramatically during the normal menstrual cycle and in pregnancy, and alterations in endometrial leucocyte populations have been described in various pathological situations, including recurrent miscarriage, recurrent implantation failure and pre-eclampsia (reviewed in Bulmer and Lash 2005; Lash and Bulmer 2011). In the 20–30 years since monoclonal antibodies allowed phenotypic identification of leucocyte populations within the endometrium, these cells have been the subject of numerous investigations, but despite considerable progress, the in vivo role of these cells is not yet fully understood.

5.2 Endometrial Leucocyte Populations

Leucocytes are present in both stratum basalis and the hormonally responsive stratum functionalis, and the populations differ between these two sites. Leucocyte populations in the stratum basalis have been less well characterised than those in the stratum functionalis.

Lymphoid aggregates are a common feature in the stratum basalis, sometimes with germinal centres. These follicles include T cells, predominantly CD8+, macrophages and B cells (Morris et al. 1985; Marshall and Jones 1988; Bulmer et al. 1988a); the aggregates have been reported to increase in size in the secretory compared with proliferative phase and to be absent in postmenopausal endometrium, suggesting possible hormonal regulation (Yeaman et al. 1997; Wira et al. 2008). The presence of CD56+ uterine natural killer (uNK) cells in the stratum basalis has not been documented in detail.

The proportion of endometrial stroma accounted for by leucocytes in the stratum functionalis varies during the menstrual cycle. Whereas leucocytes account for fewer than 10 % of stromal cells in the proliferative and early secretory phases, their numbers increase dramatically from the mid-secretory phase, increasing further in the late secretory phase and early pregnancy (Bulmer et al. 1991). Three main cell types account for the majority of endometrial stromal leucocytes in the secretory phase and early pregnancy: macrophages, T lymphocytes, including regulatory T cells, and uNK cells. Unlike other mucosal sites, including vagina, cervix and fallopian tube, B lymphocytes are uncommon in the endometrial stratum functionalis and plasma cells are not a normal feature of endometrium; their presence in non-pathological endometrium is generally considered to indicate chronic endometritis (Disep et al. 2004). Other leucocytes such as dendritic cells (Schulke et al. 2008), NK T cells (Shimada et al. 2003), eosinophils and mast cells are also present in endometrial stroma, in smaller numbers (Evans and Salamonsen 2012). Endometrial leucocyte populations also alter with the onset of menstruation, with neutrophil polymorphs becoming prominent components (Evans and Salamonsen 2012; Berbic and Fraser 2013). The mechanisms underlying the regulation of the various endometrial leucocyte populations during the menstrual cycle are not fully understood and are likely to involve complex interactions between steroid hormones, chemokine and cytokine products of endometrial stromal and epithelial cells, as well as placental trophoblast cells during pregnancy.

The relative proportion of endometrial leucocytes varies during the menstrual cycle. Whereas endometrial T lymphocytes remain relatively constant during the menstrual cycle, both macrophages and, to a greater extent, uNK cells increase in number in secretory endometrium and in early pregnancy (Bulmer et al. 1991). The prominence of uNK cells around the time of expected implantation in a fertile cycle and during early placentation has focused attention on these cells, with fewer studies of other leucocytes in human endometrium. Although uNK cells are the main focus of this chapter, it is useful to consider briefly other endometrial leucocyte populations.

5.2.1 T Lymphocytes

Although the numbers of CD3+ T cells remain relatively constant during the menstrual cycle, their proportion alters as other cells vary in number. T cells constitute 40–60 % of endometrial leucocytes in non-pregnant endometrium (Bulmer et al. 1991; Klentzeris et al. 1994) reducing to 10–20 % in early pregnancy as other cell populations become more abundant (Bulmer et al. 1991). Endometrial T lymphocyte subsets differ compared with those in peripheral blood; CD8+ cytotoxic T lymphocytes are the most abundant T cell in endometrium, with 30–45 % of CD4+ helper T cells. In vitro studies of CD8+ T cells purified from both non-pregnant endometrium and early pregnancy decidua have demonstrated that endometrial T cells are capable of cytotoxic activity (Yeaman et al. 1997; Scaife et al. 2006), although CD8+ T cells in normal pregnancy must maintain a complex balance between control of infection and tolerance towards the fetoplacental unit (Tilburgs et al. 2010; Tilburgs and Strominger 2013).

Besides CD8+ T cells, less abundant endometrial T lymphocyte populations include TCRγδ T cells, CD4-CD8-TCRαβ+ T cells and Vα24+Vβ11+ NK T cells (Shimada et al. 2003), although the function of these cell types is largely unknown. Th17+ cells are reduced in decidua compared with peripheral blood (Mjosberg et al. 2010), although Th17 cell responses in decidua are abrogated by uNK cells (Fu et al. 2013). CD4+ CD25^{bright} FOXP3+ regulatory T cells are the most studied T cell subset in pregnancy (Tilburgs et al. 2008, 2010). Although they represent a minor T cell population in human decidua, regulatory T cells may act as immune suppressors, contributing to immune homeostasis during pregnancy. They express immunosuppressive cytokines such as IL-10 and TGF-β, suppressing cytotoxic T cells (Aluvihare et al. 2004; Zenclussen et al. 2006). T regulatory cells may also play a role in vascular homeostasis at the maternal-fetal interface via production of IL-10 and TGF-β (Nevers et al. 2011).

T lymphocytes are a consistent and important component of the endometrial leucocyte population. Various subsets have been identified and even minor populations may play an important role in implantation and pregnancy, although additional studies are required to determine the role of specific T cells in both normal and pathological human pregnancy.

5.2.2 *Macrophages*

Macrophages are an important component of both non-pregnant and pregnant endometrium, although their in vivo role remains unclear. They express CD14, CD68 and class II MHC antigens as well as CD11c, CD86 and adhesion and activation markers depending on their activation state (Nagamatsu and Schust 2010; Houser 2012; Thiruchelvam et al. 2013). Macrophages have diverse functions, including remodelling of extracellular matrix, protease production, tissue regeneration and antigen presentation (Houser et al. 2011). Macrophages account for 20–25 % of the CD45+ leucocytes in non-pregnant endometrium and early pregnancy decidua, although the proportion of leucocytes accounted for by macrophages varies as uNK cell numbers increase during the menstrual cycle and pregnancy. Macrophages have been reported to increase in number in the late secretory and menstrual phases and in early pregnancy (Bulmer et al. 1991; Salamonsen et al. 2002; Thiruchelvam et al. 2013). Proteases and cytokines derived from macrophages may contribute to menstruation and clearance of menstrual debris, and macrophages may also play a role in remodelling of the stratum functionalis after menstruation (Maybin et al. 2012; Thiruchelvam et al. 2013).

In decidua, macrophages are detected in both decidua basalis and decidua parietalis, and in the former these cells are often closely associated with extravillous trophoblast (Bulmer et al. 1988b). Macrophages have been divided into two distinct subtypes, termed pro-inflammatory M1 and anti-inflammatory M2, but decidual macrophages do not conform to these subtypes; recent phenotypic studies reported two distinct populations of decidual macrophages based on CD11c expression as $CD11c^{high}$ or $CD11c^{low}$ (Houser et al. 2011). $CD11c^{high}$ macrophages exhibited increased antigen-presenting function with increased expression of CD1a, CD1c and CD1d compared with $CD11c^{low}$ cells. In addition, microarray studies suggest that decidual macrophages group closer to a M2 phenotype, but are a unique population of cells (Gustafsson et al. 2008). It has also been suggested that macrophages in decidua basalis show increased activation compared with decidua parietalis, evidenced by increased expression of HLA-DR (Repnik et al. 2008).

Despite the frequency of macrophages in endometrium and decidua, there have been relatively few studies of these cells, partly because of the difficulty in obtaining viable purified populations. Production of anti-inflammatory substances such as IL-10, IDO (Heikkinen et al. 2003) and prostaglandin E_2 (Parhar et al. 1989) raises the possibility of an immunosuppressive function in pregnancy (Nagamatsu and Schust 2010). Our recent studies of macrophages enriched from human early pregnancy decidua have demonstrated production of both IL-6 and IL-8 by macrophages in first trimester decidua, with reduced production of both chemokines in association with first trimester miscarriage (Pitman et al. 2013).

5.2.3 *Dendritic Cells*

Mature dendritic cells (DCs) are numerically a small component of the human endometrial leucocyte population, but immature DCs are seen in greater numbers in early pregnancy, although studies are at a relatively early stage. In the cycling human endometrium, mature dendritic cells (DCs) are at their peak at the late secretory phase, in contrast with early pregnancy decidua, where the majority of the DCs express CD209 (DC-SIGN), a marker of immature or inactive DC (Gardner and Moffett 2003; Kammerer et al. 2003; Rieger et al. 2004). Although it has been suggested that DCs may play a role in remodelling the cycling endometrium following menstruation, current attention is focused on their potential role in the modulation of immune responses within the pregnant uterus (Dietl et al. 2006; Blois et al. 2011; Leno-Durán et al. 2014) with increasing interest into their potential interactions with uNK cells.

5.2.4 *Uterine NK Cells*

As the most abundant leucocyte population in endometrium at the time of implantation and early placentation, uNK cells have been the most investigated of the endometrial leucocyte populations, although their in vivo function is still not fully established. Although they were originally recognised by the presence of cytoplasmic granules (Hamperl and Hellweg 1958), uNK cells are now recognised by their unusual phenotype which is distinct from most peripheral blood NK cells (Bulmer et al. 1991; Trundley and Moffett 2004; Bulmer and Lash 2005). In contrast with most peripheral blood NK cells which are $CD56^{dim}$ CD16+, uNK cells express CD56 brightly but are CD16-. Around 5 % of peripheral blood NK cells are $CD56^{bright}$ CD16-, but, in contrast with the uNK cells which show prominent cytoplasmic granules, $CD56^{bright}$ CD16- NK cells in peripheral blood are mainly non-granulated (Cooper et al. 2001). There are other notable differences in the phenotype of uNK cells compared with peripheral blood NK cells, with expression of the tetraspanin family of proteins CD9 and CD151 by uterine but not by peripheral blood NK cells. uNK cells also show expression of NK surface receptors, with differing KIR repertoires noted between uNK cells and peripheral blood NK cells from the same individual (Verma et al. 1997).

5.2.4.1 Distribution of uNK Cells

Although relatively sparse in the proliferative and early secretory phases of the menstrual cycle, uNK cells increase in number dramatically in the mid-secretory phase of the cycle, increasing further as the cycle progresses and in early pregnancy (Bulmer et al. 1991). They reduce in number after the first half of pregnancy, but a substantial number of CD56+ cells remain in the decidua at term (Williams et al. 2009a).

Early reports of a virtual absence of uNK cells at term may be explained by a reduction in the proportion of cells with perforin- and granzyme-containing cytoplasmic granules and therefore detectable using histochemical stains such as phloxine tartrazine to detect cytoplasmic granules (Bulmer et al. 2010). Electron microscope studies have suggested that the cells undergo some degranulation during pregnancy (Spornitz 1992), and this may reflect functional changes as pregnancy progresses.

uNK cells are often seen closely associated with extravillous trophoblast in the decidua basalis. Several studies have compared the distribution of uNK cells in decidua basalis and decidua parietalis with conflicting results. Williams et al. (2009a) noted no differences between uNK cell numbers in decidua basalis and decidua parietalis at various gestational ages, whereas others have suggested that uNK cell numbers are increased in decidua basalis associated with invasive EVT. Most recently, Helige et al. (2014) examined density of uNK cells in relation to EVT and noted an increased uNK cell density within 20 μm of trophoblast cells in decidua with density reducing in areas more distant from extravillous trophoblast.

uNK cells are also found adjacent to endometrial glands, as well as in aggregates adjacent to spiral arterioles and arteries (Bulmer et al. 1991). Although it was suggested that the perivascular distribution in the secretory phase may reflect influx of uNK cells from blood, the distribution of the uNK cells at this site may reflect their close association with predecidual change in the perivascular stromal cells. Both uNK cells and macrophages are closely associated with spiral arteries as pregnancy progresses, and this is likely to reflect their role in the transformation of uterine spiral arteries which is an essential feature of normal pregnancy.

The mechanisms that control the altered numbers of uNK cells during the menstrual cycle and in pregnancy are not known. uNK cells are often associated with stromal cell decidualisation; in secretory phase endometrium, they accumulate in areas of stromal predecidual change. They are also associated with stromal decidualisation at ectopic sites such as ovarian serosal or cervical decidualisation in normal pregnancy or fallopian tube mucosa in ectopic tubal pregnancy. uNK cells are also particularly common in endometrium showing pseudodecidualisation due to high-dose progesterone treatment. This association has suggested regulation by progesterone, but uNK cells do not express progesterone receptor or oestrogen receptor (ER)α (Stewart et al. 1998; Henderson et al. 2003). Regulation of uNK cells by progesterone could occur indirectly via products of decidualised endometrial stromal cells; endometrial stromal cells produce interleukin (IL)-15 in the late secretory phase of the cycle and early pregnancy, and immunostaining for IL-15 has been shown to correlate with numbers of CD56+ cells in endometrium from women with recurrent miscarriage and implantation failure (Mariee et al. 2012). uNK cells do, however, express ERβ and glucocorticoid receptor (Henderson et al. 2003). Recent studies suggest that stromal decidualisation may be associated with secretion of oestrogen (Gibson et al. 2013), raising the possibility that oestrogen may play a role in stimulating uNK cell proliferation and/or differentiation in the secretory phase of the menstrual cycle.

5.2.4.2 Increased uNK Cells: Recruitment, Differentiation and/or Proliferation?

There is no consensus regarding the origin of uNK cells; there is evidence to support both recruitment and local differentiation (Koopman et al. 2003; Bulmer and Lash 2005; Bulmer et al. 2010; Zhang et al. 2012). In mice, uNK progenitor cells have been identified in primary and secondary lymphoid tissues, and CD127 is expressed at days 10.5 and 12.5 but not at day 6.5 suggesting local differentiation and regulation (Zhang et al. 2012). However, limited extrapolation is possible from mouse studies since uNK cells are only detected after blastocyst implantation in mouse and not in non-pregnant endometrium; and human and mouse placentation show important differences.

Early studies suggested recruitment of uNK cells into endometrium in response to chemokine and cytokine secretion. For example, production of CXCL-12 by extravillous trophoblast cells was suggested to attract uNK cells into decidua in pregnancy (Wu et al. 2005), and IL-15 which is produced by secretory endometrium and decidua has been shown to have a selective chemoattractant effect on peripheral blood CD16- NK cells (Kitaya et al. 2007). Increasing evidence, however, indicates that uNK cells differentiate in situ in endometrium from haematopoietic precursor cells (HPC) or immature NK cells recruited from peripheral blood into an environment rich in growth factors, cytokines and hormones. CD34+ CD45+ HPC have been reported in non-pregnant endometrium (Lynch et al. 2007) and early pregnancy decidua (Keskin et al. 2007; Vacca et al. 2011; Szereday et al. 2012), with a frequency of 0.1–4 %. CD34+ CD45+ HPC purified from decidua and cultured in decidual stromal cell-conditioned medium or combinations of c-kit ligand (KL), IL-15, Flt3L and IL-7 produced $CD56^{bright}$ CD16- CD9+ uNK-like cells (Keskin et al. 2007; Vacca et al. 2011). Transplantation of human proliferative endometrium into nonobese diabetic/severe combined immunodeficiency/γCnull immunodeficient mice provides additional evidence for local differentiation; increased CD56+ uNK cells were detected after hormone treatment mimicking the menstrual cycle (Matsuura-Sawada et al. 2005). In contrast, Male et al. (2010) detected stage 3 NK cell precursors (CD34- CD117+ CD94-) in uterine mucosa that were able to develop into stage 4 mature (CD34- CD117+/– CD94+) NK cells in vitro but did not detect CD34+ HPC in non-pregnant endometrial pipelle samples; other studies used curettage (Lynch et al. 2007) or hysterectomy samples (Matsuura-Sawada et al. 2005) which, in contrast with pipelle samples, would consistently include the stratum basalis which is preserved after menstruation and would be a likely source of CD34+ HPC.

Although mesenchymal stem cells have been reported in endometrium (Gargett and Masuda 2010), whether there are haematopoietic stem cells (HSCs) in non-pregnant endometrium is uncertain. It could be proposed that stem cells residing in the stratum basalis, which is retained after menstruation, could differentiate into uNK cells as the cycle progresses. We have demonstrated rare CD45+ CD34+ double-labelled cells in non-pregnant endometrium with most cells within the stratum basalis. It is interesting to note that if uNK cells do develop locally within endometrium during each menstrual cycle, abnormalities which have been detected in, for example, recurrent miscarriage (see below) would point to abnormal regulation within the endometrium as the cells develop and proliferate locally or an inherent defect in the stem cell population leading to abnormal development within each individual cycle.

Locally secreted chemokines/cytokines could attract distinct mature or immature NK cell subsets to endometrium and mediate further local differentiation. Male et al. (2010) proposed migration of stage 3 immature NK cells into endometrium and differentiation in situ within the endometrium into uNK cells. In contrast, Keskin et al. (2007) reported conversion of purified peripheral blood CD16+ CD9- NK cells into CD16- CD9+ uNK-like cells after culture with decidual stromal cell-conditioned medium or TGF-β1, suggesting conversion of peripheral blood NK cells to uNK cells locally within endometrium. Further evidence that uNK cells are an immature population comes from a report that 60 % of decidual uNK cells are CD11b- CD27- (Fu et al. 2011). There is also clear evidence that uNK cells are able to proliferate within endometrium. Stromal mitotic figures which are often seen in late secretory phase endometrium are due to proliferation in uNK cells (Pace et al. 1989), and several studies have demonstrated expression of the Ki67 proliferation marker by uNK cells (Pace et al. 1989; Kämmerer et al. 1999). In a study of CD56+ cells purified from non-pregnant endometrium by immunomagnetic selection, uNK cell expression of Ki67 was highest (>40 %) in the mid- and late secretory phase, reducing to ~12 % in early pregnancy decidua (Jones et al. 1998). Using a range of different techniques, Kämmerer et al. (1999) demonstrated proliferation in CD56+ cells in decidua from 5 to 11 weeks gestational age, with 7–23.5 % CD56+ cells co-expressing Ki67. CD56+ cells in pseudodecidualised endometrium after progesterone treatment also express Ki67.

Whether there is local development from HSCs within endometrium or local modification of peripheral blood NK cells recruited into endometrium, it is clear that there is local development of uNK cells within endometrium. In addition, the phenotype of uNK cells in early pregnancy varies with gestational age, suggesting ongoing differentiation: reduced expression of KIR specific for HLA-C (KIR2DL1/S1 and KIRDDL3/L2/S2) with increasing gestational age from 6 to 12 weeks (Sharkey et al. 2008; Marlin et al. 2012) has been reported, as well as the reduction of the CD85j+NKG2D- subset and increase of the CD85j- NKG2D+ with increasing gestation from 8 to 12 weeks (Marlin et al. 2012). In addition, in double immunohistochemical labelling studies, we noted increased expression of CD122 by CD56+ cells in LH+13 endometrium compared with LH+7 endometrium (Otun et al. 2009). It is possible that the origin of uNK cells differs between non-pregnant endometrium and also different stages of pregnancy, with varying populations being recruited as their required function changes with implantation and early placentation.

5.2.4.3 Regulation of uNK Cells by Growth Factors in Endometrium

Endometrial stromal cells undergo differentiation into decidual cells in the luteal phase, and human uNK cells localise to areas of stromal decidualisation. This association suggests a role for decidualised endometrial stromal cell products in uNK cell accumulation. A range of growth factors are produced within endometrium which could play a role in the development of uNK cells.

In peripheral blood, IL-15 and/or IL-2 promote differentiation and expansion of CD34+ HPCs to CD56+ NK cells (Mrózek et al. 1996). Although IL-2 is not present in normal endometrium, IL-15 has been detected in stromal cells in luteal phase endometrium and early pregnancy decidua; secretion is stimulated by progesterone, although the regulation of this appears complex (Okada et al. 2000), involving IL-1-β (Okada et al. 2004) and IFNγ (Dunn et al. 2002). Studies of luteal phase endometrium in women with recurrent reproductive failure have shown a correlation between uNK cell number and stromal cell IL-15 levels (Mariee et al. 2012). Although no data are available for human uNK cells, IL-11 is required for mouse uNK cell maturation (Ain et al. 2004). As well as epithelial expression, IL-11 is highly expressed in decidualised endometrial stromal cells in luteal phase endometrium (Dimitriadis et al. 2005). CD117 (c-kit) is expressed by the earliest IL-15-responsive HPC throughout human NK cell differentiation (Freud et al. 2006). c-kit ligand (KL, stem cell factor, SCF) triggers haematopoiesis on binding to CD117 (Broudy 1997) and enhances the proliferative action of both IL-2 and IL-15 on human $CD56^{bright}$ NK cells (Benson et al. 2009). Endometrial expression of KL/SCF increases in early pregnancy decidua (Kauma et al. 1996; Umekage et al. 1998). Reports on other NK cell differentiation factors in endometrium are limited: IL-7 and Flt3L have recently been reported in uterine fluid (Hannan et al. 2011); and IL-18 has been proposed as a marker of endometrial function and is related to uNK cell numbers and activation of NKp46 (Petitbarat et al. 2011). Transforming growth factor beta 1 (TGF-β1) inhibits peripheral blood NK cell cytotoxic activity, cytokine production and cell proliferation (Bellone et al. 1995) and suppresses expression of NK cell surface receptors, including NKp30 and NKG2D when cultured with human peripheral blood NK cells (Castriconi et al. 2003). TGF-β1 is present in human endometrium during the menstrual cycle and in early pregnancy (Jones et al. 2006; Omwandho et al. 2010; Lash et al. 2012).

5.2.4.4 Relationship of uNK Cells in Non-pregnant Endometrium and Decidua in Pregnancy

The relationship between uNK cells in non-pregnant endometrium and decidualised endometrium in pregnancy is uncertain, at least in part because the origin of uNK cells remains controversial. Although they were initially reported as absent in proliferative endometrium based on histochemical stains to detect cytoplasmic granules, CD56+ cells are detectable, albeit in relatively low numbers (Bulmer et al. 1991). The increase in numbers is seen from day 22 onwards (Russell et al. 2011, 2013), and at least some of the increase can be accounted for by local proliferation. The phenotype of uNK cells in non-pregnant endometrium has not been investigated to the same extent as those from early pregnancy decidua, where distinct gestational age differences in phenotype and function have been reported. In a study of CD56+ cells separated from non-pregnant endometrium (>98 % CD56+) at various menstrual cycle changes, expression of various cell surface markers was investigated. There was low to no expression of CD3, CD8, CD16, HML-1, L-selectin and CD25 (IL-2 receptor-α) on CD56+ cells isolated from non-pregnant and pregnant endometrium. Expression of CD2, CD49a and CD122 increased from the proliferative to the late secretory phase of the menstrual cycle, whereas CD11a, CD69 and CD49d expression was high and did not vary with menstrual cycle phase, although CD49d levels were significantly reduced in early pregnancy. There were also differential proliferative responses: in contrast with those from early pregnancy decidua, CD56+ cells from non-pregnant endometrium did not proliferate in response to phytohaemagglutinin (Searle et al. 1999).

Manaster et al. (2008) suggested that CD56+ cells in non-pregnant endometrium are immature, non-functional cells awaiting activation in pregnancy. This contrasts with other studies that have shown both proliferative and cytotoxic activity of uNK cells from non-pregnant endometrium (Jones et al. 1997; Searle et al. 1999). More recently, microarray studies (Kopcow et al. 2010) compared CD56+ cells from non-pregnant and early pregnancy endometrium: 450 genes were differentially expressed with >twofold difference, with ~70 % over-expressed in the non-pregnant uNK cell subset, suggesting that uNK cells in non-pregnant endometrium are far from inactive and likely to play an important role in implantation and early placentation.

Studies of uNK cells in non-pregnant endometrium are limited by availability of tissues. As nonsurgical and more conservative treatments for problems associated with heavy menstrual bleeding increase in popularity, it is increasingly difficult to obtain endometrial samples for functional studies of uNK cells in non-pregnant endometrium. This difficulty makes studies of the functional relevance of the increased luteal phase uNK cells that have been reported in some women with recurrent implantation failure (RIF) and recurrent miscarriage (RM) harder to achieve.

5.2.4.5 Function of uNK Cells

The in vivo roles of uNK cells are still not clear, but in vitro studies are providing clues to their function. Because of the difficulty in obtaining sufficient non-pregnant endometrium for cell purification, the vast majority of studies have been performed using uNK cells from early pregnancy decidua, with relatively few studies performed on uNK cells from non-pregnant endometrium.

Cytotoxicity

The recognition of uNK cells as a type of natural killer cell led to early studies of their cytotoxic activity and focus on this function as a means to control trophoblast invasion in early pregnancy. Although uNK cells isolated from early pregnancy decidua exhibit cytotoxic activity against the classical NK cell target K562, this cytotoxic activity is lower than that of peripheral blood NK cells (Ritson and Bulmer 1989; Kopcow et al. 2005). Early reports suggested that uNK cells were able to lyse choriocarcinoma and normal trophoblast cells after IL-2 activation (King and Loke 1990), but the current consensus is uNK cells do not lyse normal trophoblast, due to expression of HLA-G, HLA-E and HLA-C by extravillous trophoblast and expression by uNK cells of inhibitory receptors (Chumbly et al. 1994; Rouas-Freiss et al. 1997; Chen et al. 2010; Chazara et al. 2011). Vascular endothelial growth factor (VEGF)-C also upregulates expression of TAP-1 by extravillous trophoblast; TAP-1 plays a role in peptide loading for MHC class I assembly and antigen presentation in EVT cells, thereby also potentially protecting them from uNK cell cytotoxic activity (Kalkunte et al. 2009). The cytoplasmic granules which are a characteristic feature of uNK cells express perforin and granzyme (King et al. 1993) suggesting that they are capable of effector function, although the proportion expressing these molecules reduces with increasing gestational age (Bulmer et al. 2010). However, although they possess the machinery for effector function, uNK cells have been shown to form immature synapses with K562 cells, failing to polarise their microtubule-organising centres and perforin-containing granules to the synapse (Kopcow et al. 2005). Thus, although NK cells were originally defined by their cytolytic activity, it appears that cytolytic activity is unlikely to be a primary function of uNK cells in either normal or pathological human pregnancy.

Cytokine, Growth Factor and Protease Secretion

uNK cells purified from early pregnancy decidua are a rich source of a range of cytokines and growth factors, including tumour necrosis factor (TNF)-α, IL-10, granulocyte-macrophage colony-stimulating factor (GM-CSF), IL-1-β, TGF-β1, macrophage colony-stimulating factor (M-CSF), leukaemia inhibitor factor (LIF) and interferon (IFN)γ (Saito et al. 1993; Jokhi et al. 1994; Lash et al. 2010a). Interestingly, there are gestational age differences in cytokine secretion by uNK cells: secretion of IL-1-β, GM-CSF (Lash et al. 2010a), IL-6 (Champion et al. 2012), IL-8 (De Oliveira et al. 2010) and IFNγ (Lash et al. 2006a) has been shown to increase from 8–10 to 12–14 weeks gestational age.

uNK cells are also an important source of angiogenic growth factors within early pregnancy decidua. Production of angiopoietin (Ang) 1, Ang2, VEGF-C, placental growth factor (PlGF) and TGF-β1 by uNK cells from early pregnancy decidua has been reported (Lash et al. 2006b), and uNK cells in secretory phase endometrium also produce VEGF-C, Ang1, Ang2 and PlGF (Li et al. 2001). In contrast with cytokine production, uNK cell secretion of Ang2 and VEGF-C appears to reduce from 8–10 to 12–14 weeks gestational age (Lash et al. 2006b). The results of secretion studies suggest that uNK cells at 8–10 weeks gestational age are major local intrauterine producers of angiogenic growth factors, whereas at 12–14 weeks gestational age, they have switched their secretory profiles to become major cytokine producers. As the phenotype of uNK cells alters during the first trimester of pregnancy, the change from secretion of angiogenic growth factors to cytokines as pregnancy progresses may reflect these phenotypic alterations. Indeed, it is possible that the proportion of uNK cells developing locally or trafficking from the peripheral blood varies at different stages of the menstrual cycle and pregnancy in both non-pregnant endometrium and decidua. Detailed functional investigation of phenotypically distinct uNK cell populations in accurately dated samples would be required to resolve this question. An alternative explanation is that exposure to the changing intrauterine environment induces functional changes in uNK cells. There is evidence to suggest that exposure of peripheral blood NK cells to soluble HLA-G induces a senescent phenotype that is characterised by increased secretion of IL-6 and IL-8 (reviewed in Rajagopalan 2014). It could be proposed, therefore, that the exposure of uNK cells to soluble HLA-G within the pregnant uterus could lead to alterations in their secretory profile which could impact on their in vivo function.

During early pregnancy, cytotrophoblast proliferates from the tips of the chorionic villi to form cytotrophoblast columns which extend peripherally to form a cytotrophoblast shell. Extravillous trophoblast from the cytotrophoblast shell then invades into uterine decidua and superficial myometrium. There are two main pathways of trophoblast invasion: interstitial extravillous trophoblast invades through the decidua and superficial myometrium, whereas endovascular trophoblast extends within the lumen of the spiral arteries, transiently replacing the endothelium and ultimately remodelling the uterine spiral arteries from thick-walled musculoelastic vessels into dilated tubes whose wall comprises fibrinoid material containing intramural trophoblast cells (Pijnenborg et al. 2006). Both interstitial trophoblast invasion and spiral artery remodelling require breakdown of extracellular matrix (ECM) by proteolytic enzymes. Uterine NK cells secrete metalloproteinase (MMP)-1, MMP-2, MMP-7, MMP-9, MMP-10, tissue inhibitor of metalloproteinases (TIMP)-1, TIMP-2, TIMP-3, urokinase plasminogen activator (uPA) and uPA receptor (uPAR), although not plasminogen activator inhibitor (PAI)-1 and PAI-2 (Naruse et al. 2009a, b). In contrast with cytokines and angiogenic growth factors, no gestational age differences have been detected in secretion of these proteolytic enzymes. Immunoreactivity of MMP-7 and MMP-9 by leucocytes surrounding spiral arteries during early pregnancy has also been reported (Smith et al. 2009).

Both trophoblast invasion and spiral artery remodelling are tightly controlled processes, deficient trophoblast invasion and vascular transformation being associated with pregnancy pathology such as pre-eclampsia, fetal growth restriction, late miscarriage and preterm delivery (Pijnenborg et al. 2006). By their secretion of cytokines and angiogenic growth factors, uNK cells have been suggested to play a central role in these processes in early pregnancy.

Regulation of Trophoblast Invasion by uNK Cells

Hanna et al. (2006) demonstrated that IL-15-stimulated uNK cell supernatants stimulate invasion of isolated cytotrophoblast cells in vitro, this stimulatory effect being partially abrogated by neutralising antibodies to IL-8 and IP-10 (Hanna et al. 2006). This result is supported in part by studies demonstrating that uNK cell supernatants from 12 to 14 weeks gestational age stimulated invasion of extravillous trophoblast from placental explants (Lash et al. 2010b). This uNK-mediated stimulation of EVT invasion was partially abrogated in the presence of an IL-8 neutralising antibody (De Oliveira et al. 2010). In contrast, when both uNK cells and placenta were from 8 to 10 weeks gestational age, uNK cell supernatants had no effect on extravillous trophoblast invasion (Lash et al. 2010b; De Oliveira et al. 2010). Hu et al. (2006)

demonstrated that IL-15-stimulated uNK cell supernatants inhibited migration of EVT in a two-dimensional migration assay by a mechanism dependent on IFNγ. Discrepancies between reports may, therefore, be due to variation in the gestational ages of the samples studied, as well as sources of trophoblast cells and the presence of prior activation of uNK cells by IL-15.

Role of uNK Cells in Spiral Artery Remodelling

Spiral artery remodelling is a key feature of successful human pregnancy (Pijnenborg et al. 2006). This process has often been attributed solely to the effect of extravillous trophoblast, but increasing evidence indicates that there is a 'trophoblast-independent' phase of spiral artery remodelling (Pijnenborg et al. 2006). Histological studies of the placental bed in early pregnancy demonstrated that initial stages of spiral artery remodelling, including dilatation, some fibrinoid deposition, endothelial swelling and vascular smooth muscle cell (VSMC) separation, occur in the absence of extravillous trophoblast (Craven et al. 1998; Kam et al. 1999), and more recent studies support this view.

uNK cells are often seen aggregated around the spiral arteries and arterioles in both the secretory phase of the menstrual cycle and in early human pregnancy; this distribution in pregnancy may reflect a role in mediating vascular changes in pregnancy. Increased numbers of both uNK cells and macrophages were detected within 25 μm of the vessel lumen in human decidual spiral arteries showing partial remodelling and an absence of EVT, compared with non-remodelled vessels and those with greater degrees of remodelling, including the presence of extravillous trophoblast (Smith et al. 2009). Using in vitro vessel models (either chorionic plate arteries from human term placenta or non-pregnant myometrial arteries), we demonstrated that supernatants produced by uNK cells from 8 to 10 weeks gestation were able to mediate separation of vascular smooth muscle cells within the vessels, while uNK cell supernatants from 12 to 14 weeks gestation had a greater effect on vascular smooth muscle cell dedifferentiation (Robson et al. 2012). The uNK cell-derived factors responsible for mediating the effects on uterine spiral arteries remain to be fully determined, but evidence from our in vitro studies implicates Ang1, Ang2 and VEGF-C in this process (Robson et al. 2012). It is possible that the stimulation of trophoblast invasion by uNK cells at 12–14 weeks gestational age may play a role in attracting interstitial trophoblast cells towards spiral arteries that have been partially remodelled by uNK cells for completion of the remodelling process.

5.2.4.6 Interaction of uNK Cells with Other Cell Types

uNK cells are intimately associated with a range of cell types in both non-pregnant and pregnant endometrium, including other endometrial leucocyte populations. Because of the intimate association of uNK cells with extravillous trophoblast in decidua basalis, several studies have focused on interactions between these two cell types.

Uterine NK Cell: Trophoblast Cell Interactions

Although early studies of uNK cells focused on the ability of uNK cells to lyse normal and pathological trophoblast cells, current studies are focused on their secretion of cytokines. Interactions could occur indirectly by secretion of growth factors and cytokines or directly via ligand-receptor interactions.

A key distinguishing feature of extravillous trophoblast, compared with villous cytotrophoblast and syncytiotrophoblast which do not express class I MHC antigens, is their expression of HLA-C, HLA-E and HLA-G, but not the classical, polymorphic MHC molecules HLA-A and HLA-B. HLA-C alleles are recognised by both inhibitory and activating KIRs, including KIR2DL1, KIR2DL2, KIR2DL3 and KIR2DS1. Genetic association studies have linked maternal KIR and fetal (paternal) HLA-C haplotypes and reproductive success. It has been suggested that a maternal KIR AA haplotype, with mainly inhibitory KIR in combination with a fetal HLA-C2 haplotype, is associated with pre-eclampsia and recurrent miscarriage (Hiby et al. 2004, 2010), while interaction between a fetal HLA-C2 haplotype and KIR BB, which is activating, may confer reproductive protection. However, it should be noted that the importance of this association has been disputed (Clark 2014), and it is not clear why only a proportion of pregnancies with the AA/C2 combination have compromised reproductive success.

As HLA-G expression appears to be restricted to extravillous trophoblast cells, initial assumptions were that it was important to protect these cells from lysis by uNK cells closely associated in decidua basalis. However, it is now clear that any protective effects can be attributed to interactions of HLA-G with T cells rather than NK cells (van der Meer et al. 2004, 2007), although interactions between HLA-G and uNK cells do appear to result in altered secretion of a range of soluble factors (van der Meer et al. 2004).

Several studies have investigated the effect of interactions between extravillous trophoblast and uNK cells on secretion of cytokines and angiogenic growth factors. Most studies have used either HLA-G-transfected cell lines or the choriocarcinoma cell lines JEG-3 (HLA-G positive) and JAR (HLA-G negative), and in addition, peripheral blood mononuclear cells and uterine mononuclear cells have often been used as a proxy for uNK cells. There have been few studies using primary isolates of extravillous trophoblast and uNK cells. Not surprisingly with the diverse approaches that have been used, the overall results of these studies have been variable.

We performed an extensive study of cytokine and angiogenic growth factor secretion after uNK cell coculture with either extravillous trophoblast or cytotrophoblast, with all cell types being isolated from the same patient. uNK cells and trophoblast were cultured both in direct contact or separated by a 0.4 μm pore filter (Lash et al. 2011). Secretion of both cytokines and angiogenic growth factors was reduced in cocultures, although whether this was due to uNK cells or trophoblast cells or both was not determined. In general, coculture of uNK cells with either extravillous trophoblast or cytotrophoblast produced similar results suggesting that any effects were not mediated by HLA molecules which differ between these two cell types, but by other shared molecules. Furthermore, secretion of angiogenic growth factors was altered after both direct and indirect coculture suggesting mediation by soluble factors, whereas alterations in cytokine secretion were only observed after direct coculture, suggesting involvement of a membrane-bound molecule. However, HLA-C and KIR haplotypes were not determined in this study, and recent studies indicate that secretion profiles may vary according to the HLA-C and KIR haplotypes (Xiong et al. 2013).

Xiong et al. (2013) studied the response of inhibitory KIR2DL1 or activating KIR2DS1 uNK cell subsets to coculture with 721.221 parent cells or those transfected with HLA-C2 (221-C2) by microarray analysis. uNK cell subsets were KIR2DS1+ KIR2DL1- (KIR2DS1 single positive (sp)), KIR2DS1- KIR2DL1+ (KIR2DL1sp), KIR2DS1+ KIR2DL1+ (double positive (dp)) or KIR2DS1- KIR2DL1- (double negative (dn)). Distinct differences in all four uNK cell subsets were demonstrated using cluster analysis; 45 transcripts were altered in the KIR2DL1sp group, 378 in the KIR2DS1sp group, 289 in the dp group and 3 in the dn group. There was little overlap between the 3 positive groups with only 24 altered transcripts common in the KIR2DS1sp and dp groups, despite both expressing KIR2DS1. Furthermore, there were only six common transcripts between the KIR2DL1sp and dp groups, as well as between the two different sp subsets. It is important to note that each individual has different subsets of uNK cells suggesting that differential local production of cytokines and growth factors may occur throughout the placental bed where uNK cells and EVT interact. In addition, the described study specifically concentrated on the effect of HLA-C2 on uNK cell transcription, whereas extravillous trophoblast cells express a range of ligands that would simultaneously interact with uNK cell-expressed receptors.

The functional consequence of interactions between uNK cells and trophoblast remains unclear. Although specific cytokines and growth factors have been implicated in these processes, it is likely that many different cytokines and chemokines are involved in the control of trophoblast invasion and spiral artery remodelling. Close association of extravillous trophoblast cells with uNK cells leads to reduced cytokine secretion, and it is possible that after recruitment of extravillous trophoblast or uNK cells to specific sites in the placental bed by higher cytokine levels, this serves to inhibit their onward passage. Extravillous trophoblasts are naturally highly invasive cells (Lash et al. 2006c), yet cease to invade in the inner third of the myometrium. Fusion of mononuclear trophoblast cells to form non-invasive multinucleate trophoblast giant cells may play a role in the control of trophoblast invasion. In women with focal placenta accreta, the numbers of myometrial multinucleate trophoblast giant cells are highly reduced in areas lacking decidua compared with areas with intact decidua, suggesting that decidual derived factors mediate or initiate this fusion process as the cells move towards the myometrium (Hannon et al. 2012). Furthermore, in placenta adhesiva, the mildest form of adherent placenta in which decidua basalis is present (van Beekhuizen et al. 2009), numbers of uNK cells were markedly reduced (van Beekhuizen et al. 2010). It is tempting to speculate that uNK cells may play a role in the fusion of mononuclear to multinucleate trophoblast.

Interaction of uNK Cells with Other Endometrial Cells

Many different cell types are present in human endometrium, and potentially uNK cells could interact with any of these cells. There is considerable interest in potential interactions between uNK cells and CD209+ (DC-SIGN) dendritic cells (DCs) in early pregnancy decidua. Although many studies have focused on mouse (Blois et al. 2011), there are a few studies that point to important interactions in human pregnancy. Using immunohistochemistry, Kammerer et al. (2003) noted intimate contact between DCs and uNK cells within human decidua, and this was confirmed in studies of decidual leucocyte suspensions in which it was suggested that a proportion of the DC-SIGN+ cells associated with uNK cells were apoptotic (Tirado-González et al. 2012). Investigation of the DC/uNK cell interaction in human pregnancy is at a very early stage, and technical issues in in vitro studies using a decidual source for both cell populations are challenging. However, studies in mouse highlight the potential importance, with evidence suggesting direct control of stromal cell proliferation, angiogenesis and homing and maturation of uNK cell precursors in the pregnant uterus (Blois et al. 2011).

5.2.4.7 uNK Cells in Pregnancy Pathology

As the predominant endometrial leucocyte population in the secretory phase and early pregnancy, when the seeds are sewn for pregnancy problems that may not manifest until much later in gestation, uNK cells have been a focus for studies of pathological pregnancy.

Miscarriage

Miscarriage is a common gynaecological problem; it has been estimated that 11–20 % of all clinically recognised pregnancies are lost before the 20th week of gestation (Everett 1997). Miscarriages are often separated into early (≤12 completed weeks from LMP) and late (≥13 weeks): early miscarriage accounts for the majority, with ~50 % being associated with aneuploidy, whereas late miscarriage affects only 1–2 % of pregnancies (Regan and Rai 2000) and is less likely to be associated with chromosomal abnormalities. Whereas some studies have focused on uNK cell populations in secretory phase endometrium, studies of decidua in miscarriage are hampered by the potential criticism that since clinical presentation may occur days after demise of the pregnancy, any alteration in leucocyte populations may simply reflect the result rather than the cause of the miscarriage.

Having been referred to by several names, such as Kornchenzellen, K cells and endometrial/decidual granulated/granular lymphocytes, the terminology of 'uterine natural killer cell' was adopted following the recognition that these cells are a type of NK cell, expressing NK receptors. Unfortunately, the term may suggest potential for these cells to 'kill' a pregnancy if activated, although there is no evidence for this and activation of uNK cells in pregnancy is considered beneficial. The focus on peripheral blood NK cells and uNK cells in recurrent miscarriage and implantation failure has led to patient demand for testing and treatment. At present, although some associations with NK cells have been reported, the functional consequences and importance of these are not known.

Sporadic Miscarriage Increased uNK cells have been reported in immunohistochemical studies of decidua from women with sporadic miscarriage compared with controls (Zenclussen et al. 2001; Plaisier et al. 2009), although this was not reproduced in an immunohistochemical study of placental bed biopsies from 8 to 20 weeks gestation (Scaife et al. 2004). A recent study of chromosomally normal and abnormal miscarriage reported an increased incidence of NK cell aggregates in both euploid and aneuploid miscarriage compared with elective abortion (Lee et al. 2015). In contrast, Yamamoto et al. (1999) reported reduced CD56+ uNK cells in a flow cytometric study of decidua of women with sporadic miscarriage. Vassiliadou and Bulmer (1996) noted increased numbers of CD57+ NK cells in a subgroup of cases of sporadic miscarriage. Thus, whether numbers or phenotype of uNK cells are altered in sporadic miscarriage remains uncertain.

Altered NK cell function is likely to be more important than any change in NK cell numbers within decidua. Early suggestions that activation of NK cells in decidua could lead to pregnancy loss by cytotoxicity against fetoplacental tissues have not been supported by studies in human pregnancy. Compared with controls, Vassiliadou and Bulmer (1998) detected reduced cytotoxic activity against K562 cells in uNK cells isolated from decidua of women with early sporadic miscarriage. Yamada et al. (2005) reported an increased proportion of perforin-positive uNK cells in decidua associated with sporadic miscarriage compared with controls. In contrast, Nakashima et al. (2008) reported no difference in the proportion of uNK cells expressing perforin and granzyme B in sporadic miscarriage compared with controls, although the proportion of granulysin-positive uNK cells in decidua was increased and this was linked to an increase in EVT apoptosis.

Hence, there are several studies that suggest that uNK cells could play a role in sporadic early pregnancy loss. However, results are conflicting and studies are complicated by the potential of inflammation secondary to pregnancy demise. Furthermore, despite many studies of uNK cell function in the first trimester of pregnancy, understanding of their function around implantation and in very early pregnancy remains very limited.

Recurrent Miscarriage Recurrent miscarriage (RM) is defined as three or more consecutive first trimester miscarriages and in 50 % of cases the cause remains unknown (Quenby and Farquharson 1993; Rai and Regan 2006). Several immunohistochemical studies from different groups have reported increased numbers of uNK cells in mid-secretory phase endometrium from women with a history of RM (Clifford et al. 1999; Quenby et al. 1999; Tuckerman et al. 2007). Quenby et al. (1999) noted significantly higher uNK cell numbers in women who went on to a further miscarriage compared with those who had a live birth, although a subsequent larger study failed to detect such an association (Tuckerman et al. 2007). In contrast, using flow cytometry, others have not detected altered numbers of endometrial uNK cell numbers in women with RM (Lachapelle et al. 1996; Shimada et al. 2004), although Lachapelle et al. (1996) did report reduced CD56^{bright} CD16- and increased CD56^{dim}CD16+ subsets. Michimata et al. (2002) also did not detect any differences in endometrial uNK cell numbers, although RM in this study was defined as two rather than three or more consecutive miscarriages. Although there are discrepancies in results, immunohistochemical studies have consistently shown increased uNK cell numbers in mid-secretory phase endometrium of women with a well-defined history of RM.

It is difficult to interpret studies of uNK cell numbers in the miscarried decidua of women with RM which have differing results; this may reflect the use of immunohistochemistry versus flow cytometry, but in addition whether any changes noted represent 'cause' or 'effect' is uncertain. Chao et al. (1995) reported similar numbers but increased NK activity in decidua associated with RM. Emmer et al. (2002) reported expression of CD16 by CD56+ cells in decidua from RM samples, while Kwak et al. (1995) noted increased expression of CD57+ cells in 29.6 % of RM cases. In contrast, Quack et al. (2001) reported reduced CD56+ cells in decidua from RM, although there were significantly increased numbers of CD25+ cells, suggesting increased leucocyte activation. Yamamoto et al. (1999) also reported reduced numbers of CD56+ cells in RM compared with controls, but an increased proportion expressed CD16.

In summary, reports of increased uNK cells in secretory phase endometrium from women with RM are consistent, but other approaches, including investigation of decidua, have not yielded consistent results. The functional importance of this is largely unknown. Quenby et al. (2009) linked increased uNK cells to altered blood flow in endometrium from women with both RM and RIF. However, little is known of uNK cell function in non-pregnant endometrium, and further studies of normal endometrium are required before the functional significance of these increased uNK cells is known.

Recurrent Implantation Failure and Infertility

There are several studies that have suggested that uNK cells may be implicated in both infertility and RIF, but results are variable. Although, using flow cytometry, Matteo et al. (2007) reported no difference in uNK cells in late secretory phase endometrium in women with RIF compared with normal fertile controls, others have reported increased uNK cells in pre-pregnancy endometrium in association with RIF. Tuckerman et al. (2010) reported a dramatic increase in uNK cells as a proportion of stromal cells in the mid-secretory phase in women with RIF after IVF compared with normal fertile controls, with uNK cell numbers correlating with stromal expression of IL-15 (Mariee et al. 2012). Similarly, Ledee-Bataille et al. (2004) demonstrated elevated uNK cell numbers in women with RIF in association with increased endometrial IL-12 or IL-18. This cohort also showed abnormal uterine artery Doppler compared with normal controls or RIF women without elevated uNK cells and cytokines. Quenby et al. (2009) also reported that elevated uNK cell numbers in women with RIF correlated with abnormal uterine artery Doppler.

Women with unexplained infertility have been less well studied. Klentzeris et al. (1994) reported reduced CD56+ cells in timed luteal phase biopsies from women with unexplained infertility compared with fertile controls. In contrast, using flow cytometry, Fukui et al. (1999) did not demonstrate alteration in overall CD56+ cell numbers but reported increased CD16+ $CD56^{dim}$ NK cells and reduced $CD56^{bright}$ CD16- NK cells in endometrium from women who subsequently went on to have failed IVF compared with those who had a successful pregnancy. McGrath et al. (2009) subsequently reported that while uNK cell numbers were not altered in infertility, uNK cell expression of CD94, CD158a and CD158a was increased, suggesting an altered phenotype in this cell type.

Infertility may be associated with uterine pathology, and although there are few studies, this may also be associated with altered uNK cells. Tremellen and Russell (2012) studied infertile women with adenomyosis and demonstrated increased uNK cells and macrophages in late secretory phase endometrium in women with diffuse or 'adenomyoma'-type adenomyosis compared with focal adenomyosis or no disease. Leucocytes have also been investigated in the endometrium of women with leiomyomata (Kitaya and Yasuo 2010). Compared with controls, uNK cell numbers in proliferative and mid- to late secretory phase endometrium were lower in association with leiomyoma, with reduced uNK cells seen in 'non-near nodule' (i.e. distant from the leiomyoma) endometrium with a further reduction in 'near nodule' endometrium.

Clinical Applications of uNK Assessment in Recurrent Reproductive Failure

The association of altered uNK cells in endometrium has led to testing of uNK cells in endometrium from women suffering RM and RIF. Although increased uNK cell numbers have consistently been detected in a proportion of women suffering RM (Clifford et al. 1999; Quenby et al. 1999; Tuckerman et al. 2007), assessment differs between different laboratories and definitions of 'abnormal' uNK cell levels differs. Since there is variation in uNK cell numbers at different levels in endometrium (e.g. between zona spongiosum and zona compactum), an initial step should be to establish consistent methodology for quantification of uNK cell numbers in endometrium (Lash et al. 2014). Furthermore, the date of assessment may vary from LH+6 to LH+9 in different studies, reflecting a cycle stage when uNK cell numbers can vary dramatically (Russell et al. 2011, 2013). It is essential that sampling is on a specific day of the cycle or that 'normal' ranges for uNK cell numbers are established in large subject groups.

At present, the importance of increased uNK cells in recurrent reproductive failure remains unknown, and whether there is a correlation between high uNK cell levels and pregnancy outcome is uncertain and unclear. A recent meta-analysis did not demonstrate any difference in uNK cell numbers in RM subjects compared with controls, although percentages and numbers of peripheral blood NK cells were increased in women with RM (Seshadri and Sunkara 2014). Tang et al. (2011) conducted a systematic review and concluded that studies of uNK cells were not sufficiently large to determine whether high levels of uNK cells in the luteal phase predicted subsequent miscarriage.

Fetal Growth Restriction and Pre-eclampsia

Both fetal growth restriction (FGR) and pre-eclampsia (PE) are associated with failed spiral artery transformation and trophoblast invasion (Pijnenborg et al. 2006). Since uNK cells have been implicated in the control and stimulation of trophoblast invasion as well as in early stages of spiral artery transformation, it is feasible that altered uNK cell numbers or function may contribute to the aetiology of these pathological pregnancies. Furthermore, as already highlighted, there are genetic association studies that suggest that pre-eclampsia and recurrent miscarriage may be associated with specific maternal KIR haplotypes expressed by uNK cells interacting with paternal HLA-C haplotypes expressed by extravillous trophoblast (Moffett and Colucci 2014).

In common with RM and RIF, reports of uNK cells in FGR and PE are variable. Some groups have reported increased numbers of $CD56^+$ uNK cells in decidua from women with pre-eclampsia compared with age-matched controls (Stallmach et al. 1999; Wilczynski et al. 2003; Bachmayer et al. 2006). However, in an immunohistochemical study of placental bed biopsies, we demonstrated reduced CD56+ uNK cells in both pre-eclampsia and FGR (Williams et al. 2009b). These results are similar to another report of reduced decidual $CD56^+$ uNK cells in women with severe FGR with and without pre-eclampsia, although there were no differences in pre-eclampsia not associated with FGR (Eide et al. 2006). A flow cytometry study of decidual curettings reported no difference in the proportion of CD45+ cells that were CD56+ CD16- in pre-eclampsia compared with controls, although the proportion of CD45+ cells that were CD56+ CD16+ was reduced in pre-eclampsia (Rieger et al. 2009).

Studies of both pre-eclampsia and fetal growth restriction are limited by the fact that their clinical presentation occurs in the second half of pregnancy, whereas the pathogenetic lesions are likely to be established at a much earlier gestational age. Recent studies have investigated uNK cell function in early pregnancy decidua from pregnancies screened prior to pregnancy termination using uterine artery Doppler (Fraser et al. 2012; Wallace et al. 2013, 2014, 2015). Pregnancies were separated into those showing normal or high uterine artery Doppler resistance index as those at least risk (<1 %) and at most risk (21 %) of developing pre-eclampsia. Compared with uNK cells from low-risk pregnancies, those from pregnancies showing a high resistance index were less able to induce apoptosis of vascular smooth muscle cells and produce factors able to stimulate trophoblast invasion (Fraser et al. 2012). The cells were also less chemoattractant for trophoblast and less able to stimulate outgrowth from placental villous explants (Wallace et al. 2013), and uNK cell secretion of various angiogenic growth factors was increased (Wallace et al. 2014).

Perhaps most interesting in the context of reports of the importance of paternal HLA-C/maternal KIR interactions in the pathogenesis of pre-eclampsia, a reduced proportion of uNK cells from the high-risk pregnancies expressed KIR2DL/S1, 3 and 5 and LILRB1 (Wallace et al. 2015), raising the possibility of an altered interaction with extravillous trophoblast via class I MHC antigens. Whether these results will be confirmed by others is not yet known, and it is notable that only a proportion of those pregnancies with a high uterine artery Doppler resistance index will go on to develop pre-eclampsia. Furthermore, the studies are based on vascular smooth muscle cell and extravillous trophoblast cell lines which may not be representative of the primary cells. However, these studies represent an exciting new approach to investigation of pre-eclampsia and potentially other pregnancy disorders that are associated with deficient trophoblast invasion.

Could uNK Cells Interact with Seminal Fluid Components?

Investigation of the role of uNK cells in pregnancy disorders has mainly focused on their potential role in mediating successful implantation and early placentation. However, uNK cells are potentially exposed to components within the seminal fluid, and this is an alternative mechanism whereby their function and/or differentiation and proliferation may be disturbed, thereby contributing to reproductive failure. Despite several studies, many in animals, that have indicated that seminal fluid contributes to reproductive success by signalling to elicit functional adaptations in the female (Robertson 2007; Schjenken and Robertson 2014), no studies have considered the potential effect of seminal fluid on uNK cell function.

Seminal fluid contains cytokines and prostaglandins that are synthesised in the male accessory glands which are transferred to the female at insemination. Early studies in humans demonstrated an influx of leucocytes, predominantly neutrophil polymorphs, into the cervix following insemination (Thompson et al. 1992), and more recent studies have demonstrated that the seminal fluid induces production of GM-CSF, IL-6, IL-8, MCP-1, MIP-3-α and IL-1-α (Sharkey et al. 2012). Evidence in mice suggests that seminal fluid induces T regulatory cells (Guerin et al. 2011), and there is some evidence for a similar effect in humans, although evidence to date is not definitive (Schjenken and Robertson 2014). Molecules within seminal fluid which may play a role in signalling are TGF-β, HLA-G5 and PGE_2 (Kelly and Critchley 1997; Hutter and Dohr 1998; Robertson et al. 2002; Schjenken and Robertson 2014), and it is feasible that these could affect the function of uNK cells within endometrium. TGF-β affects the differentiation of NK cells, and incubation of peripheral blood NK cells in TGF-β1 has been shown to result in formation of cells with uNK-like phenotype (Keskin et al. 2007). TGF-β1 also affects uNK cell function by altering expression of cytokines (Eriksson et al. 2004). Soluble HLA-G has been detected in seminal fluid (Larsen et al. 2011); although effects on uNK cells have not been studied directly, it has been shown that soluble HLA-G induces a senescent phenotype in peripheral blood NK cells, altering their cytokine secretion profile (reviewed in Rajagopalan 2014).

Several pregnancy complications are commoner in the first pregnancy or when pregnancy occurs with a new partner in multiparous women, emphasising the importance of a paternal component in successful reproduction. Furthermore, women who have had recurrent miscarriages with one partner may go on to have successful pregnancy with a different partner. Although the focus has been on interactions of endometrial leucocytes with placental trophoblast cells as the pathogenetic mechanism in problems such as recurrent miscarriage and pre-eclampsia, the possibility that problems arise much earlier in the process should be considered. Given the prominence of uNK cells in endometrium around the time of implantation, as well as the known effects on NK cells in general, and in some cases on uNK cells, of the various components of seminal fluid, it is surprising that attention has not focused on the potential effect of seminal fluid on uNK cell function. This may be a promising area for future investigation.

5.3 Conclusions

Uterine NK cells have been the focus of many studies since it was appreciated that they are a major leucocytic component of the endometrial stroma at the time of implantation and in early pregnancy. There have been considerable advances in knowledge, but studies of very early pregnancy are limited by availability of tissues from humans and the limitations of extrapolating from animal (predominantly mouse) models. Functional studies of uNK cells from early pregnancy have highlighted their functions, with particular focus on secretion of cytokines and angiogenic growth factors pointing to roles in facilitation and control of trophoblast invasion and spiral artery remodelling in early pregnancy. Despite studies suggesting altered uNK cell numbers in recurrent reproductive failure, the functional consequences remain uncertain. It is important that the mechanisms that underlie the increased numbers of uNK cells as well as the functional consequences are determined before it will be possible to develop effective diagnostic and therapeutic approaches. Furthermore, the possible contribution of male factors such as components within seminal fluid should be considered in the context of uNK cell differentiation and function in recurrent reproductive failure.

References

Ain R, Trinh ML, Soares MJ (2004) Interleukin-11 signaling is required for the differentiation of natural killer cells at the maternal-fetal interface. Dev Dyn 231:700–708

Aluvihare VR, Kallikourdis M, Betz AG (2004) Regulatory T cells mediate maternal tolerance to the fetus. Nat Immunol 5:266–271

Bachmayer N, Rafik Hamad R, Liszka L, Bremme K, Sverremark-Ekstrom E (2006) Aberrant uterine natural killer (NK)-cell expression and altered placental and serum levels of the NK-cell promoting cytokine interleukin-12 in pre-eclampsia. Am J Reprod Immunol 56:292–301

Bellone G, Aste-Amezaga M, Trinchieri G, Rodeck U (1995) Regulation of NK cell functions by TGF-beta 1. J Immunol 155:1066–1073
Benson DM Jr, Yu J, Beckneil B, Wei M, Freud AG, Ferketich AK, Trotta R, Perrotti D, Briesewitz R, Caligiuri MA (2009) Stem cell factor and interleukin-2/15 combine to enhance MAPK-mediated proliferation of human natural killer cells. Blood 113:2706–2714
Berbic M, Fraser IS (2013) Immunology of normal and abnormal menstruation. Womens Health (Lond Engl) 9:387–395
Blois SM, Klapp BF, Barrientos G (2011) Decidualization and angiogenesis in early pregnancy: unravelling the functions of DC and NK cells. J Reprod Immunol 88:86–92
Broudy VC (1997) Stem cell factor and hematopoiesis. Blood 90:1345–1364
Bulmer JN, Lash GE (2005) Human uterine natural killer cells: a reappraisal. Mol Immunol 42:511–521
Bulmer JN, Lunny DP, Hagin SV (1988a) Immunohistochemical characterization of stromal leucocytes in nonpregnant human endometrium. Am J Reprod Immunol Microbiol 17:83–90
Bulmer JN, Smith J, Morrison L, Wells M (1988b) Maternal and fetal cellular relationships in the human placental basal plate. Placenta 9:237–246
Bulmer JN, Morrison L, Longfellow M, Ritson A, Pace D (1991) Granulated lymphocytes in human endometrium: histochemical and immunohistochemical studies. Hum Reprod 6:791–798
Bulmer JN, Williams PJ, Lash GE (2010) Immune cells in the placental bed. Int J Dev Biol 54:281–294
Castriconi R, Cantoni C, Della Chiesa M, Vitale M, Marcenaro E, Conte R, Biassoni R, Bottino C, Moretta L, Moretta A (2003) Transforming growth factor beta 1 inhibits expression of NKp30 and NKG2D receptors: consequences for the NK-mediated killing of dendritic cells. Proc Natl Acad Sci USA 100:4120–4125
Champion H, Innes BA, Robson SC, Lash GE, Bulmer JN (2012) Effects of interleukin-6 on extravillous trophoblast invasion in early human pregnancy. Mol Hum Reprod 18:391–400
Chao KH, Yang YS, Ho HN, Chen SU, Chen HF, Dai HJ, Huang SC, Gill TJ 3rd (1995) Decidual natural killer cytotoxicity decreased in normal pregnancy but not in anembryonic pregnancy and recurrent spontaneous abortion. Am J Reprod Immunol 34:274–280
Chazara O, Xiong S, Moffett A (2011) Maternal KIR and fetal HLA-C: a fine balance. J Leukoc Biol 90:703–716
Chen LJ, Han ZQ, Zhou H, Zou L, Zou P (2010) Inhibition of HLA-G expression via RNAi abolishes resistance of Extravillous trophoblast cell line TEV-1 to NK lysis. Placenta 31:519–527
Chumbly G, King A, Robertson K, Holmes N, Loke YW (1994) Resistance of HLA-G and HLA-A2 transfectants to lysis by decidual NK cells. Cell Immunol 155:312–322
Clark DA (2014) Popular myths in reproductive immunology. J Reprod Immunol 104–105:54–62
Clifford K, Flanagan AM, Regan L (1999) Endometrial CD56+ natural killer cells in women with recurrent miscarriage: a histomorphometric study. Hum Reprod 14:2727–2730
Cooper MA, Fehniger TA, Caligiuri MA (2001) The biology of human natural killer-cell subsets. Trends Immunol 22:633–640
Craven CM, Morgan T, Ward K (1998) Decidual spiral artery remodelling begins before cellular interaction with cytotrophoblasts. Placenta 19:241–252
De Oliveira LG, Lash GE, Murray-Dunning C, Bulmer JN, Innes BA, Searle RF, Sass N, Robson SC (2010) Role of interleukin 8 in uterine natural killer cell regulation of extravillous trophoblast cell invasion. Placenta 31:595–601
Dietl J, Hönig A, Kämmerer U, Rieger L (2006) Natural killer cells and dendritic cells at the human feto-maternal interface: an effective cooperation? Placenta 27:341–347
Dimitriadis E, White CA, Jones RL, Salamonsen LA (2005) Cytokines, chemokines and growth factors in endometrium related to implantation. Hum Reprod Update 11:613–630
Disep B, Innes BA, Cochrane HR, Tijani S, Bulmer JN (2004) Immunohistochemical characterization of endometrial leucocytes in endometritis. Histopathology 45:625–632
Dunn CL, Critchley HO, Kelly RW (2002) IL-15 regulation in human endometrial stromal cells. J Clin Endocrinol Metab 87:1898–1901

Eide IP, Rolfseng T, Isaksen CV, Mecsei R, Roald B, Lydersen S, Salvesen KA, Harsem NK, Austgulen R (2006) Serious foetal growth restriction is associated with reduced proportions of natural killer cells in decidua basalis. Virchows Arch 448:269–276

Emmer PM, Steegers EA, Kerstens HM, Bulten J, Nelen WL, Boer K, Joosten I (2002) Altered phenotype of HLA-G expressing trophoblast and decidual natural killer cells in pathological pregnancies. Hum Reprod 17:1072–1080

Eriksson M, Meadows SK, Wira CR, Sentman CL (2004) Unique phenotype of human uterine NK cells and their regulation by endogenous TGF-beta. J Leukoc Biol 76:667–675

Evans J, Salamonsen LA (2012) Inflammation, leukocytes and menstruation. Rev Endocr Metab Disord 13:277–288

Everett C (1997) Incidence and outcome of bleeding before the 20th week of pregnancy: prospective study from general practice. BMJ 315:32–34

Fraser R, Whitley GS, Johnstone AP et al (2012) Impaired decidual natural killer cell regulation of vascular remodelling in early human pregnancies with high uterine artery resistance. J Pathol 228:322–332

Freud AG, Yokohama A, Becknell B, Lee MT, Mao HC, Ferketich AK, Caligiuri MA (2006) Evidence for discrete stages of human natural killer cell differentiation in vivo. J Exp Med 203:1033–1043

Fu B, Wang F, Sun R, Ling B, Tian Z, Wei H (2011) CD11b and CD27 reflect distinct population and functional specialization in human natural killer cells. Immunology 133:350–359

Fu B, Li X, Sun R, Tong X, Ling B, Tian Z, Wei H (2013) Natural killer cells promote immune tolerance by regulating inflammatory TH17 cells at the human maternal-fetal interface. Proc Natl Acad Sci USA 110:E231–240

Fukui A, Fujii S, Yamaguchi E, Kimura H, Sato S, Saito Y (1999) Natural killer cell subpopulations and cytotoxicity for infertile patients undergoing in vitro fertilization. Am J Reprod Immunol 41:413–422

Gardner L, Moffett A (2003) Dendritic cells in the human decidua. Biol Reprod 69:1438–1446

Gargett CE, Masuda H (2010) Adult stem cells in the endometrium. Mol Hum Reprod 16:818–834

Gibson DA, McInnes KJ, Critchley HO, Saunders PT (2013) Endometrial Intracrinology—generation of an estrogen-dominated microenvironment in the secretory phase of women. J Clin Endocrinol Metab 98:E1802–1806

Guerin LR, Moldenhauer LM, Prins JR, Bromfield JJ, Hayball JD, Robertson SA (2011) Seminal fluid regulates accumulation of FOXP3+ regulatory T cells in the preimplantation mouse uterus through expanding the FOXP3+ cell pool and CCL19-mediated recruitment. Biol Reprod 85:397–408

Gustafsson C, Mjösberg J, Matussek A, Geffers R, Matthiesen L, Berg G, Sharma S, Buer J, Ernerudh J (2008) Gene expression profiling of human decidual macrophages: evidence for immunosuppressive phenotype. PLoS One 3, e2078

Hamperl H, Hellweg G (1958) Granular endometrial stroma cells. Obstet Gynecol 11:379–387

Hanna J, Goldman-Wohl D, Hamani Y, Avraham I, Greenfield C, Natanson-Yaron S, Prus D, Cohen-Daniel L, Arnon TI, Manaster I, Gazit R, Yutkin V, Benharroch D, Porgador A, Keshet E, Yagel S, Mandelboim O (2006) Decidual NK cells regulate key developmental processes at the human fetal-maternal interface. Nat Med 12:1065–1074

Hannan NJ, Paiva P, Meehan KL, Rombauts LJ, Gardner DK, Salamonsen LA (2011) Analysis of fertility-related soluble mediators in human uterine fluid identifies VEGF as a key regulator of embryo implantation. Endocrinology 152:4948–4956

Hannon T, Innes BA, Lash GE, Bulmer JN, Robson SC (2012) Effects of local decidua on trophoblast invasion and spiral artery remodeling in focal placenta creta—an immunohistochemical study. Placenta 33:998–1004

Heikkinen J, Mottonen M, Komi J, Alanen A, Lassila O (2003) Phenotypic characterization of human decidual macrophages. Clin Exp Immunol 131:498–505

Helige C, Ahammer H, Moser G, Hammer A, Dohr G, Huppertz B, Sedlmayr P (2014) Distribution of decidual natural killer cells and macrophages in the neighbourhood of the trophoblast invasion front: a quantitative evaluation. Hum Reprod 29:8–17

Henderson TA, Saunders PTK, Moffett-King A, Groome NP, Critchley HOD (2003) Steroid receptor expression in uterine natural killer cells. J Clin Endocrinol Metab 88:440–449
Hiby SE, Walker JJ, O'shaughnessy KM, Redman CW, Carrington M, Trowsdale J, Moffett A (2004) Combinations of maternal KIR and fetal HLA-C genes influence the risk of preeclampsia and reproductive success. J Exp Med 200:957–965
Hiby SE, Apps R, Sharkey AM, Farrell LE, Gardner L, Mulder A, Claas FH, Walker JJ, Redman CC, Morgan L, Tower C, Regan L, Moore GE, Carrington M, Moffett A (2010) Maternal activating KIRs protect against human reproductive failure mediated by fetal HLA-C2. J Clin Invest 120:4102–4110
Houser BL (2012) Decidual macrophages and their roles at the maternal-fetal interface. Yale J Biol Med 85:105–118
Houser BL, Tilburgs T, Hill J, Nicotra ML, Strominger JL (2011) Two unique human decidual macrophage populations. J Immunol 186:2633–2642
Hu Y, Dutz JP, Maccalman CD, Yong P, Tan R, Von Dadelszen P (2006) Decidual NK cells alter in vitro first trimester extravillous cytotrophoblast migration: a role for IFN-gamma. J Immunol 177:8522–8530
Hutter H, Dohr G (1998) HLA expression on immature and mature human germ cells. J Reprod Immunol 38:101–122
Jokhi PP, King A, Loke YW (1994) Production of granulocyte-macrophage colony-stimulating factor by human trophoblast cells and by decidual large granular lymphocytes. Hum Reprod 9:1660–1669
Jones RK, Bulmer JN, Searle RF (1997) Cytotoxic activity of endometrial granulated lymphocytes during the menstrual cycle in humans. Biol Reprod 57:1217–1222
Jones RK, Searle RF, Stewart JA, Turner S, Bulmer JN (1998) Apoptosis, bcl-2 expression, and proliferative activity in human endometrial stroma and endometrial granulated lymphocytes. Biol Reprod 58:995–1002
Jones RL, Stoikos C, Findlay JK, Salamonsen LA (2006) TGF-beta superfamily expression and actions in the endometrium and placenta. Reproduction 132:217–232
Kalkunte SS, Mselle TF, Norris WE, Wira CR, Sentman CL, Sharma S (2009) Vascular endothelial growth factor C facilitates immune tolerance and endovascular activity of human uterine NK cells at the maternal-fetal interface. J Immunol 182:4085–4092
Kam EP, Gardner L, Loke YW, King A (1999) The role of trophoblast in the physiological change in decidual spiral arteries. Hum Reprod 14:2131–2138
Kämmerer U, Marzusch K, Kröber S, Ruck P, Handgretinger R, Dietl J (1999) A subset of CD56+ large granular lymphocytes in first-trimester human decidua are proliferating cells. Fertil Steril 71:74–79
Kammerer U, Eggert AO, Kapp M, McLellan AD, Geijtenbeek TB, Dietl J, van Kooyk Y, Kampgen E (2003) Unique appearance of proliferating antigen-presenting cells expressing DC-SIGN (CD209) in the decidua of early human pregnancy. Am J Pathol 162:887–896
Kauma S, Huff T, Krystal G, Ryan J, Takacs P, Turner T (1996) The expression of stem cell factor and its receptor, c-kit in human endometrium and placental tissues during pregnancy. J Clin Endocrinol Metab 81:1261–1266
Kelly RW, Critchley HO (1997) Immunomodulation by human seminal plasma: a benefit for spermatozoon and pathogen? Hum Reprod 12:2200–2207
Keskin DB, Allan DSJ, Rybalov B, Andzelm MM, Stern JNH, Kopcow HD, Koopman LA, Strominger JL (2007) TGFβ promotes conversion of CD16+ peripheral blood NK cells into CD16- NK cells with similarities to decidual NK cells. Proc Natl Acad Sci USA 104:3378–3383
King A, Loke YW (1990) Human trophoblast and JEG choriocarcinoma cells are sensitive to lysis by IL-2-stimulated decidual NK cells. Cell Immunol 129:435–448
King A, Wooding P, Gardner L, Loke YW (1993) Expression of perforin, granzyme A and TIA-1 by human uterine CD56+ NK cells implies they are activated and capable of effector functions. Hum Reprod 8:2061–2067
Kitaya K, Yasuo T (2010) Leukocyte density and composition in human cycling endometrium with uterine fibroids. Hum Immunol 71:158–163

Kitaya K, Yamaguchi T, Yasuo T, Okubo T, Honjo H (2007) Post-ovulatory rise of endometrial CD16(−) natural killer cells: in situ proliferation of residual cells or selective recruitment from circulating peripheral blood? J Reprod Immunol 76:45–53
Klentzeris LD, Bulmer JN, Warren MA, Morrison L, Li TC, Cooke ID (1994) Lymphoid tissue in the endometrium of women with unexplained infertility: morphometric and immunohistochemical aspects. Hum Reprod 9:646–652
Koopman LA, Kopcow HD, Rybalov B, Boyson JE, Orange JS, Schatz F, Masch R, Lockwood CJ, Schachter AD, Park PJ, Strominger JL (2003) Human decidual natural killer cells are a unique NK cell subset with immunomodulatory potential. J Exp Med 198:1201–1212
Kopcow HD, Allan DS, Chen X, Rybalov B, Andzelm MM, Ge B, Strominger JL (2005) Human decidual NK cells form immature activating synapses and are not cytotoxic. Proc Natl Acad Sci USA 102:15563–15568
Kopcow HD, Eriksson M, Mselle TF, Damrauer SM, Wira CR, Sentman CL, Strominger JL (2010) Human decidual uNK cells from gravid uteri and NK cells from cycling endometrium are distinct NK cell subsets. Placenta 31:334–338
Kwak JY, Beaman KD, Gilman-Sachs A, Ruiz JE, Schewitz D, Beer AE (1995) Up-regulated expression of CD56+, CD56+/CD16+, and CD19+ cells in peripheral blood lymphocytes in pregnant women with recurrent pregnancy losses. Am J Reprod Immunol 34:93–99
Lachapelle MH, Miron P, Hemmings R, Roy DC (1996) Endometrial T, B, and NK cells in patients with recurrent spontaneous abortion. Altered profile and pregnancy outcome. J Immunol 156:4027–4034
Larsen MH, Bzorek M, Pass MB, Larsen LG, Nielsen MW, Svendsen SG, Lindhard A, Hviid TV (2011) Human leukocyte antigen-G in the male reproductive system and in seminal plasma. Mol Hum Reprod 17:727–738
Lash GE, Bulmer JN (2011) Do uterine natural killer (uNK) cells contribute to female reproductive disorders? J Reprod Immunol 88:156–164
Lash GE, Otun HA, Innes BA, Kirkley M, De Oliveira L, Searle RF, Robson SC, Bulmer JN (2006a) Interferon-gamma inhibits extravillous trophoblast cell invasion by a mechanism that involves both changes in apoptosis and protease levels. FASEB J 20:2512–2518
Lash GE, Schiessl B, Kirkley M, Innes BA, Cooper A, Searle RF, Robson SC, Bulmer JN (2006b) Expression of angiogenic growth factors by uterine natural killer cells during early pregnancy. J Leukoc Biol 80:572–580
Lash GE, Otun HA, Innes BA, Bulmer JN, Searle RF, Robson SC (2006c) Low oxygen concentrations inhibit trophoblast cell invasion from early gestation placental explants via alterations in levels of the urokinase plasminogen activator system. Biol Reprod 74:403–409
Lash GE, Robson SC, Bulmer JN (2010a) Review: functional role of uterine natural killer (uNK) cells in human early pregnancy decidua. Placenta 31:S87–S92
Lash GE, Otun HA, Innes BA, Percival K, Searle RF, Robson SC, Bulmer JN (2010b) Regulation of extravillous trophoblast invasion by uterine natural killer cells is dependent on gestational age. Hum Reprod 25:1137–1145
Lash GE, Naruse K, Robson A, Innes BA, Searle RF, Robson SC, Bulmer JN (2011) Interaction between uterine natural killer cells and extravillous trophoblast cells: effect on cytokine and angiogenic growth factor production. Hum Reprod 26:2289–2295
Lash GE, Innes BA, Drury JA, Robson SC, Quenby S, Bulmer JN (2012) Localization of angiogenic growth factors and their receptors in the human endometrium throughout the menstrual cycle and in recurrent miscarriage. Hum Reprod 27:183–195
Lash GE, Laird SM, Li TC, Innes BA, Mariee N, Patel N, Quenby S, Bulmer JN (2014) Standardisation of uterine natural killer (uNK) cell measurements in endometrium of women with recurrent miscarriage. ESHRE Annual Conference Abstract
Ledee-Bataille N, Dubanchet S, Coulomb-L'hermine A, Durand-Gasselin I, Frydman R, Chaouat G (2004) A new role for natural killer cells, interleukin (IL)-12, and IL-18 in repeated implantation failure after in vitro fertilization. Fertil Steril 81:59–65

Lee JY, Kim EN, Hong JS, Shim JY, Hwang D, Kim KC, Lee J, Kim CJ (2015) Diffuse decidual leucocytoclastic necrosis and NK cell aggregates: two distinct features of miscarriage. J Clin Pathol 68:119–124

Leno-Durán E, Muñoz-Fernández R, Olivares EG, Tirado-González I (2014) Liaison between natural killer cells and dendritic cells in human gestation. Cell Mol Immunol 11:449–455

Li XF, Charnock-Jones DS, Zhang E, Hiby S, Malik S, Day K, Licence D, Bowen JM, Gardner L, King A, Loke YW, Smith SK (2001) Angiogenic growth factor messenger ribonucleic acids in uterine natural killer cells. J Clin Endocrinol Metab 86:1823–1834

Lynch L, Golden-Mason L, Eogan M, O'Herlihy C, O'Farrelly C (2007) Cells with haematopoietic stem cell phenotype in adult human endometrium: relevance to infertility? Hum Reprod 22:919–926

Male V, Hughes T, McClory S, Colucci F, Caligiuri MA, Moffett A (2010) Immature NK cells, capable of producing IL-22, are present in human uterine mucosa. J Immunol 185:3913–3918

Manaster I, Mizrahi S, Goldman-Wohl D, Sela HY, Stern-Ginossar N, Lankry D, Gruda R, Hurwitz A, Bdolah Y, Haimov-Kochman R, Yagel S, Mandelboim O (2008) Endometrial NK cells are special immature cells that await pregnancy. J Immunol 181:1869–1876

Mariee N, Li TC, Laird SM (2012) Expression of leukaemia inhibitory factor and interleukin 15 in endometrium of women with recurrent implantation failure after IVF; correlation with the number of endometrial natural killer cells. Hum Reprod 27:1946–1954

Marlin R, Duriez M, Berkane N, de Truchis C, Madec Y, Rey-Cuille MA, Cummings JS, Cannou C, Quillay H, Barré-Sinoussi F, Nugeyre MT, Menu E (2012) Dynamic shift from CD85j/ILT-2 to NKG2D NK receptor expression pattern on human decidual NK during the first trimester of pregnancy. PLoS One 7, e30017

Marshall RJ, Jones DB (1988) An immunohistochemical study of lymphoid tissue in human endometrium. Int J Gynecol Pathol 3:225–235

Matsuura-Sawada R, Murakami T, Ozawa Y, Nabeshima H, Akahira J, Sato Y, Koyanagi Y, Ito M, Terada Y, Okamura K (2005) Reproduction of menstrual changes in transplanted human endometrial tissue in immunodeficient mice. Hum Reprod 20:1477–1484

Matteo MG, Greco P, Rosenberg P, Mestice A, Baldini D, Falagario T, Martino V, Santodirocco M, Massenzio F, Castellana L, Specchia G, Liso A (2007) Normal percentage of CD56bright natural killer cells in young patients with a history of repeated unexplained implantation failure after in vitro fertilization cycles. Fertil Steril 88:990–993

Maybin JA, Barcroft J, Thiruchelvam U, Hirani N, Jabbour HN, Critchley HO (2012) The presence and regulation of connective tissue growth factor in the human endometrium. Hum Reprod 27:1112–1121

McGrath E, Ryan EJ, Lynch L, Golden-Mason L, Mooney E, Eogan M, O'Herlihy C, O'Farrelly C (2009) Changes in endometrial natural killer cell expression of CD94, CD158a and CD158b are associated with infertility. Am J Reprod Immunol 61:265–276

Michimata T, Ogasawara MS, Tsuda H, Suzumori K, Aoki K, Sakai M, Fujimura M, Nagata K, Nakamura M, Saito S (2002) Distributions of endometrial NK cells, B cells, T cells, and Th2/Tc2 cells fail to predict pregnancy outcome following recurrent abortion. Am J Reprod Immunol 47:196–202

Mjosberg J, Berg G, Jenmalm MC, Ernerudh J (2010) FOXP3+ regulatory T cells and T helper 1, T helper 2, and T helper 17 cells in human early pregnancy decidua. Biol Reprod 82: 698–705

Moffett A, Colucci F (2014) Uterine NK cells: active regulators at the maternal-fetal interface. J Clin Invest 124:1872–1879

Morris H, Edwards J, Tiltman A, Emms M (1985) Endometrial lymphoid tissue: an immunohistological study. J Clin Pathol 38:644–652

Mrózek E, Anderson P, Caligiuri MA (1996) Role of interleukin-15 in the development of human CD56+ natural killer cells from CD34+ hematopoietic progenitor cells. Blood 87:2632–2640

Nagamatsu T, Schust DJ (2010) The contribution of macrophages to normal and pathological pregnancies. Am J Reprod Immunol 63:460–471

Nakashima A, Shiozaki A, Myojo S, Ito M, Tatematsu M, Sakai M, Takamori Y, Ogawa K, Nagata K, Saito S (2008) Granulysin produced by uterine natural killer cells induces apoptosis of extravillous trophoblasts in spontaneous abortion. Am J Pathol 173:653–664
Naruse K, Lash GE, Innes BA, Otun HA, Searle RF, Robson SC, Bulmer JN (2009a) Localization of matrix metalloproteinase (MMP)-2, MMP-9 and tissue inhibitors for MMPs (TIMPs) in uterine natural killer cells in early human pregnancy. Hum Reprod 24:553–561
Naruse K, Lash GE, Bulmer JN, Innes BA, Otun HA, Searle RF, Robson SC (2009b) The urokinase plasminogen activator (uPA) system in uterine natural killer cells in the placental bed during early pregnancy. Placenta 30:398–404
Nevers T, Kalkunte S, Sharma S (2011) Uterine Regulatory T cells, IL-10 and hypertension. Am J Reprod Immunol 66:88–92
Okada S, Okada H, Sanezumi M, Nakajima T, Yasuda K, Kanzaki H (2000) Expression of interleukin-15 in human endometrium and decidua. Mol Hum Reprod 6:75–80
Okada H, Nakajima T, Yasuda K, Kanzaki H (2004) Interleukin-1 inhibits interleukin-15 production by progesterone during in vitro decidualization in human. J Reprod Immunol 61:3–12
Omwandho CO, Konrad L, Halis G, Oehmke F, Tinneberg HR (2010) Role of TGF-betas in normal human endometrium and endometriosis. Hum Reprod 25:101–109
Otun HA, Innes BA, Lash GE, Robson SC, Bulmer JN (2009) Immunohistochemical characterization of uterine NK cell differentiation markers in endometrium and decidua. Reprod Sci 16(Suppl):182A
Pace D, Morrison L, Bulmer JN (1989) Proliferative activity in endometrial stromal granulocytes throughout menstrual cycle and early pregnancy. J Clin Pathol 42:35–39
Pace D, Longfellow M, Bulmer JN (1991) Characterization of intraepithelial lymphocytes in human endometrium. J Reprod Fertil 91:165–174
Parhar RS, Yagel S, Lala PK (1989) PGE2-mediated immunosuppression by first trimester human decidual cells blocks activation of maternal leukocytes in the decidua with potential anti-trophoblast activity. Cell Immunol 120:61–74
Petitbarat M, Rahmati M, Sérazin V, Dubanchet S, Morvan C, Wainer R, de Mazancourt P, Chaouat G, Foidart JM, Munaut C, Lédée N (2011) TWEAK appears as a modulator of endometrial IL-18 related cytotoxic activity of uterine natural killers. PLoS One 6, e14497
Pijnenborg R, Vercruysse L, Hanssens M (2006) The uterine spiral arteries in human pregnancy: facts and controversies. Placenta 27:939–958
Pitman H, Innes BA, Robson SC, Bulmer JN, Lash GE (2013) Altered expression of interleukin-6, interleukin-8 and their receptors in decidua of women with sporadic miscarriage. Hum Reprod 28:2075–2086
Plaisier M, Dennert I, Rost E, Koolwijk P, van Hinsbergh VW, Helmerhorst FM (2009) Decidual vascularization and the expression of angiogenic growth factors and proteases in first trimester spontaneous abortions. Hum Reprod 24:185–197
Quack KC, Vassiliadou N, Pudney J, Anderson DJ, Hill JA (2001) Leukocyte activation in the decidua of chromosomally normal and abnormal fetuses from women with recurrent abortion. Hum Reprod 16:949–955
Quenby SM, Farquharson RG (1993) Predicting recurring miscarriage: what is important? Obstet Gynecol 82:132–138
Quenby S, Bates M, Doig T, Brewster J, Lewis-Jones DI, Johnson PM, Vince G (1999) Pre-implantation endometrial leukocytes in women with recurrent miscarriage. Hum Reprod 14:2386–2391
Quenby S, Nik H, Innes B, Lash G, Turner M, Drury J, Bulmer J (2009) Uterine natural killer cells and angiogenesis in recurrent reproductive failure. Hum Reprod 24:45–54
Rai R, Regan L (2006) Recurrent miscarriage. Lancet 368:601–611
Rajagopalan S (2014) HLA-G-mediated NK cell senescence promotes vascular remodeling: implications for reproduction. Cell Mol Immunol 11:460–466
Regan L, Rai R (2000) Epidemiology and the medical causes of miscarriage. Baillieres Best Pract Res Clin Obstet Gynaecol 14:839–854

Repnik U, Tilburgs T, Roelen DL, van der Mast BJ, Kanhai HH, Scherjon S, Claas FH (2008) Comparison of macrophage phenotype between decidua basalis and decidua parietalis by flow cytometry. Placenta 29:405–412
Rieger L, Honig A, Sutterlin M, Kapp M, Dietl J, Ruck P, Kammerer U (2004) Antigen-presenting cells in human endometrium during the menstrual cycle compared to early pregnancy. J Soc Gynecol Investig 11:488–493
Rieger L, Segerer S, Bernar T, Kapp M, Majic M, Morr AK, Dietl J, Kammerer U (2009) Specific subsets of immune cells in human decidua differ between normal pregnancy and preeclampsia—a prospective observational study. Reprod Biol Endocrinol 7:132
Ritson A, Bulmer JN (1989) Isolation and functional studies of granulated lymphocytes in first trimester human decidua. Clin Exp Immunol 77:263–268
Robertson SA (2007) Seminal fluid signaling in the female reproductive tract: lessons from rodents and pigs. J Anim Sci 85(Suppl):E36–44
Robertson SA, Ingman WV, O'Leary S, Sharkey DJ, Tremellen KP (2002) Transforming growth factor beta–a mediator of immune deviation in seminal plasma. J Reprod Immunol 57:109–128
Robson A, Harris LK, Innes BA, Lash GE, Aljunaidy MM, Aplin JD, Baker PN, Robson SC, Bulmer JN (2012) Uterine natural killer cells initiate spiral artery remodeling in human pregnancy. FASEB J 26:4876–4885
Rouas-Freiss N, Goncalves RM, Menier C, Dausset J, Carosella ED (1997) Direct evidence to support the role of HLA-G in protecting the fetus from maternal uterine natural killer cytolysis. Proc Natl Acad Sci USA 94:11520–11525
Russell P, Anderson L, Lieberman D, Tremellen K, Yilmaz H, Cheerala B, Sacks G (2011) The distribution of immune cells and macrophages in the endometrium of women with recurrent reproductive failure I: Techniques. J Reprod Immunol 91:90–102
Russell P, Sacks G, Tremellen K, Gee A (2013) The distribution of immune cells and macrophages in the endometrium of women with recurrent reproductive failure. III: Further observations and reference ranges. Pathology 45:393–401
Saito S, Nishikawa K, Morii T, Enomoto M, Narita N, Motooshi K, Ichijo M (1993) Cytokine production by $CD16^{-}CD56^{bright}$ natural killer cells in the human early pregnancy decidua. Int Immunol 5:559–563
Salamonsen LA, Zhang J, Brasted M (2002) Leukocyte networks and human endometrial remodelling. J Reprod Immunol 57:95–108
Scaife PJ, Searle RF, Robson SC, Innes BA, Bulmer JN (2004) Comparison of decidual leucocyte populations in spontaneous early pregnancy loss. J Soc Gynecol Investig 11:285A
Scaife PJ, Bulmer JN, Robson SC, Innes BA, Searle RF (2006) Effector activity of decidual CD8+ T lymphocytes in early human pregnancy. Biol Reprod 75:562–567
Schjenken JE, Robertson SA (2014) Seminal fluid and immune adaptation for pregnancy–comparative biology in mammalian species. Reprod Domest Anim 49(Suppl 3):27–36
Schulke L, Manconi F, Markham R, Fraser IS (2008) Endometrial dendritic cell populations during the normal menstrual cycle. Hum Reprod 23:1574–1580
Searle RF, Jones RK, Bulmer JN (1999) Phenotypic analysis and proliferative responses of human endometrial granulated lymphocytes during the menstrual cycle. Biol Reprod 60:871–878
Seshadri S, Sunkara SK (2014) Natural killer cells in female infertility and recurrent miscarriage: a systematic review and meta-analysis. Hum Reprod Update 20:429–438
Sharkey AM, Gardner L, Hiby S, Farrell L, Apps R, Masters L, Goodridge J, Lathbury L, Stewart CA, Verma S, Moffett A (2008) Killer Ig-like receptor expression in uterine NK cells is biased toward recognition of HLA-C and alters with gestational age. J Immunol 181:39–46
Sharkey DJ, Tremellen KP, Jasper MJ, Gemzell-Danielsson K, Robertson SA (2012) Seminal fluid induces leukocyte recruitment and cytokine and chemokine mRNA expression in the human cervix after coitus. J Immunol 188:2445–2454
Shimada S, Iwabuchi K, Kato EH, Morikawa M, Sakuragi N, Onoé K, Minakami H, Yamada H (2003) No difference in natural-killer-T cell population, but Th2/Tc2 predominance in peripheral blood of recurrent aborters. Am J Reprod Immunol 50:334–339

Shimada S, Kato EH, Morikawa M, Iwabuchi K, Nishida R, Kishi R, Onoe K, Minakami H, Yamada H (2004) No difference in natural killer or natural killer T-cell population, but aberrant T-helper cell population in the endometrium of women with repeated miscarriage. Hum Reprod 19:1018–1024

Smith SD, Dunk CE, Aplin JD, Harris LK, Jones RL (2009) Evidence for immune cell involvement in decidual spiral arteriole remodeling in early human pregnancy. Am J Pathol 174:1959–1971

Spornitz UM (1992) The functional morphology of the human endometrium and decidua. Adv Anat Embryol Cell Biol 124:1–99

Stallmach T, Hebisch G, Orban P, Lu X (1999) Aberrant positioning of trophoblast and lymphocytes in the feto-maternal interface with pre-eclampsia. Virchows Arch 434:207–211

Stewart JA, Bulmer JN, Murdoch AP (1998) Endometrial leucocytes: expression of steroid hormone receptors. J Clin Pathol 51:121–126

Szereday L, Miko E, Meggyes M, Barakonyi A, Farkas B, Varnagy A, Bodis J, Lynch L, O'Farrelly C, Szekeres-Bartho J (2012) Commitment of decidual haematopoietic progenitor cells in first trimester pregnancy. Am J Reprod Immunol 67:9–16

Tang AW, Alfirevic Z, Quenby S (2011) Natural killer cells and pregnancy outcomes in women with recurrent miscarriage and infertility: a systematic review. Hum Reprod 26:1971–1980

Thiruchelvam U, Dransfield I, Saunders PT, Critchley HO (2013) The importance of the macrophage within the human endometrium. J Leukoc Biol 93:217–225

Thompson LA, Barratt CL, Bolton AE, Cooke ID (1992) The leukocytic reaction of the human uterine cervix. Am J Reprod Immunol 28:85–89

Tilburgs T, Strominger JL (2013) CD8+ effector T cells at the fetal-maternal interface, balancing fetal tolerance and antiviral immunity. Am J Reprod Immunol 69:395–407

Tilburgs T, Roelen DL, van der Mast BJ, de Groot-Swings GM, Kleijburg C, Scherjon SA, Claas FH (2008) Evidence for a selective migration of fetus-specific CD4+CD25^{bright} regulatory T cells from the peripheral blood to the decidua in human pregnancy. J Immunol 180:5737–5745

Tilburgs T, Claas FH, Scherjon SA (2010) Elsevier Trophoblast Research Award Lecture: Unique properties of decidual T cells and their role in immune regulation during human pregnancy. Placenta 31:S82–S86

Tirado-González I, Muñoz-Fernández R, Prados A, Leno-Durán E, Martin F, Abadía-Molina AC, Olivares EG (2012) Apoptotic DC-SIGN+ cells in normal human decidua. Placenta 33:257–263

Tremellen KP, Russell P (2012) The distribution of immune cells and macrophages in the endometrium of women with recurrent reproductive failure. II: adenomyosis and macrophages. J Reprod Immunol 93:58–63

Trundley A, Moffett A (2004) Human uterine leukocytes and pregnancy. Tissue Antigens 63:1–12

Tuckerman E, Laird SM, Prakash A, Li TC (2007) Prognostic value of the measurement of uterine natural killer cells in the endometrium of women with recurrent miscarriage. Hum Reprod 22:2208–2213

Tuckerman E, Mariee N, Prakash A, Li TC, Laird S (2010) Uterine natural killer cells in peri-implantation endometrium from women with repeated implantation failure after IVF. J Reprod Immunol 87:60–66

Umekage H, Saito S, Morikawa H (1998) Enhancement by stem cell factor of interleukin-2 (IL-2)-induced DNA synthesis in human decidual CD16- CD56bright natural killer cells mediated by increased expression of the IL-2 receptor alpha chain. J Reprod Immunol 40:1–24

Vacca P, Vitale C, Montaldo E, Conte R, Cantoni C, Fulcheri E, Darretta V, Moretta L, Mingari MC (2011) CD34+ hematopoietic precursors are present in human decidua and differentiate into natural killer cells upon interaction with stromal cells. Proc Natl Acad Sci USA 108: 2402–2407

van Beekhuizen HJ, Joosten I, de Groot AN, Lotgering FK, van der Laak J, Bulten J (2009) The number of multinucleated trophoblastic giant cells in the basal decidua is decreased in retained placenta. J Clin Pathol 62:794–797 van Beekhuizen HJ, Joosten I, Lotgering FK, Bulten J, van Kempen LC (2010) Natural killer cells and HLA-G expression in the basal decidua of human placenta adhesiva. Placenta 31: 1078–1084

van der Meer A, Lukassen HGM, van Lierop MJC, Wijnands F, Mosselman S, Braat DDM, Joosten I (2004) Membrane-bound HLA-G activates proliferation and interferon-γ production by uterine natural killer cells. Mol Hum Reprod 10:189–195

van der Meer A, Lukassen HG, van Cranenbroek B, Weiss EH, Braat DD, van Lierop MJ, Joosten I (2007) Soluble HLA-G promotes Th1-type cytokine production by cytokine-activated uterine and peripheral natural killer cells. Mol Hum Reprod 13:123–133

Vassiliadou N, Bulmer JN (1996) Immunohistochemical evidence for increased numbers of 'classic' CD57+ natural killer cells in the endometrium of women suffering spontaneous early pregnancy loss. Hum Reprod 11:1569–1574

Vassiliadou N, Bulmer JN (1998) Functional studies of human decidua in spontaneous early pregnancy loss: effect of soluble factors and purified CD56+ lymphocytes on killing of natural killer- and lymphokine-activated killer-sensitive targets. Biol Reprod 58:982–987

Verma S, King A, Loke YW (1997) Expression of killer cell inhibitory receptors on human uterine natural killer cells. Eur J Immunol 27:979–983

Wallace AE, Host AJ, Whitley GS, Cartwright JE (2013) Decidual natural killer cell interactions with trophoblasts are impaired in pregnancies at increased risk of preeclampsia. Am J Pathol 183:1853–186

Wallace AE, Fraser R, Gurung S, Goulwara SS, Whitley GS, Johnstone AP, Cartwright JE (2014) Increased angiogenic factor secretion by decidual natural killer cells from pregnancies with high uterine artery resistance alters trophoblast function. Hum Reprod 29:652–60

Wallace AE, Whitley GS, Thilaganathan B, Cartwright JE (2015) Decidual natural killer cell receptor expression is altered in pregnancies with impaired vascular remodeling and a higher risk of pre-eclampsia. J Leukoc Biol 97:79–86

Wilczynski JR, Tchorzewski H, Banasik M, Glowacka E, Wieczorek A, Lewkowicz P, Malinowski A, Szpakowski M, Wilczynski J (2003) Lymphocyte subset distribution and cytokine secretion in third trimester decidua in normal pregnancy and preeclampsia. Eur J Obstet Gynecol Reprod Biol 109:8–15

Williams PJ, Searle RF, Robson SC, Innes BA, Bulmer JN (2009a) Decidual leucocyte populations in early to late gestation normal human pregnancy. J Reprod Immunol 82:24–31

Williams PJ, Bulmer JN, Searle RF, Innes BA, Robson SC (2009b) Altered decidual leucocyte populations in the placental bed in pre-eclampsia and foetal growth restriction: a comparison with late normal pregnancy. Reproduction 138:177–184

Wira CR, Fahey JV, Schaefer TM, Pioli PA, Sentman CL, Shen L (2008) Innate and adaptive immunity in the human female reproductive tract:influence of the menstrual cycle and menopause on the mucosal immune system in the uterus. The endometrium: molecular, cellular and clinical perspectives, 2nd edn. Informa Healthcare, New York

Wu X, Jin LP, Yuan MM, Zhu Y, Wang MY, Li DJ (2005) Human first-trimester trophoblast cells recruit CD56brightCD16- NK cells into decidua by way of expressing and secreting of CXCL12/stromal cell-derived factor 1. J Immunol 175:61–68

Xiong S, Sharkey AM, Kennedy PR, Gardner L, Farrell LE, Chazara O, Bauer J, Hiby SE, Colucci F, Moffett A (2013) Maternal uterine NK cell-activating receptor KIR2DS1 enhances placentation. J Clin Invest 123:4264–4272

Yamada H, Shimada S, Morikawa M, Iwabuchi K, Kishi R, Onoe K, Minakami H (2005) Divergence of natural killer cell receptor and related molecule in the decidua from sporadic miscarriage with normal chromosome karyotype. Mol Hum Reprod 11:451–457

Yamamoto T, Takahashi Y, Kase N, Mori H (1999) Decidual natural killer cells in recurrent spontaneous abortion with normal chromosomal content. Am J Reprod Immunol 41:337–342

Yeaman GR, Guyre PM, Fanger MW, Collins JE, White HD, Rathbun W, Orndorff KA, Gonzalez J, Stern JE, Wira CR (1997) Unique CD8+ T cell-rich lymphoid aggregates in human uterine endometrium. J Leukoc Biol 61:427–435

Zenclussen AC, Fest S, Sehmsdorf US, Hagen E, Klapp BF, Arck PC (2001) Upregulation of decidual P-selectin expression is associated with an increased number of Th1 cell populations in patients suffering from spontaneous abortions. Cell Immunol 213:94–103
Zenclussen AC, Gerlof K, Zenclussen ML, Ritschel S, Zambon Bertoja A, Fest S, Hontsu S, Ueha S, Matsushima K, Leber J, Volk HD (2006) Regulatory T cells induce a privileged tolerant microenvironment at the fetal-maternal interface. Eur J Immunol 36:82–94
Zhang J, Chen Z, Fritz JH, Rochman Y, Leonard WJ, Gommerman JL, Plumb AW, Abraham N, Croy BA (2012) Unusual timing of CD127 expression by mouse uterine natural killer cells. J Leukoc Biol 91:417–426

Chapter 6
Seminal Fluid Signalling in the Female Reproductive Tract: Implications for Reproductive Success and Offspring Health

John E. Schjenken and Sarah A. Robertson

Abstract Carriage of sperm is not the only function of seminal fluid in mammals. Studies in mice show that at conception, seminal fluid interacts with the female reproductive tract to induce responses which influence whether or not pregnancy will occur, and to set in train effects that help shape subsequent fetal development. In particular, seminal fluid initiates female immune adaptation processes required to tolerate male transplantation antigens present in seminal fluid and inherited by the conceptus. A tolerogenic immune environment to facilitate pregnancy depends on regulatory T cells (Treg cells), which recognise male antigens and function to suppress inflammation and immune rejection responses. The female response to seminal fluid stimulates the generation of Treg cells that protect the conceptus from inflammatory damage, to support implantation and placental development. Seminal fluid also elicits molecular and cellular changes in the oviduct and endometrium that directly promote embryo development and implantation competence. The plasma fraction of seminal fluid plays a key role in this process with soluble factors, including TGFB, prostaglandin-E, and TLR4 ligands, demonstrated to contribute to the peri-conception immune environment. Recent studies show that conception in the absence of seminal plasma in mice impairs embryo development and alters fetal development to impact the phenotype of offspring, with adverse effects on adult metabolic function particularly in males. This review summarises our current understanding of the molecular responses to seminal fluid and how this contributes to the establishment of pregnancy, generation of an immune-regulatory environment and programming long-term offspring health.

Keywords Seminal fluid • Prostaglandin E • Transforming growth factor-beta • Natural killer cells • Toll-like receptors • Regulatory T cells • Cytokines • Maternal immune tolerance

---

J.E. Schjenken • S.A. Robertson (✉)
Robinson Research Institute, University of Adelaide, Adelaide, SA 5005, Australia

School of Paediatrics and Reproductive Health, University of Adelaide, Adelaide, SA 5005, Australia
e-mail: sarah.robertson@adelaide.edu.au

R. Bronson (ed.), *The Male Role in Pregnancy Loss and Embryo Implantation Failure*, Advances in Experimental Medicine and Biology 868,
DOI 10.1007/978-3-319-18881-2_6

Abbreviations

AI	Artificial Insemination
AMP	Antimicrobial peptide
APC	Antigen-presenting cells
ART	Assisted reproductive technologies
ASG	Accessory sex glands
BSP	Bovine seminal plasma protein
CCL	C-C motif chemokine ligand
COX2	Cyclooxygenase 2
CRISP3	Cysteine-rich secretory protein-3
CSF	Colony-stimulating factor
CXCL	C-X-C motif chemokine ligand
DAMP	Danger-associated molecular patterns
FOXP3	Forkhead box P3
GCSF	Granulocyte colony-stimulating factor
GMCSF	Granulocyte-macrophage colony-stimulating factor
GRO/KC	Growth regulated alpha
IFNG	Interferon gamma
IL	Interleukin
IVF	In vitro fertilisation
JAK/STAT	Janus kinase/Signal transducer and activator of transcription
LIF	Leukaemia inhibitory factor
MAPK	Mitogen-activated protein kinase
MCP1	Monocyte chemotactic protein 1
MHC	Major histocompatibility complex
MIP	Macrophage inflammatory protein
MMP	Matrix metalloproteinase
NK cells	Natural killer cells
OIF	Ovulation-inducing factor
P13K-AKT	Phosphatidylinositol-3 kinase-protein kinase B
PGE	Prostaglandin E
PSP	Porcine sperm adhesion proteins
sFlt1	Soluble fms-like tyrosine kinase-1
SVX	Seminal vesicle deficient
TGFB	Transforming growth factor beta
TIMP	Tissue inhibitor of matrix metalloproteinase
TLR	Toll-like receptor
TNF	Tumor Necrosis Factor
Treg cells	Regulatory T cells
VEGF	Vascular endothelial growth factor

6.1 Introduction

Over the last 20 years, it has become clear that the soluble fraction of the male ejaculate, known as seminal plasma, provides more than simply a transport medium for sperm. Compelling reports in a range of mammalian species demonstrate that seminal fluid interacts with female reproductive tract tissues to induce cellular and molecular changes that impact reproductive function. The post-copulatory environment activated by seminal fluid not only ensures fertilisation but also activates female responses that facilitate conception and sustain progression to pregnancy. Within seminal plasma, a range of soluble signalling factors including hormones and cytokines have been identified (Aumuller and Riva 1992; Maegawa et al. 2002; Mann 1964) and interfering with their function compromises early embryo development and affects the longer-term health of offspring (Bromfield et al. 2014; Poon et al. 2009). Recent advances in male–female seminal fluid signalling mainly focus on the mouse and human, but there is evidence of seminal fluid effects in every mammalian species examined, with considerable variation in the physiological origin, fate, and consequences of seminal fluid constituents between species (Schjenken and Robertson 2014; Robertson 2005). Remarkably, the biological mechanisms and effects have parallels with invertebrate species such as Drosophila, where male seminal fluid influences ovulation and egg storage, the production of antimicrobial proteins, female tract remodelling, and female receptivity to mating and related behaviours (Wolfner 2002; Chapman and Davies 2004). This conservation across taxa, with the tremendous variation across species in signalling proteins and their rapid evolution, implies there is substantial biological value in seminal fluid signalling that is driven by sexual selection with implications at a population level (McGraw et al. 2015).

In this review, we describe our current understanding of how seminal fluid acts in the female reproductive tract after coitus to promote pregnancy and reproductive success. We discuss the active components of seminal fluid and recent findings demonstrating that seminal fluid affects not only the peri-conception environment but also the course of gestation, such that alterations to seminal fluid composition impact fetal growth and offspring development. The majority of relevant studies concern the mouse and human, but since a comparative approach provides insight on mechanisms conserved across species, we note commonalities and differences between the signalling properties of seminal fluid in other mammalian and non-mammalian species. Finally, the clinical implications of these findings will be discussed with a particular focus on the use of seminal fluid or active components of seminal fluid in assisted reproductive technologies (ART), where the frequent absence of seminal plasma may be part of the explanation for their limited success.

6.2 Seminal Fluid Signalling in Invertebrate and Avian Species

Our knowledge of the effects of male accessory gland fluids on female reproductive physiology originated in invertebrates [reviewed in Wolfner (2002) and Gillott (2003)]. In flies, crickets, and other insects where intromission results in seminal fluid delivery to the female tract, the male accessory gland fluids have multiple functions, ultimately improving the likelihood of the male siring offspring. Studies in *Drosophila* are particularly informative and show that accessory gland proteins contribute to the induction of female refractoriness to subsequent mating, support sperm function and storage, induce ovulation and oviposition, regulate egg development, and alter behaviour, metabolism, immune function, and ultimately lifespan of the female. Successful fertilisation and transmission of the male germ line to the next generation depend on the male's capacity to induce an adequate response in the female. Since most intromitting insects are polyandrous (females regularly mate with different males each reproductive cycle), the female tract response is part of a process through which different males compete to sire offspring, in so-called 'sexual conflict'. The female has a high threshold for recognition and sufficient responsiveness to signals in seminal fluid—unless the seminal fluid elicits a sufficient response in the female reproductive tract, ovulation and progression of the reproductive cycle will not occur. Through this process of 'cryptic female choice', females interrogate the reproductive fitness of the prospective male partner and invest reproductive resources accordingly (Roldan et al. 1992; Eberhard 2009). This ensures optimal female investment of reproductive resources and ensures maximal progeny fitness since the next generation are sired only by the most competitive males.

Accessory gland proteins have been identified that contribute to various aspects of the female response. In *Drosophila*, a protein named Acp26Aa (otherwise known as ovulin) acts to trigger ovulation (Heifetz et al. 2000), while Acp36DE contributes to the formation of the anterior plug and facilitates sperm storage (Lung and Wolfner 2001; Chapman et al. 2000). There is a multi-step network of interactions between different seminal fluid proteins to induce behavioural responses lasting several days after mating (Ram and Wolfner 2009). Some male seminal fluid proteins utilise female reproductive tract proteases to achieve full activity, indicating a dynamic interaction between male and female reproductive biology for male effects to be fully exerted (Laflamme et al. 2014). Interestingly, Toll pathway and Imd pathway (which akin to mammalian Toll like receptors (TLRs), are activated by endogenous and pathogen-associated molecular structures) act to mediate the effects of seminal fluid signals to induce an innate immune response mediated by antimicrobial peptides (AMPs) in females (Peng et al. 2005). It is speculated that this response protects the female from injury and potential pathogen transfer at mating (Peng et al. 2005). Other studies show that despite the acute effect on AMP synthesis, females exposed to sperm and seminal fluid proteins have compromised humoral immune system activity and impaired immune defence against bacterial infection (Short et al. 2012). This trade-off between reproductive success and immune defence is common to many species of fruit flies, damselflies, beetles, crickets, and ant queens and is thought to lie at the heart of the well-known cost to longevity of reproductive success in insects (Short et al. 2012).

In birds, which are generally also polyandrous despite often being socially monogamous, there is also evidence of sperm competition and seminal plasma proteins interacting with the immune system may contribute to this. Fresh semen instilled into the chicken oviduct acts to induce local expression of proinflammatory cytokines (Das et al. 2009) and to recruit immune cells (Zheng et al. 2001). Since sperm not involved in fertilisation in the fowl oviduct are phagocytosed by immune cells (Koyanagi and Nishiyama 1981), the immune response induced by seminal fluid is speculated to contribute to the fate and survivability of sperm in the oviduct, as well as defence from infectious microorganisms (Das et al. 2008).

6.3 Seminal Fluid Signalling in Mammals

In mammalian species, various effects of male seminal fluid in the female reproductive tract are reported in humans, rodents, domestic and livestock animals, and all mammals examined to date (Schjenken and Robertson 2014; Robertson 2005). The first evidence of a female response to seminal fluid in mammals was reported in rabbits in 1952 (McDonald et al. 1952), where an influx of leukocytes into the female tract following semen exposure was observed. Since then, a leukocytic response to seminal fluid has been reported in pigs, sheep, cow, horses, donkeys and dogs (Schjenken and Robertson 2014). Typically the response is characterised by an influx of immune cells including macrophages and neutrophils into the cervical or uterine lumen, as well as changes in the leukocyte content of stromal tissues accompanied by the induction of expression of immune-regulatory genes (Lovell and Getty 1968; Troedsson et al. 2001; Mattner 1968; England et al. 2012).

The inflammatory response to seminal fluid is likely to have a central role in female tract processing of seminal material and recovery of tissue homeostasis after mating, which is common to all species. However the extent to which seminal fluid influences subsequent events to impact reproductive success presumably relates to extraordinary variation in reproductive strategies, the timing and nature of implantation and placentation, and the operation of different molecular pathways to induce and maintain pregnancy. Despite patchiness in the depth of investigation in different species, it is clear that the precise nature of the female response to seminal fluid—its tissue site, cellular composition, and the pattern of gene expression induced—varies between species according to the different reproductive physiology and anatomy of each, particularly the site of seminal fluid deposition and its access to different parts of the female tract.

In many livestock species, the success of artificial insemination (AI) with diluted semen suggests that seminal constituents other than sperm are not mandatory for pregnancy. Depending on the species, reproductive success and quality of the pregnancy can be compromised if females are not exposed to seminal plasma.

In cattle, AI with diluted semen routinely achieves pregnancy at rates approaching natural service, despite very limited carry-over of seminal plasma (Lima et al. 2009). In other species such as rodents and pigs, effects ranging from reduced fertilisation and embryo implantation to altered growth of the placenta and fetus are observed when seminal signalling is perturbed. Moreover, given that some effects of seminal fluid manifest in offspring phenotype, there is the prospect that ignoring the effects of seminal plasma is potentially contributing to reduced fertility or declining reproductive fitness as seen in dairy cows (Lucy 2001), or the limited success of human in vitro fertilisation (IVF) programmes (Schieve 2002; Maher et al. 2003; Ceelen et al. 2007, 2008) (see later). Furthermore, long-held views that post-mating responses are detrimental to reproductive success in some species may be inappropriately founded. For example in horses, inflammation following seminal fluid exposure is linked with a pathological condition known as mating-induced endometritis, but recent studies contend that contrary to this, inflammation induced by seminal fluid facilitates resolution of the endometrium to a receptive state (Katila 2012).

Studies in rodents show clearly that seminal plasma increases the likelihood of conception and embryo implantation, through influencing sperm survival and competence, development of the pre-implantation embryo, and receptivity of the uterine endometrium to embryo implantation. Experiments in which the seminal vesicle, prostate, or coagulating glands are surgically removed from mice, rats, and hamsters prior to mating each show that seminal vesicle fluid is the most vital non-sperm component of the ejaculate (Pang et al. 1979; O et al. 1988; Queen et al. 1981; Peitz and Olds 1986), and in hamsters, this is accompanied by a slower cleavage rate in pre-implantation embryos and higher fetal loss after implantation (O et al. 1988). In mice, fetal loss and abnormality is much greater when embryos are transferred into recipients not exposed to male fluids (Watson et al. 1983). In rats, implantation rates and fetal growth are similarly impaired unless females are inseminated prior to embryo transfer (Carp et al. 1984).

Research in pigs also indicates that pregnancies conceived in the presence of seminal plasma can have better outcomes. Comparisons made when AI was first implemented suggested that conception rates and litter sizes were compromised by AI compared with natural mating (Claus 1990; Skjervold 1975). However with current AI practice in modern, large-scale facilities, farrowing rates, and litter sizes are comparable to natural service. This suggests that seminal fluid stimulation of the female tract is not essential, particularly when insemination occurs close to the time of ovulation, and when sperm are deposited high in the reproductive tract (Vazquez et al. 2005). However, in herds with poorer reproductive performance, seminal plasma can improve reproductive outcomes. Increases in farrowing rate from 70 % to 81 % were reported when seminal plasma was administered prior to natural service, and seminal plasma treatment increased both farrowing rate and litter size when given together with AI plus natural service (Flowers and Esbenshade 1993). Uterine infusion with heat-killed semen in the previous estrus can also increase litter size and improve farrowing rate (Murray et al. 1983, 1986), and similar effects result from mating with vasectomised boars in previous estrous periods (Flowers and Esbenshade 1993), suggesting that some benefit persists into subsequent cycles. Similarly when fertility is impaired by less than optimal female condition, addition of seminal plasma to sperm at AI may promote both conception rate and farrowing rate (Rozeboom et al. 2000).

6.4 Seminal Fluid and the Peri-conception Inflammatory Response

The molecular basis of the female response to seminal fluid mostly comes from laboratory studies using the mouse as a model (Robertson 2005; Robertson et al. 1996; Schjenken et al. 2015). At coitus in the mouse, seminal fluid is deposited in the uterus and disseminates throughout the female reproductive tract, where it interacts with epithelial cells of the uterine lumen to activate gene expression and synthesis of several proinflammatory cytokines and chemokines. Early studies focused on granulocyte-macrophage colony-stimulating factor (GMCSF) as a key element of the female response. More recently, several other cytokines and chemokines including interleukin-6 (IL6), tumor necrosis factor (TNF), C-X-C motif chemokine ligand-1 (CXCL1, also known as growth regulated alpha protein, GRO or KC), CXCL2 (macrophage inflammatory protein-2, MIP2), C-C motif chemokine ligand-3 (CCL3, macrophage inflammatory protein-1 alpha, MIP1A), and granulocyte colony-stimulating factor (GCSF) have been identified as elevated in the mouse uterus during the immediate post-coital phase (Schjenken et al. 2015; Sanford et al. 1992; Robertson et al. 1998; Pollard et al. 1998; Johansson et al. 2004). Seminal plasma is required for much of this response as coitus with males rendered seminal fluid deficient, by surgery to remove seminal vesicles and vasectomy, results in either no induction or substantially reduced cytokine induction (Schjenken et al. 2015). In contrast, delivery of seminal plasma by vasectomised males induces a response comparable to that induced by intact males (Robertson et al. 1996).

In response to the seminal fluid-induced cytokine and chemokine influx, immune cells including macrophages, granulocytes, and dendritic cells are recruited into the endometrial stroma (Robertson et al. 1992, 1996, 1998). Large populations of neutrophils are also observed to migrate into the luminal cavity (De et al. 1991). This leukocytic response is transient and is resolved prior to embryo implantation, when inflammatory cytokine release declines due to rising levels of progesterone. The effects of seminal fluid extend the full depth of the reproductive tract to the oviduct (Jasper et al. 2005) and to the ovary (Gangnuss et al. 2004) where cytokine synthesis and leukocyte numbers also increase.

Similar cytokine responses are observed in the human, where the cervix is the site of seminal fluid deposition (Pudney et al. 2005). In vitro and in vivo studies of the effect of seminal fluid on the cervix and vagina have shown an induction of cytokines and chemokines including GMCSF, IL1A, IL6, IL8, CCL2 (monocyte chemotactic protein-1 (MCP1)), and CCL20 (MIP3A) (Sharkey et al. 2007, 2012a, b). As with the mouse, the seminal plasma fraction appears to play a critical role in this response (Sharkey et al. 2007, 2012a), as no response is observed when barrier contraceptives are used (Sharkey et al. 2012b) while a significant induction mimicking the in vivo response is observed in cervical cells exposed to seminal fluid in vitro (Sharkey et al. 2007). Seminal fluid-induced cytokines initiate the recruitment of leukocytes into the epithelial layers and deeper stromal tissues with macrophages, dendritic cells, and memory T cells also being detected (Sharkey et al. 2012b). Even amongst proven fertile men, the cytokine response elicited by seminal plasma is highly variable (Sharkey et al. 2007) and the seminal fluid composition can vary greatly between individual men depending on fertility status or infection (Owen and Katz 2005). Thus there is the real prospect that coitus induces quite variable cytokine responses in women depending on seminal fluid composition, but how this might relate to fertility or subfertility remains to be determined.

In women, seminal fluid is largely retained in the cervix but effects of seminal fluid constituents may extend to the higher reproductive tract. In vitro experiments show that human endometrial epithelial cells can respond to seminal plasma with upregulated expression of *IL1B*, *IL6*, and leukaemia inhibitory factor (*LIF*) (Gutsche et al. 2003). It has been demonstrated that moieties in seminal plasma can bind to the sperm surface (Chu et al. 1996) and in the subcellular or sperm-bound form would be transported into the higher reproductive tract by uterine peristalsis, as demonstrated using hysterosalpingo scintigraphy which indicates rapid and extensive transport of particulate material from the cervix to the higher tract in women (Kunz et al. 1996). The ‘first uterine pass effect’, which is the vaginal to uterine transport of molecules (Bulletti et al. 1997), also supports a uterine response to semen exposure.

Studies in the pig, horse, and sheep all demonstrate the induction of endometrial cytokine expression following seminal plasma exposure with GMCSF, IL6, MCP1, and IL10 being induced in the pig (O’Leary et al. 2004), IL1B, IL6, TNF, and cyclooxygenase 2 (COX2) in the horse (Palm et al. 2008), and GMCSF and IL8 in the sheep (Scott et al. 2009). Studies in the cat and camel provide evidence for a similar function for seminal fluid as in other mammalian species (Lockett et al. 2010; Li and Zhao 2004).

6.5 Function of Seminal Fluid Response at Coitus

The consequences of seminal fluid-induced cytokine production and leukocyte recruitment in the female reproductive tract in turn have a range of effects on conception and pregnancy. In different species, effects include clearing the uterine cavity of microorganisms introduced at mating (Robertson et al. 1999), induction of ovulation and promoting corpus luteum formation (Gangnuss et al. 2004), supporting development of the pre-implantation embryo (Robertson et al. 2001), inducing endometrial expression of embryo attachment molecules and angiogenic factors to promote uterine receptivity (Jasper et al. 2011), and priming the female immune response to paternal antigens (Robertson and Sharkey 2001; Moldenhauer et al.

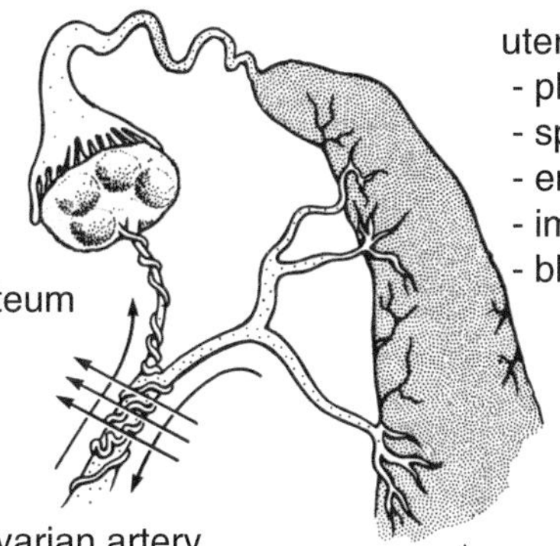

Fig. 6.1 The function of seminal fluid in the female reproductive tract following coitus. Seminal fluid exposure at coitus induces female responses that have consequences for conception and pregnancy. In the uterus, seminal fluid mediated effects include the phagocytic clearance of superfluous sperm and microorganisms introduced at mating, sperm selection, preparation of a receptive endometrium, promotion of blastocyst development, and regulation of the maternal immune response towards paternal antigens. In addition to effects in the uterus, seminal fluid has also been demonstrated to have effects in the higher reproductive tract. In the oviduct, components of seminal fluid influence sperm storage as well as promote the development of the embryo through generation of embryotrophic cytokines. Further effects are seen in the ovary, where seminal fluid exposure contributes to the induction of ovulation and the promotion of corpus luteum formation, potentially via a countercurrent mechanism between the uterine vein and ovarian artery. Figure is updated from (O'Leary et al. 2002)

2009) to promote T cell-mediated immune tolerance (Moldenhauer et al. 2009) (Fig. 6.1). When seminal fluid signals operate optimally, the result is to facilitate fertilisation, embryo development, and implantation. But depending on the balance of cytokines produced and immune cells recruited, the female response to seminal fluid may also induce mechanisms that constrain embryo implantation and progression of the reproductive cycle (Robertson 2010). In this way, it seems biologically plausible that the mammalian response to seminal fluid provides a mechanism of post-copulatory selection, or cryptic female choice, analogous to that described in invertebrate species (Eberhard 2009).

6.6 Clearance of Microorganisms and Superfluous Sperm

A major component of the female immune response to seminal fluid is the recruitment of neutrophils, which are prevalent in the uterine luminal cavity during the initial acute inflammatory phase following mating (De et al. 1991; Pandya and Cohen 1985). These neutrophils are believed to influence the capacity of sperm to reach the oocyte by removing superfluous sperm, microorganisms, and seminal debris (Robertson 2007). At coitus, commensal microorganisms from male or female tissues or sexually transmitted infections have the opportunity to enter the normally sterile higher reproductive tract where they can be detected in the immediate period post-coitus with sterility being recovered within 24 h (Robertson et al. 1999).

In addition to a function for neutrophils and other phagocytes in the rapid clearance of microorganisms introduced at coitus, there is also evidence that these cells may respond to seminal fluid exposure and act to phagocytose non-fertilising sperm. This process was originally postulated to act to filter out morphologically abnormal spermatozoa (Tomlinson et al. 1992). However, both abnormal and apparently viable and morphologically normal spermatozoa can undergo phagocytosis, suggesting that sperm selection may occur on the basis of morphological or antigenic parameters other than the ability to fertilise (Robertson 2005), and conferring female immune cells with the capacity to influence sperm selection through the recognition of surface markers that indicate faulty or damaged sperm (Sutovsky and Lovercamp 2010). The means by which phagocytes could discriminate between gametes remains to be determined, but if a biologically plausible mechanism were identified, this could explain observations of active female tract selection of male gametes (Robertson 2007).

6.7 Endometrial Receptivity and Embryo Implantation

The immune cells recruited into the endometrium in response to seminal fluid contribute to restructuring the endometrial environment to facilitate embryo implantation and support the first phase of placental development (Robertson 2005). Of the leukocytes present in the female reproductive tract following coitus, macrophages are postulated to play a critical role in the tissue remodelling process through the regulation of angiogenesis. Macrophages are a potent source of vascular endothelial growth factor (VEGF) and other key angiogenic factors and vascular permeability agents (Yoshida et al. 1997). The expression of these angiogenic factors is tightly regulated through the estrous cycle and early pregnancy (Ma et al. 2001) with a reduction of VEGF mRNA expression in hamsters after coitus occurring when seminal fluid signalling is perturbed by mating with accessory gland-deficient males (Chow et al. 2003). Evidence of a role for seminal fluid in angiogenesis is also demonstrated by studies in the horse and pig which show an increase in uterine vasodilation and oedema in the days following coitus (O'Leary et al. 2004; Bollwein et al. 2003).

Macrophage secreted products may also target the extracellular matrix of the endometrial stroma, which is remodelled during the decidualisation period (Aplin 2002). In particular, macrophage-secreted matrix metalloproteinases (MMPs) and tissue inhibitors of MMPs (TIMPs) coordinate their expression to contribute to this remodelling process (Robertson 2005). In rodents, MMPs and TIMPs are induced during the pre-implantation period (Das et al. 1997; Feng et al. 1998), with seminal plasma postulated to contribute to their induction as the absence of accessory gland fluid at coitus in the hamster is associated with a reduction in MMP2 at the implantation site (Chow et al. 2003).

In addition to macrophages, seminal fluid-induced dendritic cells may contribute to tissue remodelling in preparation for implantation. Uterine dendritic cells have been demonstrated to be crucial in decidua formation and maternal receptivity. Depletion of uterine dendritic cells around the period of implantation leads to impaired decidual formation through perturbed angiogenesis. This process was postulated to be regulated by dendritic cell soluble fms-like tyrosine kinase-1 (sFlt1) and transforming growth factor beta (TGFB) which promote coordinated blood vessel maturation (Plaks et al. 2008).

Embryo attachment and implantation is a process requiring specific changes in the expression of integrins and mucins which function to allow close apposition and then adhesion between the blastocyst and implantation site (Aplin 1997). Fucosyltransferases, which modulate the expression of fucosylated structures involved in the attachment and implantation process, are deficient when seminal plasma signalling is perturbed by surgical removal of the seminal vesicle from males (Jasper et al. 2011), suggesting that seminal plasma exposure contributes to this process.

Macrophages are implicated as cellular mediators transmitting the effects of seminal fluid on endometrial receptivity. They are postulated to secrete molecules that target luminal epithelial cells to induce modifications to proteins involved in embryo attachment (Robertson 2005). In support of this, mouse macrophage-derived LIF and IL1B induce fucosyltransferases *Fut2* and *Fut4* required for embryo attachment and implantation (Jasper et al. 2011), and activity of fucosyltransferases is impaired when females are mated with seminal fluid-deficient males (Jasper et al. 2011). Studies in human tissues indicate a similar mechanism, where leukocytes influence the adhesive properties of uterine epithelial cells in vitro (Kosaka et al. 2003). In the cow, intrauterine administration of leukocytes may operate through a similar pathway to increase embryo implantation rates (Ideta et al. 2010).

6.8 Synthesis of Embryotrophic Cytokines

A key role of the female tract response to seminal fluid is the synthesis of embryotrophic cytokines that control the development of the pre-implantation embryo. Alterations to embryo programming may have a major influence on long-term offspring health as small perturbations in blastomere number and inner cell mass/trophectoderm allocation in the blastocyst are associated with an altered growth trajectory of the fetus and resulting offspring (Thompson et al. 2002). Embryos express cytokine receptors from conception until implantation with seminal fluid-induced cytokines GMCSF, IL6, LIF, and MCSF all exerting direct effects on blastocyst cell number and viability, gene expression, and developmental competence (Robertson 2005). Prominent cytokine receptor pathways demonstrated to be involved in pre-implantation embryo development include the phosphatidylinositol-3 kinase-protein kinase B (PI3K-AKT), Janus kinase/signal transducer and activator of transcription (JAK/STAT), and mitogen-activated protein kinase (MAPK) pathways. Depending on the cytokine, ligation may affect one or more of these signal transduction pathways to modify gene expression, metabolism, and differentiation, alter the epigenome, suppress the cell stress response, and ultimately impact blastomere survival (Robertson et al. 2011). Cytokine effects have profound impact on embryo developmental competence and programming future developmental trajectory, providing a pathway whereby the peri-conception environment affects the future growth of the fetus and phenotype of offspring after birth (Robertson et al. 2011; Sjoblom et al. 2005).

GMCSF, CSF1, IL6, and LIF are well-described examples of key embryotrophic cytokines induced in the uterus and oviduct by seminal fluid contact (Bromfield et al. 2014). In mice, GMCSF targets the pre-implantation embryo to promote blastocyst formation and increase the number of viable blastomeres by inhibiting apoptosis and facilitating glucose uptake (Robertson et al. 2001) and GMCSF deficiency compromises blastocyst development by inducing stress response and apoptotic gene pathways (Robertson et al. 2001; Chin et al. 2009). In cattle embryos, GMCSF regulates genes involved in de novo methylation to influence epigenetic reprogramming (Loureiro et al. 2009). IL6 regulates STAT3-dependent miRNAs that function to protect embryos from apoptosis (Shen et al. 2009) while neutralisation of LIF affects embryo implantation rates and fetoplacental development (Mitchell et al. 2002). Addition of CSF1 to embryo culture media improves embryo development (Pampfer et al. 1991) and reduced fertility in CSF1-deficient mice may be due to adverse effects in blastocysts (Cohen et al. 1997a). The contribution of seminal plasma to embryo development is illustrated by findings that coitus with seminal plasma-deficient males is associated with reduced cytokine synthesis and this is linked with a significant reduction in the rate of cleavage to the 2-cell stage in zygotes and reduction in the development of blastocysts from 2-cell embryos (Bromfield et al. 2014).

In contrast to cytokines and growth factors that exert a positive influence on embryonic development, factors that have profound inhibitory effects on embryo development have also been identified. Increased levels of apoptosis-inducing cytokines TNF, interferon gamma (IFNG), and TNF-related apoptosis-inducing ligand (TRAIL) in the female reproductive tract following coitus are emerging as factors that cause pre-implantation embryonic loss or impair embryo development. TNF acts in vitro to increase the percentage of apoptotic blastomeres in the mouse (Fabian et al. 2007), while IFNG inhibits the embryotrophic factor GMCSF and also exerts direct inhibitory effects in mouse embryo culture (Robertson et al. 1992; Haimovici et al. 1988). TRAIL induces apoptosis in the pre-implantation embryo (Riley et al. 2004). Like the embryotrophic cytokines, these cytokines are regulated by seminal fluid and particularly oviduct TRAIL expression is suppressed in response to seminal fluid factors (Bromfield et al. 2014).

In the human, the embryotrophic cytokines GMCSF, IL6, and LIF are induced in cervical and endometrial epithelial cells in response to seminal plasma exposure (Sharkey et al. 2012a, b; Gutsche et al. 2003), but to date whether this occurs in the higher reproductive tract in vivo has not been confirmed. These cytokines are predicted to have a similar role in the human as in vitro culture of human embryos in GMCSF improves developmental outcomes through an increase in embryos reaching blastocyst stage, with earlier blastulation and increased inner cell mass and trophectoderm cell number (Sjoblom et al. 1999), followed by improved implantation rates and progression to healthy birth after IVF treatment (Ziebe et al. 2013). Due to the challenges in accessing appropriate tissues, it has not been possible to ascertain the extent to which expression of embryotrophic and embryotoxic cytokines responds to coital activity to influence human embryo development and implantation after natural conception.

The capacity for seminal fluid to alter the balance of embryotrophic and embryotoxic cytokines provides a mechanism through which the female reproductive tract can reflect signals from the male, to either support or impair progression of preimplantation embryo development, and impart longer-term programming. The capacity to liberate embryotoxic cytokines such as TRAIL in the absence of sufficient seminal fluid suppression raises the prospect of maternal tract 'quality control'. If proven to occur, this would be relevant in couples where altered seminal fluid composition provides insufficient signalling factors (Sharkey et al. 2007) or where sexually transmitted infection alters the balance of seminal fluid signals to induce a proinflammatory phenotype (Rasmussen et al. 1997). These changes may affect progression of these embryos as well as imparting effects on offspring phenotype and health in adult life (Bromfield et al. 2014; Sjoblom et al. 2005).

6.9 Maternal Immune Tolerance

A critical component of the female tract response to seminal fluid which is essential for ongoing pregnancy success is programming of the maternal immune system to respond to the presence of the genetically disparate fetus. Immune tolerance must exist from the very earliest time the embryo contacts the maternal tissues at implantation. A combination of strategies contributes to the activation of maternal immune tolerance, with seminal fluid providing antigens and cytokines that, in the correct immune environment, drive production of regulatory T cells (Trowsdale and Betz 2006; Robertson et al. 2009a).

Seminal fluid contains several antigens specific to individual males including class Ia, Ib, and II major histocompatibility complex (MHC) (Hutter and Dohr 1998) that can be presented by antigen-presenting cells (APCs) such as macrophages and dendritic cells recruited into the endometrium following exposure to seminal fluid (Robertson et al. 1996; McMaster et al. 1992). These dendritic cells and macrophages in the presence of seminal fluid factors prostaglandin E (PGE) and TGFB differentiate into cells that mediate tolerogenic immune responses (Blois et al. 2007; Jaiswal et al. 2012). However, not all APCs are tolerogenic as the immune environment at this stage of pregnancy is highly varied with both tolerogenic and immunogenic cells present, demonstrating differing roles for these cells during peri-conception (Jaiswal et al. 2012). The APCs are thought to take up paternal antigens and either traffic to the uterine draining lymph nodes or interact locally with resident uterine T cells driving activation and expansion of clonal subsets of Treg cells which recognise and appropriately respond to paternal/fetal antigens (Robertson et al. 2009a). An antigen-specific T cell response has been demonstrated using a T cell receptor transgenic model with OVA as a model paternal antigen (Moldenhauer et al. 2009). The concept of seminal fluid-induced tolerance of paternal antigens is supported by experiments demonstrating prolonged survival of skin grafts of paternal origin in mated female mice (Lengerova and Vojtiskova 1966) and studies demonstrating tolerance towards tumor cells expressing the same MHC as the donor male (Robertson et al. 2009b).

Treg cells operate as potent suppressors of inflammation and cell-mediated immunity (Rudensky 2011). These cells act via a variety of mechanisms, generally involving the suppression of cytokine production and effector function in T cells, B cells, natural killer (NK) cells, dendritic cells, and macrophages (Sakaguchi 2000; Shevach 2002). In pregnancy, Treg cells play a critical role in mediating immune tolerance required for embryo implantation, with several studies demonstrating in the mouse that allogenic mating leads to fetal rejection unless sufficient Treg cells are present in the endometrium in the pre- and peri-implantation phase (Aluvihare et al. 2004). In humans, Treg cells play a similar regulatory role, with an increase in pregnant women of circulating CD4+ CD25+ cells highly enriched for the signature Treg transcription factor forkhead box P3 (FOXP3) in early pregnancy. The number of these regulatory cells peaks during the second trimester before declining at term (Somerset et al. 2004).

The presence of TGFB and PGE in seminal fluid confers immune-deviating activity that drives immune cells into tolerogenic phenotypes. Both factors have been linked with the induction of naïve T cells into suppressor T cells expressing *Foxp3* (Chen et al. 2003; Baratelli et al. 2005) and exogenous TGFB delivery at conception is shown to boost vaginal Treg cell numbers which acts to reduce fetal loss in the abortion prone CBA/J x DBA/2 J model (Clark et al. 2008). Because of poor availability of suitable reagents for Treg cells, they have not been extensively studied in other species, other than in the cat where there is evidence that an increase in Treg cells in early pregnancy occurs after seminal fluid exposure (Lockett et al. 2010).

6.10 Ovulation and Corpus Luteum Formation

As well as the endometrium, seminal fluid exposure at coitus has effects in the ovary. This effect was first described in camelids, where effects of seminal plasma on induction of ovulation are well known (Li and Zhao 2004). It is challenging to conceive how seminal fluid moieties could reach the ovary via the oviduct. An alternative pathway for seminal fluid effects on ovarian function may be due to a unique countercurrent exchange mechanism through which prostaglandins and other small molecules can be transmitted from the uterine vein to the ovarian artery (Krzymowski et al. 1989). In the presence of a mature follicle, ovulation can be induced by natural coitus, artificial insemination, or intramuscular injection of seminal plasma, but not washed sperm. The ovulation-inducing factor in camelid seminal plasma has been identified as the highly conserved beta-nerve growth factor, which acts via an endocrine mechanism targeting the hypothalamic–pituitary axis to induce LH release and ovulation (Ratto et al. 2012). This so-called ovulation-inducing factor (OIF) is present in other species demonstrating induced ovulation (koalas and rabbits). Recent studies indicate that OIF in seminal plasma is conserved among species, including cattle, horses, pigs, mice, and other species considered to be spontaneous ovulators (Bogle et al. 2011; Ratto et al. 2006). Additional studies showing that OIF can induce ovulation in prepubertal mice (Bogle et al. 2011) and alter ovarian follicular wave dynamics in cows (Tanco et al. 2012) imply this factor may have broader relevance than originally thought (Ratto et al. 2012). Seminal plasma-mediated effects on ovarian function are clearly demonstrated in pigs, where a reduction in the interval between LH surge and ovulation, followed by elevated plasma progesterone concentration over the pre-implantation phase, occurs when seminal plasma is instilled into the uterus just prior to ovulation (Waberski et al. 1997; O'Leary et al. 2001).

As well as endocrine actions, the effects of seminal fluid on ovarian function in rodents and pigs may be mediated through local mechanisms involving stimulation of macrophage populations. Ovarian macrophages contribute to regulation of follicle development, ovulation, and post-ovulatory tissue remodelling associated with the conversion of the ruptured follicle into the corpus luteum (Cohen et al. 1997b; Care et al. 2013). Ovarian macrophages are present in the theca of developing follicles and their numbers increase at ovulation (Brannstrom et al. 1993). These macrophages support follicle growth, ovulation, and vascular homeostasis (Brannstrom et al. 1993; Wu et al. 2004; Van der Hoek et al. 2000; Turner et al. 2011). In both rodents and humans, macrophages migrate into the developing corpora lutea immediately after ovulation (Cohen et al. 1997b; Hellberg et al. 1991; Brannstrom et al. 1994a, b) where they influence steroidogenic function of luteal cells as well as tissue remodelling after luteal regression (Kirsch et al. 1981; Nelson et al. 1992).

In mice, seminal plasma influences macrophage recruitment into the ovary, as mating with seminal plasma-deficient males results in fewer macrophages in the corpus luteum compared to mating with seminal plasma sufficient males (Gangnuss et al. 2004). Cytokines that are induced by seminal fluid in the uterus such as MCP1, MCSF, and GMCSF (Robertson 2005) also act as chemoattractants to recruit macrophages into the ovary (Cohen et al. 1997b; Jasper et al. 2000; Townson et al. 1996), although whether these cytokines are induced in the ovary by seminal fluid remains to be examined. Serum progesterone levels on days 1, 2, and 4 of pregnancy in mice are not different with seminal plasma deficiency (Gangnuss et al. 2004). However, a threshold level of macrophage activity is essential for normal corpus luteum development as macrophage depletion disrupts the luteal microvascular network and alters expression of VEGFs required to support luteal angiogenesis in the peri-implantation period (Care et al. 2013). The result is increased inflammatory and apoptotic gene expression, decreased steroidogenic gene expression, and infertility which can be rescued by exogenous progesterone administration (Care et al. 2013).

6.11 Active Factors in Seminal Fluid

To further understand the function of seminal fluid at coitus, studies have aimed to identify the components of the ejaculate that induce the female tract response. Early studies in the mouse demonstrated through removal of the male accessory glands that the seminal plasma fraction contains many of the active factors, as its absence resulted in a reduced production of GMCSF and reduced leukocyte infiltration in the endometrium (Robertson et al. 1996). Further analysis of the high molecular weight component of mouse seminal fluid that induced these changes was able to identify TGFB as the principal trigger (Tremellen et al. 1998). These findings have since been confirmed in the human, where high levels of TGFB in seminal plasma contribute to the peri-conception inflammatory response by inducing GMCSF, IL1B, IL6, and LIF in human cervical epithelial and endometrial epithelial cells in vitro (Sharkey et al. 2012a; Gutsche et al. 2003). Secreted TGFB is initially produced in a precursor dimeric form where it is activated in the female tract after insemination as a consequence of the low pH, enzymatic activation with plasmin or interaction of the latent TGFB with thrombospondin-1 or avb6 integrins (Robertson 2005). In the mouse there is approximately 30 ng/ml TGFB1 in the ejaculate (Robertson et al. 2002) while in the human the TGFB concentrations are 219 ng/ml TGFB1, 5 ng/ml TGFB2, and 172 ng/ml TGFB3 (Sharkey et al. 2012a). TGFB is also detected in the seminal plasma of pigs (O'Leary et al. 2011) and sheep (Scott et al. 2006), with the pig carrying similar concentrations [150 ng/ml TGFB1 (O'Leary et al. 2011)] to that of the human and mouse. There is no correlation between TGFB and fertility (Loras et al. 1999), but studies in the seminal plasma of fertile men show that there are substantial differences in TGFB concentrations between individuals (Sharkey et al. 2012a).

In addition to TGFB, we have recently identified that ligands for TLR4 contribute to the peri-conception inflammatory response in the mouse. In these studies, the induction of colony-stimulating factor 3 (*Csf3*), *Cxcl1*, and *Cxcl2* was shown to require the presence of TLR4 while the TLR4 signalling pathway may also contribute to the induction of *Il1a*, *Tnf*, and *Csf2* (Schjenken et al. 2015). While it has been speculated that bacterial products including endotoxin may contribute to the capacity of seminal fluid to interact with the female reproductive tract (Schaefer et al. 2004), the amount of endotoxin detectable in the female reproductive tract of mice following coitus (Schjenken et al. 2015) and in human seminal fluid (Sharkey, in preparation) is insufficient to explain the induction of TLR4 regulated cytokines. This suggests that other TLR4 ligands such as the endogenous danger-associated molecular patterns (DAMPs) contribute to eliciting the female response; however, these studies are yet to be completed. DAMPS have previously been identified in seminal fluid with beta-defensin-2 (Narciandi et al. 2014), fibronectin (Lilja et al. 1987), heat shock proteins (Rego et al. 2014; Pilch and Mann 2006), hyaluronidase (Geipel et al. 1992; Shimada et al. 2008), and S100 proteins (Rego et al. 2014; Donato et al. 2013) all being present.

In line with the functions of seminal fluid, activation of TLR4 signalling may contribute to tolerogenic immune responses (Conroy et al. 2008), as studies have demonstrated that CD45RB^{low} CD25^{+} regulatory T cells can express TLR4 and enhance suppressive function following TLR4 ligation (Caramalho et al. 2003), dendritic cells from TLR4-deficient mice have reduced capacity to produce IL10 in response to TLR4 ligation and have impaired expansion of Treg cells (Higgins et al. 2003), and Treg cells cultured in the presence of dendritic cells can induce tryptophan catabolism which enhances tolerogenic dendritic cell production (Fallarino et al. 2015). More detailed studies are required to understand how disruption of TLR4 signalling may affect the female response to seminal fluid in the mouse and other species and how in turn offspring health might be affected.

In humans, PGE in seminal fluid are postulated to act as signalling agents, most notably those in the 19-hydroxy (19-OH) form (Templeton et al. 1978). While PGE is detectable in equine seminal plasma (Claus et al. 1992), it is undetectable in rodent and porcine samples. In the human, extremely high levels of PGE are detected in seminal fluid, with 300 μM of the 19-OH form detectable (Templeton et al. 1978). Due to the immunosuppressive nature of 19-OH PGE, it was postulated that seminal prostaglandins may protect sperm from immunological damage in the male and protect the female against sensitisation to sperm antigens (James and Hargreave 1984; Alexander and Anderson 1987). In vitro experiments using human cervical explants stimulated with 19-OH PGE demonstrated an induction of IL8 and suppression of the anti-inflammatory molecule, secretory leukocyte protease inhibitor (Denison et al. 1999). Seminal plasma PGE2 has also been linked to facilitate tumorigenesis-angiogenesis in reproductive tract cells through activation of fibroblast growth factor 2, COX2, and VEGF expression and E series prostanoid-2 and -4 receptor, EGF receptor, and ERK1/2 signalling pathways (Battersby et al. 2007; Muller et al. 2006). Further, PGE has been demonstrated in the human to induce a regulatory phenotype in naïve CD4^{+}CD25^{-} T cells, enhance the in vitro inhibitory function of Treg cells, and induce expression of Foxp3 (Baratelli et al. 2005) which is consistent with the female tract response to seminal fluid.

Other factors postulated to contribute to seminal fluid signalling include IL8, which is present in seminal plasma of healthy human donors at 1456 pg/ml. Treatment of human endometrial epithelial cells in vitro with recombinant IL8 at physiological levels can induce *IL1B*, *IL6*, and *LIF* (Gutsche et al. 2003). Outside of mice and humans, other novel seminal fluid signalling molecules that have been examined are porcine sperm adhesion proteins (PSP)1 and PSP2, which contribute in part to the influx of neutrophils in pigs and can preserve sperm viability, motility, and mitochondrial activity (Rodriguez-Martinez et al. 2010; Caballero et al. 2006), and cysteine-rich secretory protein-3 (CRISP3) in horses, which mediates the interaction between sperm and neutrophils (Doty et al. 2011). Positive correlations between specific seminal plasma proteins and fertility have been observed in the horse (Brandon et al. 1999) and bull (Killian et al. 1993).

Despite the evidence showing an integral role for seminal plasma in the signalling process, it is becoming apparent that not all seminal fluid effects can be accounted for by the soluble fraction, raising the prospect that sperm also communicate with female reproductive tract cells. In domestic animal species, there is extensive evidence that sperm is critical to the inflammatory response and may be required to interact with seminal plasma to regulate the influx and function of immune cells (Schjenken and Robertson 2014). In the human and mouse, while the literature has focused on seminal plasma, there is some evidence for a sperm signalling component. In the mouse, sperm has been shown to contribute to the peri-conception immune environment as *Foxp3* expression requires the presence of both sperm and seminal plasma for its complete induction (Guerin et al. 2011) while in the human, AI of human sperm into the cervix results in an influx of leukocytes (Pandya and Cohen 1985; Thompson et al. 1992). A function for sperm in the signalling process is further supported by studies in the mouse showing that sperm can form intimate associations with cells of the female reproductive tract following coitus (Reid 1965). Non-fertilising sperm can be taken up by female tissues, potentially by phagocytosis, where sperm-associated transcripts can be identified in the uterus and other distal sites, including the uterine draining lymph nodes for several days post-coitus (Watson et al. 1983). Similar interactions between sperm and the female reproductive tract can be seen in other species, including human, pigs, cow, dog, cats, bats, lizards, and even some marsupials (Suarez and Pacey 2006; Murakami et al. 1985; Rasweiler 1987; Nogueira et al. 2011; Pacey et al. 1995; Taylor et al. 2008; Rijsselaere et al. 2004; Ignotz et al. 2001; Breed et al. 1989). Seminal plasma proteins contribute to these interactions as bovine seminal plasma (BSP) proteins coat bull sperm and mediate interactions between sperm and oviductal epithelial cells to assist in the formation of sperm storage reservoirs, which function to maintain fertility of sperm until ovulation (Suarez 2008). In addition to the formation of storage reservoirs, these interactions assist in the clearance of sperm at coitus. Additionally, these interactions may potentially be a mechanism by which sperm communicates with the female reproductive tract to transmit antigen and facilitate pregnancy success; however, this remains to be determined.

In contrast to signalling molecules that promote successful pregnancy, there is also evidence that seminal fluid may contain factors that inhibit seminal fluid signalling. The type 1 cytokine IFNG can be detected in the seminal fluid of both mouse (~50 pg/ml) (Gopichandran et al. 2006) and human (range: 0–130 pg/ml) (Politch et al. 2007) and in the mouse has been shown to inhibit the tolerance inducing properties of TGFB during early pregnancy (Glynn et al. 2004). Whether IFNG levels are altered in the seminal fluid of infertile or subfertile men remains to be determined.

While several active factors that contribute to seminal fluid signalling have been identified, future studies will be required to understand in more detail how these factors interact to regulate the female response. In addition, if sperm contain signalling factors as has been postulated, studies will be required to elucidate these factors and their function. Additional factors beyond those discussed here no doubt exist. Exciting candidates include CD52, transglutaminase 4, and seminal vesicle secretion 2. CD52 is present in soluble form in seminal plasma and on the sperm surface (Koyama et al. 2009), which is recently reported to be a potent immune-regulatory signal driving induction of antigen-activated suppressor T cells, distinct from FOXP3^{+} Treg cells, from CD4 precursors (Bandala-Sanchez et al. 2013). Transglutaminase 4 and seminal vesicle secretion 2 are proteins which are known to play a critical role in copulatory plug formation (Dean 2013; Kawano et al. 2014). Interestingly, the absence of these seminal proteins disrupts not only plug formation but also fertility. Sperm from ejaculates lacking transglutaminase 4 fertilised at normal rates but were significantly less likely to give birth to a litter (Dean 2013). Sperm from seminal vesicle secretion 2-deficient mice are disrupted by uterine-derived cytotoxic factors (Kawano et al. 2014), suggesting that seminal plasma plays a protective role in sperm function.

Consideration of species-specific differences is integral as it is highly likely, due to the different sites of semen deposition, that different species will utilise different active factors. Within a species, fluctuations in the seminal fluid profile between individuals, or even within an individual, may influence the female response leading to altered outcomes for offspring depending on male health and environmental exposures.

6.12 Seminal Fluid and Programming of Offspring Health

The studies detailed above document a complex role for seminal fluid in preparing the female reproductive tract for pregnancy, particularly through regulating the maternal immune response to tolerate exposure to paternal antigens. These changes not only influence the period of early pregnancy but also have the potential to influence fetal development and offspring health. It is well documented that the in utero environment is a major determinant in programming how adult individuals respond to stressors and challenge, and the risk of disease later in life (Godfrey et al. 2010). Increasingly the peri-conception period is identified as the most sensitive for imprinting later life consequences, and the health and experiences of both the mother and father even prior to conception contribute through effects transmitted by the male and female gametes (Lane et al. 2014). There is evidence that altered conditions at conception can induce adaptations to protect the fetus from immediate effects, but a consequence of these changes is an increased risk of later metabolic disease (Barker and Clark 1997; Hanson and Gluckman 2005). Studies on the paternal contribution towards health and disease have identified smoking, age, environmental exposures, and obesity being linked to later offspring illness (Fullston et al. 2012), with sperm transmission of epigenetic mechanisms including non-coding RNAs (Fullston et al. 2012; Liu et al. 2012) being postulated to contribute to these changes.

As well as a contribution from sperm, there is the prospect that seminal fluid can contribute at fertilisation to influence the peri-conception environment in turn affecting later development and offspring health (Lane et al. 2014). Some evidence supporting a role for seminal plasma through its influence over the female reproductive tract at coitus is emerging. In mice, Bromfield et al. (Bromfield et al. 2014) demonstrate that conception in the absence of seminal plasma using mice surgically rendered seminal vesicle deficient (SVX) leads to a reduction in fecundity and altered fetal and neonatal outcomes. Mice mated with SVX males had a significant reduction in progression to pregnancy and fewer implanting embryos compared to intact mated females. Changes in development were observed as early as the eight-cell stage, where most embryos from SVX fathers were developmentally delayed. Prior to birth, changes in phenotype were evident as placental hypertrophy. Male offspring were observed to have altered growth trajectory and metabolic parameters with evidence of obesity, distorted metabolic hormones, reduced glucose tolerance, and hypertension. Embryo transfer experiments confirmed the role of seminal plasma in programming altered phenotype as increased adiposity was also observed in adult male progeny when normal 2-cell embryos were transferred to females mated with SVX males (Bromfield et al. 2014).

Similar studies have been conducted in the golden hamster, where ablation of the accessory sex glands (ASG), which produce seminal plasma, delayed entry of the zygote into the first cell cycle (Ying and Chow 1998), reduced pre-implantation embryonic cell number (Chan et al. 2001), resulted in an early transit of the embryo into the uterus from the oviduct (Wong et al. 2008), decreased the rate of implantation, and ultimately resulted in embryo death (Jiang et al. 2001). The offspring of fathers deficient in seminal plasma exhibited altered postnatal growth and elevated anxiety (Wong et al. 2007). It is postulated that these changes may be associated with epigenetic mechanisms as cleavage stage embryos from females mated with ASG-deficient males showed reduced acetylation and altered methylation kinetics which were associated with dysregulated expression of the paternally expressed *Igf2* and *Dlk1* (Jiang et al. 2001).

In the human, there is no clear direct evidence implicating altered seminal fluid signalling activity with long-term health of offspring. However, there are clinical observations that are consistent with a similar function (discussed below). This means that alterations to the content of seminal fluid outside of the normal range may not manifest as impaired fertility, but could potentially alter fetal development and impart long-term consequences for offspring health. Carefully designed studies will be required to evaluate this, as effects on offspring can be subtle and require large populations to detect. Additionally, it will be important to distinguish effects of seminal fluid factors exerted via the female reproductive tract from those exerted through effects on sperm fertilising capacity and epigenetic status.

6.13 Clinical Implications of Altered Seminal Fluid Signalling

These considerations raise the question of the extent to which seminal fluid exposure may influence pregnancy outcomes in humans. The success of IVF even in women without male partners demonstrates that seminal plasma is not an absolute requirement for pregnancy. Rather, evidence suggests that rather than being an absolute prerequisite, seminal plasma influences the success and quality of pregnancy outcome (Robertson 2005). There are few definitive studies in humans focussed on evaluating effects of seminal fluid, but clinical studies point to a consistent beneficial effect in improving the chance of conception and protecting from immune-mediated gestational disorders such as preeclampsia.

Preeclampsia is characterised by a significant reduction in peripheral blood $CD4^{+}CD25^{high}$ T regulatory cells and decidual Foxp3-positive cells within the $CD3^{+}$ population (Sasaki et al. 2007). The altered Treg cell numbers may in part be due to altered seminal fluid function, with extensive evidence pointing to an important role for TGFB in regulating human Treg cell reactivity and autoimmunity (Rubtsov and Rudensky 2007). Clinical studies show that preeclampsia is more common in situations where limited exposure to the conceiving partners seminal fluid has occurred—including first conception, after a short period of cohabitation, when barrier contraception has been used, or when multiparous mothers conceive with a new male partner [reviewed in Schjenken and Robertson (2014)]. Treg cell numbers and function are found to be altered in the peripheral blood and decidua of pregnant women who have spontaneous abortions compared to induced abortions (Sasaki et al. 2004; Arruvito et al. 2007), while *FOXP3* expression, used as a surrogate measure of Treg cells, is reduced in the endometrium of women with primary unexplained infertility (Jasper et al. 2006).

Data from ART further support a protective function for seminal fluid. The absence of seminal plasma in the in vivo setting has been associated with higher rates of implantation failure and decreases in embryo quality [reviewed in Schjenken and Robertson (2014)]. Exposure to semen around the time of embryo transfer is demonstrated to improve embryo viability 6–8 weeks post-transfer (Tremellen et al. 2000). Given the emerging recognition that IVF is linked with reduced birth weights and impaired metabolic health outcomes for IVF children (Hart and Norman 2013), and the overlap between these changes and the phenotypes of offspring conceived without seminal fluid in mouse studies (Bromfield et al. 2014), it would seem prudent to more formally investigate the impact of seminal fluid on perinatal outcome and offspring health in human IVF. One difficulty with advising couples to have coitus at the time of embryo transfer is the small but real prospect of multiple pregnancy due to natural conception, but this would be expected to be a small risk in couples experiencing long-term infertility.

6.14 Conclusions

Seminal fluid signalling appears to play an integral role in eliciting inflammatory changes in the female reproductive tract following coitus in all species examined so far, despite differences in sites of semen deposition and reproductive physiology. This implies that the capacity for seminal fluid to exert effects in the female reproductive tract has biological value beyond the support of sperm. The enormous variation between mammalian species in seminal fluid volume, tissue origin and composition, and actions in the female implies rapid evolution that is driven by sexual selection, analogous to that which occurs in invertebrate species. For sexual selection to occur there must be variation in the female response that impacts the likelihood of progression to pregnancy. The array of effects induced by seminal fluid that impact conception, embryo development, implantation, and fetal growth together impact fertility and fecundity and ultimately influence offspring survival and health. Signalling factors within seminal fluid, in particular in the seminal plasma fraction, have been identified, but there are clearly other, as yet unknown factors that contribute to signalling in the female tract, potentially including sperm.

Despite the advances that have been made in understanding the contribution of seminal fluid, there is still much to learn. In particular, the contribution of seminal fluid to offspring phenotype requires further study and the knowledge to be gained will have major implications in human and animal ART, where seminal fluid signalling is often overlooked or avoided. In humans there are substantial variations in seminal fluid profiles, but the implications and causes of this are not known. Whether these variations, or the presence of conditions such as reproductive tract infection, affect the capacity of seminal fluid to induce the appropriate response and ultimately affect male fertility, pregnancy health, or even offspring phenotype remains to be understood. Ultimately, a better understanding of seminal fluid signalling may improve success rates and outcomes of human IVF, and yield novel therapies for infertility and pathologies of pregnancy.

References

Alexander NJ, Anderson DJ (1987) Immunology of semen. Fertil Steril 47:192–204

Aluvihare VR, Kallikourdis M, Betz AG (2004) Regulatory T cells mediate maternal tolerance to the fetus. Nat Immunol 5:266–271

Aplin JD (1997) Adhesion molecules in implantation. Rev Reprod 2:84–93

Aplin JD (2002) Endometrial extracellular matrix. In: Glasser SR, Aplin J, Guidice LC, Tabibzadeh S (eds) The endometrium. Taylor and Francis, London and New York, pp 294–307

Arruvito L, Sanz M, Banham AH, Fainboim L (2007) Expansion of CD4+CD25+and FOXP3+ regulatory T cells during the follicular phase of the menstrual cycle: implications for human reproduction. J Immunol 178:2572–2578

Aumuller G, Riva A (1992) Morphology and functions of the human seminal vesicle. Andrologia 24:183–196

Bandala-Sanchez E, Zhang Y, Reinwald S, Dromey JA, Lee BH, Qian J, Bohmer RM, Harrison LC (2013) T cell regulation mediated by interaction of soluble CD52 with the inhibitory receptor Siglec-10. Nat Immunol 14:741–748

Baratelli F, Lin Y, Zhu L, Yang S-C, Heuze-Vourc'h N, Zeng G, Reckamp K, Dohadwala M, Sharma S, Dubinett SM (2005) Prostaglandin E2 induces FOXP3 gene expression and T regulatory cell function in human CD4+ T cells. J Immunol 175:1483–1490

Barker DJ, Clark PM (1997) Fetal undernutrition and disease in later life. Rev Reprod 2:105–112

Battersby S, Sales KJ, Williams AR, Anderson RA, Gardner S, Jabbour HN (2007) Seminal plasma and prostaglandin E2 up-regulate fibroblast growth factor 2 expression in endometrial adenocarcinoma cells via E-series prostanoid-2 receptor-mediated transactivation of the epidermal growth factor receptor and extracellular signal-regulated kinase pathway. Hum Reprod 22:36–44

Blois SM, Kammerer U, Alba Soto C, Tometten MC, Shaikly V, Barrientos G, Jurd R, Rukavina D, Thomson AW, Klapp BF, Fernandez N, Arck PC (2007) Dendritic cells: key to fetal tolerance? Biol Reprod 77:590–598

Bogle OA, Ratto MH, Adams GP (2011) Evidence for the conservation of biological activity of ovulation-inducing factor in seminal plasma. Reproduction 142:277–283

Bollwein H, Sowade C, Stolla R (2003) The effect of semen extender, seminal plasma and raw semen on uterine and ovarian blood flow in mares. Theriogenology 60:607–616

Brandon CI, Heusner GL, Caudle AB, Fayrer-Hosken RA (1999) Two-dimensional polyacrylamide gel electrophoresis of equine seminal plasma proteins and their correlation with fertility. Theriogenology 52:863–873

Brannstrom M, Mayrhofer G, Robertson SA (1993) Localization of leukocyte subsets in the rat ovary during the periovulatory period. Biol Reprod 48:277–286

Brannstrom M, Giesecke L, Moore IC, van den Heuvel CJ, Robertson SA (1994a) Leukocyte subpopulations in the rat corpus luteum during pregnancy and pseudopregnancy. Biol Reprod 50:1161–1167

Brannstrom M, Pascoe V, Norman RJ, McClure N (1994b) Localization of leukocyte subsets in the follicle wall and in the corpus luteum throughout the human menstrual cycle. Fertil Steril 61:488–495

Breed WG, Leigh CM, Bennett JH (1989) Sperm morphology and storage in the female reproductive tract of the fat-tailed dunnart, Sminthopsis crassicaudata (Marsupialia: Dasyuridae). Gamete Res 23:61–75

Bromfield JJ, Schjenken JE, Chin PY, Care AS, Jasper MJ, Robertson SA (2014) Maternal tract factors contribute to paternal seminal fluid impact on metabolic phenotype in offspring. Proc Natl Acad Sci USA 111:2200–2205

Bulletti C, De Ziegler D, Giacomucci E, Polli V, Rossi S, Alfieri S, Flamigni C (1997) Vaginal drug delivery: the first uterine pass effect. Ann N Y Acad Sci 828:285–290

Caballero I, Vazquez JM, Garcia EM, Roca J, Martinez EA, Calvete JJ, Sanz L, Ekwall H, Rodriguez-Martinez H (2006) Immunolocalization and possible functional role of PSP-I/PSP-II heterodimer in highly extended boar spermatozoa. J Androl 27:766–773

Caramalho I, Lopes-Carvalho T, Ostler D, Zelenay S, Haury M, Demengeot J (2003) Regulatory T cells selectively express toll-like receptors and are activated by lipopolysaccharide. J Exp Med 197:403–411

Care AS, Diener KR, Jasper MJ, Brown HM, Ingman WV, Robertson SA (2013) Macrophages regulate corpus luteum development during embryo implantation in mice. J Clin Invest 123:3472–3487

Carp HJ, Serr DM, Mashiach S, Nebel L (1984) Influence of insemination on the implantation of transferred rat blastocysts. Gynecol Obstet Invest 18:194–198

Ceelen M, van Weissenbruch MM, Roos JC, Vermeiden JP, van Leeuwen FE, Delemarre-van de Waal HA (2007) Body composition in children and adolescents born after in vitro fertilization or spontaneous conception. J Clin Endocrinol Metab 92:3417–3423

Ceelen M, van Weissenbruch MM, Vermeiden JP, van Leeuwen FE, Delemarre-van de Waal HA (2008) Cardiometabolic differences in children born after in vitro fertilization: follow-up study. J Clin Endocrinol Metab 93:1682–1688

Chan OC, Chow PH, O WS (2001) Total ablation of paternal accessory sex glands curtails developmental potential in preimplantation embryos in the golden hamster. Anat Embryol (Berl) 204:117–122

Chapman T, Davies SJ (2004) Functions and analysis of the seminal fluid proteins of male Drosophila melanogaster fruit flies. Peptides 25:1477–1490

Chapman T, Neubaum DM, Wolfner MF, Partridge L (2000) The role of male accessory gland protein Acp36DE in sperm competition in Drosophila melanogaster. Proc Biol Sci 267:1097–1105

Chen W, Jin W, Hardegen N, Lei KJ, Li L, Marinos N, McGrady G, Wahl SM (2003) Conversion of peripheral CD4+CD25- naive T cells to CD4+CD25+ regulatory T cells by TGF-beta induction of transcription factor Foxp3. J Exp Med 198:1875–1886

Chin PY, Macpherson AM, Thompson JG, Lane M, Robertson SA (2009) Stress response genes are suppressed in mouse preimplantation embryos by granulocyte-macrophage colony-stimulating factor (GM-CSF). Hum Reprod 24:2997–3009

Chow PH, Jiang HY, Poon HK, Lee KH, O WS (2003) Embryos sired by males without accessory sex glands induce failure of uterine support: a study of VEGF, MMP and TGF expression in the golden hamster. Anat Embryol 206:203–213
Chu TM, Nocera MA, Flanders KC, Kawinski E (1996) Localization of seminal plasma transforming growth factor-beta1 on human spermatozoa: an immunocytochemical study. Fertil Steril 66:327–330
Clark DA, Fernandes J, Banwatt D (2008) Prevention of spontaneous abortion in the CBA x DBA/2 mouse model by intravaginal TGF-beta and local recruitment of CD4+8+ FOXP3+ cells. Am J Reprod Immunol 59:525–534
Claus R (1990) Physiological role of seminal components in the reproductive tract of the female pig. J Reprod Fertil Suppl 40:117–131
Claus R, Dimmick MA, Gimenez T, Hudson LW (1992) Estrogens and prostaglandin F2alpha in the semen and blood plasma of stallions. Theriogenology 38:687–693
Cohen PE, Hardy MP, Pollard JW (1997a) Colony-stimulating factor-1 plays a major role in the development of reproductive function in male mice. Mol Endocrinol 11:1636–1650
Cohen PE, Zhu L, Pollard JW (1997b) Absence of colony stimulating factor-1 in osteopetrotic (csfmop/csfmop) mice disrupts estrous cycles and ovulation. Biol Reprod 56:110–118
Conroy H, Marshall NA, Mills KH (2008) TLR ligand suppression or enhancement of Treg cells? A double-edged sword in immunity to tumours. Oncogene 27:168–180
Das SK, Yano S, Wang J, Edwards DR, Nagase H, Dey SK (1997) Expression of matrix metalloproteinases and tissue inhibitors of metalloproteinases in the mouse uterus during the periimplantation period. Dev Genet 21:44–54
Das SC, Isobe N, Yoshimura Y (2008) Mechanism of prolonged sperm storage and sperm survivability in hen oviduct: a review. Am J Reprod Immunol 60:477–481
Das SC, Isobe N, Yoshimura Y (2009) Changes in the expression of interleukin-1beta and lipopolysaccharide-induced TNF factor in the oviduct of laying hens in response to artificial insemination. Reproduction 137:527–536
De M, Choudhuri R, Wood GW (1991) Determination of the number and distribution of macrophages, lymphocytes, and granulocytes in the mouse uterus from mating through implantation. J Leukoc Biol 50:252–262
Dean MD (2013) Genetic disruption of the copulatory plug in mice leads to severely reduced fertility. PLoS Genet 9:e1003185
Denison FC, Grant VE, Calder AA, Kelly RW (1999) Seminal plasma components stimulate interleukin-8 and interleukin-10 release. Mol Hum Reprod 5:220–226
Donato R, Cannon BR, Sorci G, Riuzzi F, Hsu K, Weber DJ, Geczy CL (2013) Functions of S100 proteins. Curr Mol Med 13:24–57
Doty A, Buhi WC, Benson S, Scoggin KE, Pozor M, Macpherson M, Mutz M, Troedsson MH (2011) Equine CRISP3 modulates interaction between spermatozoa and polymorphonuclear neutrophils. Biol Reprod 85:157–164
Eberhard WG (2009) Postcopulatory sexual selection: Darwin's omission and its consequences. Proc Natl Acad Sci USA 106(Suppl 1):10025–10032
England GC, Russo M, Freeman SL (2012) The bitch uterine response to semen deposition and its modification by male accessory gland secretions. Vet J 195:179–84
Fabian D, Juhas S, Il'kova G, Koppel J (2007) Dose- and time-dependent effects of TNFalpha and actinomycin D on cell death incidence and embryo growth in mouse blastocysts. Zygote 15:241–249
Fallarino F, Pallotta MT, Matino D, Gargaro M, Orabona C, Vacca C, Mondanelli G, Allegrucci M, Boon L, Romani R, Talesa VN, Puccetti P et al (2015) LPS-conditioned dendritic cells confer endotoxin tolerance contingent on tryptophan catabolism. Immunobiology 220:315–21
Feng J, Wocssncr JF Jr, Zhu C (1998) Matrilysin activity in the rat uterus during the oestrous cycle and implantation. J Reprod Fertil 114:347–350
Flowers WL, Esbenshade KL (1993) Optimizing management of natural and artificial matings in swine. J Reprod Fertil Suppl 48:217–228

Fullston T, Palmer NO, Owens JA, Mitchell M, Bakos HW, Lane M (2012) Diet-induced paternal obesity in the absence of diabetes diminishes the reproductive health of two subsequent generations of mice. Hum Reprod 27:1391–1400

Gangnuss S, Sutton-McDowall ML, Robertson SA, Armstrong DT (2004) Seminal plasma regulates corpora lutea macrophage populations during early pregnancy in mice. Biol Reprod 71:1135–1141

Geipel U, Kropf J, Krause W, Gressner AM (1992) The concentration pattern of laminin, hyaluronan, and aminoterminal propeptide of type III procollagen in seminal fluid. Andrologia 24:205–211

Gillott C (2003) Male accessory gland secretions: modulators of female reproductive physiology and behavior. Annu Rev Entomol 48:163–184

Glynn DJ, Sharkey DJ, Robertson SA (2004) Interferon-gamma inhibits female reproductive tract responsiveness to seminal plasma, 37th Annual Meeting of the Society for the Study of Reproduction. Biol Reprod 70:242

Godfrey KM, Gluckman PD, Hanson MA (2010) Developmental origins of metabolic disease: life course and intergenerational perspectives. Trends Endocrinol Metab 21:199–205

Gopichandran N, Ekbote UV, Walker JJ, Brooke D, Orsi NM (2006) Multiplex determination of murine seminal fluid cytokine profiles. Reproduction 131:613–621

Guerin LR, Moldenhauer LM, Prins JR, Bromfield JJ, Hayball JD, Robertson SA (2011) Seminal fluid regulates accumulation of FOXP3(+) regulatory T cells in the preimplantation mouse uterus through expanding the FOXP3(+) cell pool and CCL19-mediated recruitment. Biol Reprod 85:397–408

Gutsche S, von Wolff M, Strowitzki T, Thaler CJ (2003) Seminal plasma induces mRNA expression of IL-1beta, IL-6 and LIF in endometrial epithelial cells in vitro. Mol Hum Reprod 9:785–791

Haimovici F, Hill JA, Anderson DJ (1988) Variables affecting toxicity of human sera in mouse embryo cultures. J In Vitro Fert Embryo Transf 5:202–206

Hanson MA, Gluckman PD (2005) Developmental processes and the induction of cardiovascular function: conceptual aspects. J Physiol 565:27–34

Hart R, Norman RJ (2013) The longer-term health outcomes for children born as a result of IVF treatment: part I–general health outcomes. Hum Reprod Update 19:232–243

Heifetz Y, Lung O, Frongillo EA Jr, Wolfner MF (2000) The Drosophila seminal fluid protein Acp26Aa stimulates release of oocytes by the ovary. Curr Biol 10:99–102

Hellberg P, Thomsen P, Janson PO, Brannstrom M (1991) Leukocyte supplementation increases the luteinizing hormone-induced ovulation rate in the in vitro-perfused rat ovary. Biol Reprod 44:791–797

Higgins SC, Lavelle EC, McCann C, Keogh B, McNeela E, Byrne P, O'Gorman B, Jarnicki A, McGuirk P, Mills KH (2003) Toll-like receptor 4-mediated innate IL-10 activates antigen-specific regulatory T cells and confers resistance to Bordetella pertussis by inhibiting inflammatory pathology. J Immunol 171:3119–3127

Hutter H, Dohr G (1998) HLA expression on immature and mature human germ cells. J Reprod Immunol 38:101–122

Ideta A, Sakai S, Nakamura Y, Urakawa M, Hayama K, Tsuchiya K, Fujiwara H, Aoyagi Y (2010) Administration of peripheral blood mononuclear cells into the uterine horn to improve pregnancy rate following bovine embryo transfer. Anim Reprod Sci 117:18–23

Ignotz GG, Lo MC, Perez CL, Gwathmey TM, Suarez SS (2001) Characterization of a fucose-binding protein from bull sperm and seminal plasma that may be responsible for formation of the oviductal sperm reservoir. Biol Reprod 64:1806–1811

Jaiswal MK, Mallers TM, Larsen B, Kwak-Kim J, Chaouat G, Gilman-Sachs A, Beaman KD (2012) V-ATPase upregulation during early pregnancy: a possible link to establishment of an inflammatory response during preimplantation period of pregnancy. Reproduction 143: 713–725

James K, Hargreave TB (1984) Immunosuppression by seminal plasma and its possible clinical significance. Immunol Today 5:357–363

Jasper MJ, Robertson SA, Van der Hoek KH, Bonello N, Brannstrom M, Norman RJ (2000) Characterization of ovarian function in granulocyte-macrophage colony-stimulating factor-deficient mice. Biol Reprod 62:704–713
Jasper MJ, Bromfield JJ, Skinner R, Robertson SA (2005) 242. LIF expression is induced in the mouse oviduct following activation by seminal factors. Reprod Fertil Dev 17:96–96
Jasper MJ, Tremellen KP, Robertson SA (2006) Primary unexplained infertility is associated with reduced expression of the T-regulatory cell transcription factor Foxp3 in endometrial tissue. Mol Hum Reprod 12:301–308
Jasper MJ, Care AS, Sullivan B, Ingman WV, Aplin JD, Robertson SA (2011) Macrophage-derived LIF and IL1B regulate alpha(1,2)fucosyltransferase 2 (Fut2) expression in mouse uterine epithelial cells during early pregnancy. Biol Reprod 84:179–188
Jiang HY, O WS, Lee KH, Tang PL, Chow PH (2001) Ablation of paternal accessory sex glands is detrimental to embryo development during implantation. Anat Embryol (Berl) 203:255–263
Johansson M, Bromfield JJ, Jasper MJ, Robertson SA (2004) Semen activates the female immune response during early pregnancy in mice. Immunology 112:290–300
Katila T (2012) Post-mating inflammatory responses of the uterus. Reprod Domest Anim 47(Suppl 5): 31–41
Kawano N, Araki N, Yoshida K, Hibino T, Ohnami N, Makino M, Kanai S, Hasuwa H, Yoshida M, Miyado K, Umezawa A (2014) Seminal vesicle protein SVS2 is required for sperm survival in the uterus. Proc Natl Acad Sci USA 111:4145–4150
Killian GJ, Chapman DA, Rogowski LA (1993) Fertility-associated proteins in Holstein bull seminal plasma. Biol Reprod 49:1202–1207
Kirsch TM, Friedman AC, Vogel RL, Flickinger GL (1981) Macrophages in corpora lutea of mice: characterization and effects on steroid secretion. Biol Reprod 25:629–638
Kosaka K, Fujiwara H, Tatsumi K, Yoshioka S, Higuchi T, Sato Y, Nakayama T, Fujii S (2003) Human peripheral blood mononuclear cells enhance cell-cell interaction between human endometrial epithelial cells and BeWo-cell spheroids. Hum Reprod 18:19–25
Koyama K, Hasegawa A, Komori S (2009) Functional aspects of CD52 in reproduction. J Reprod Immunol 83:56–59
Koyanagi F, Nishiyama H (1981) Fate of spermatozoa that do not participate in fertilization in the domestic fowl. Cell Tissue Res 214:89–95
Krzymowski T, Stefanczyk-Krzymowska S, Koziorowski M (1989) Counter current transfer of PGF2 alpha in the mesometrial vessels as a mechanism for prevention of luteal regression in early pregnancy. Acta Physiol Pol 40:23–34
Kunz G, Beil D, Deininger H, Wildt L, Leyendecker G (1996) The dynamics of rapid sperm transport through the female genital tract: evidence from vaginal sonography of uterine peristalsis and hysterosalpingoscintigraphy. Hum Reprod 11:627–632
Laflamme BA, Avila FW, Michalski K, Wolfner MF (2014) A Drosophila protease cascade member, seminal metalloprotease-1, is activated stepwise by male factors and requires female factors for full activity. Genetics 196:1117–1129
Lane M, Robker RL, Robertson SA (2014) Parenting from before conception. Science 345: 756–760
Lengerova A, Vojtiskova M (1966) Prolonged survival of syngenic male skin grafts in parous C57 B1 mice. Folia Biol 8:21
Li X, Zhao X (2004) Separation and purification of ovulation-inducing factors in the seminal plasma of the Bactrian camel (Camelus bactrianus). Vet Res Commun 28:235–245
Lilja H, Oldbring J, Rannevik G, Laurell CB (1987) Seminal vesicle-secreted proteins and their reactions during gelation and liquefaction of human semen. J Clin Invest 80:281–285
Lima FS, Risco CA, Thatcher MJ, Benzaquen ME, Archbald LF, Santos JE, Thatcher WW (2009) Comparison of reproductive performance in lactating dairy cows bred by natural service or timed artificial insemination. J Dairy Sci 92:5456–5466
Liu WM, Pang RT, Chiu PC, Wong BP, Lao K, Lee KF, Yeung WS (2012) Sperm-borne microRNA-34c is required for the first cleavage division in mouse. Proc Natl Acad Sci USA 109:490–494

Lockett NN, Scott VL, Boudreaux CE, Clay BT, Pruett SB, Ryan PL, Coats KS (2010) Expression of regulatory T cell (Treg) activation markers in endometrial tissues from early and late pregnancy in the feline immunodeficiency virus (FIV)-infected cat. Placenta 31:796–802
Loras B, Vetele F, El Malki A, Rollet J, Soufir JC, Benahmed M (1999) Seminal transforming growth factor-beta in normal and infertile men. Hum Reprod 14:1534–1539
Loureiro B, Bonilla L, Block J, Fear JM, Bonilla AQ, Hansen PJ (2009) Colony-stimulating factor 2 (CSF-2) improves development and posttransfer survival of bovine embryos produced in vitro. Endocrinology 150:5046–5054
Lovell JW, Getty R (1968) Fate of semen in the uterus of the sow: histologic study of endometrium during the 27 hours after natural service. Am J Vet Res 29:609–625
Lucy MC (2001) Reproductive loss in high-producing dairy cattle: where will it end? J Dairy Sci 84:1277–1293
Lung O, Wolfner MF (2001) Identification and characterization of the major Drosophila melanogaster mating plug protein. Insect Biochem Mol Biol 31:543–551
Ma W, Tan J, Matsumoto H, Robert B, Abrahamson DR, Das SK, Dey SK (2001) Adult tissue angiogenesis: evidence for negative regulation by estrogen in the uterus. Mol Endocrinol 15:1983–1992
Maegawa M, Kamada M, Irahara M, Yamamoto S, Yoshikawa S, Kasai Y, Ohmoto Y, Gima H, Thaler CJ, Aono T (2002) A repertoire of cytokines in human seminal plasma. J Reprod Immunol 54:33–42
Maher ER, Afnan M, Barratt CL (2003) Epigenetic risks related to assisted reproductive technologies: epigenetics, imprinting, ART and icebergs? Hum Reprod 18:2508–2511
Mann T (1964) The biochemistry of semen and the male reproductive tract. Wiley, New York
Mattner PE (1968) The distribution of spermatozoa and leucocytes in the female genital tract in goats and cattle. J Reprod Fertil 17:253–261
McDonald LE, Black WG, McNutt SH, Casida LE (1952) The response of the rabbit uterus to instillation of semen at different phases of the estrous cycle. Am J Vet Res 13:419–424
McGraw LA, Suarez SS, Wolfner MF (2015) On a matter of seminal importance. Bioessays 37:142–147
McMaster M, Newton R, Dey S, Andrews G (1992) Activation and distribution of inflammatory cells in the mouse uterus during the preimplantation period. J Immunol 148:1699–1705
Mitchell MH, Swanson RJ, Oehninger S (2002) In vivo effect of leukemia inhibitory factor (LIF) and an anti-LIF polyclonal antibody on murine embryo and fetal development following exposure at the time of transcervical blastocyst transfer. Biol Reprod 67:460–464
Moldenhauer LM, Diener KR, Thring DM, Brown MP, Hayball JD, Robertson SA (2009) Cross-presentation of male seminal fluid antigens elicits T cell activation to initiate the female immune response to pregnancy. J Immunol 182:8080–8093
Muller M, Sales KJ, Katz AA, Jabbour HN (2006) Seminal plasma promotes the expression of tumorigenic and angiogenic genes in cervical adenocarcinoma cells via the E-series prostanoid 4 receptor. Endocrinology 147:3356–3365
Murakami M, Nishida T, Shiromoto M, Iwanaga S (1985) Phagocytosis of spermatozoa and latex beads by the epithelial cell of the cat oviduct: combined SEM and TEM study. Arch Histol Jpn 48:519–526
Murray FA, Grifo P, Parker CF (1983) Increased litter size in gilts by intrauterine infusion of seminal and sperm antigens before mating. J Anim Sci 56:895–900
Murray FA, Grifo P, Parker CF (1986) Increased litter size in gilts by presensitization with boar semen. J Anim Sci 196(Suppl):189
Narciandi F, Lloyd A, Meade KG, O'Farrelly C (2014) A novel subclass of bovine β-defensins links reproduction and immunology. Reprod Fertil Dev 26:769–77
Nelson SE, McLean MP, Jayatilak PG, Gibori G (1992) Isolation, characterization, and culture of cell subpopulations forming the pregnant rat corpus luteum. Endocrinology 130:954–966
Nogueira KD, Araujo VA, Rodrigues Sartori SS, Neves CA (2011) Phagocytosis of spermatozoa by epithelial cells in the vagina of the lizard Hemidactylus mabouia (Reptilia, Squamata). Micron 42:377–80

O WS, Chen HQ, Chow PH (1988) Effects of male accessory sex gland secretions on early embryonic development in the golden hamster. J Reprod Fertil 84:341–344
O'Leary S, Robertson SA, Armstrong DT (2001) Seminal plasma treatment stimulates progesterone synthesis during early pregnancy in the pig. In: Proceedings of the 32nd annual conference of the society for reproductive biology, Gold Coast, QLD
O'Leary S, Robertson SA, Armstrong DT (2002) The influence of seminal plasma on ovarian function in pigs—a novel inflammatory mechanism? J Reprod Immunol 57:225–238
O'Leary S, Armstrong DT, Robertson SA (2011) Transforming growth factor-beta (TGFbeta) in porcine seminal plasma. Reprod Fertil Dev 23:748–758
O'Leary S, Jasper MJ, Warnes GM, Armstrong DT, Robertson SA (2004) Seminal plasma regulates endometrial cytokine expression, leukocyte recruitment and embryo development in the pig. Reproduction 128:237–247
Owen DH, Katz DF (2005) A review of the physical and chemical properties of human semen and the formulation of a semen simulant. J Androl 26:459–469
Pacey AA, Hill CJ, Scudamore IW, Warren MA, Barratt CL, Cooke ID (1995) The interaction in vitro of human spermatozoa with epithelial cells from the human uterine (fallopian) tube. Hum Reprod 10:360–366
Palm F, Walter I, Budik S, Kolodziejek J, Nowotny N, Aurich C (2008) Influence of different semen extenders and seminal plasma on PMN migration and on expression of IL-1beta, IL-6, TNF-alpha and COX-2 mRNA in the equine endometrium. Theriogenology 70:843–851
Pampfer S, Arceci RJ, Pollard JW (1991) Role of colony stimulating factor-1 (CSF-1) and other lympho-hematopoietic growth factors in mouse pre-implantation development. Bioessays 13:535–540
Pandya IJ, Cohen J (1985) The leukocytic reaction of the human uterine cervix to spermatozoa. Fertil Steril 43:417–421
Pang SF, Chow PH, Wong TM (1979) The role of the seminal vesicle, coagulating glands and prostate glands on the fertility and fecundity of mice. J Reprod Fertil 56:129–132
Peitz B, Olds CP (1986) Effects of seminal vesicle removal on fertility and uterine sperm motility in the house mouse. Biol Reprod 35:608–617
Peng J, Zipperlen P, Kubli E (2005) Drosophila sex-peptide stimulates female innate immune system after mating via the Toll and Imd pathways. Curr Biol 15:1690–1694
Pilch B, Mann M (2006) Large-scale and high-confidence proteomic analysis of human seminal plasma. Genome Biol 7:R40
Plaks V, Birnberg T, Berkutzki T, Sela S, BenYashar A, Kalchenko V, Mor G, Keshet E, Dekel N, Neeman M, Jung S (2008) Uterine DCs are crucial for decidua formation during embryo implantation in mice. J Clin Invest 118:3954–3965
Politch JA, Tucker L, Bowman FP, Anderson DJ (2007) Concentrations and significance of cytokines and other immunologic factors in semen of healthy fertile men. Hum Reprod 22:2928–2935
Pollard JW, Lin EY, Zhu L (1998) Complexity in uterine macrophage responses to cytokines in mice. Biol Reprod 58:1469–1475
Poon HK, Lee KH, Wong CL, O WS, Chow PH (2009) Absence of paternal accessory sex gland secretions disturbs epigenetic reprogramming and expression of Igf2 and Dlk1 in golden hamster embryos. Theriogenology 71:1367–1380
Pudney J, Quayle AJ, Anderson DJ (2005) Immunological microenvironments in the human vagina and cervix: mediators of cellular immunity are concentrated in the cervical transformation zone. Biol Reprod 73:1253–1263
Queen F, Dhabuwala CB, Pierrepoint CG (1981) The effect of removal of the various accessory sex glands on the fertility of male rats. J Reprod Fertil 62:423–436
Ram KR, Wolfner MF (2009) A network of interactions among seminal proteins underlies the long-term postmating response in Drosophila. Proc Natl Acad Sci USA 106:15384–15389
Rasmussen SJ, Eckmann L, Quayle AJ, Shen L, Zhang YX, Anderson DJ, Fierer J, Stephens RS, Kagnoff MF (1997) Secretion of proinflammatory cytokines by epithelial cells in response to

Chlamydia infection suggests a central role for epithelial cells in chlamydial pathogenesis. J Clin Invest 99:77–87
Rasweiler JJ (1987) Prolonged receptivity to the male and the fate of spermatozoa in the female black mastiff bat, Molossus ater. J Reprod Fertil 79:643–654
Ratto M, Huanca W, Singh J, Adams GP (2006) Comparison of the effect of natural mating, LH, and GnRH on interval to ovulation and luteal function in llamas. Anim Reprod Sci 91:299–306
Ratto MH, Leduc YA, Valderrama XP, van Straaten KE, Delbaere LT, Pierson RA, Adams GP (2012) The nerve of ovulation-inducing factor in semen. Proc Natl Acad Sci USA 109:15042–15047
Rego JP, Crisp JM, Moura AA, Nouwens AS, Li Y, Venus B, Corbet NJ, Corbet DH, Burns BM, Boe-Hansen GB, McGowan MR (2014) Seminal plasma proteome of electroejaculated Bos indicus bulls. Anim Reprod Sci 148:1–17
Reid BL (1965) Interaction between homologous sperm and somatic cells of the uterus and peritoneum in the mouse. Exp Cell Res 40:679–683
Rijsselaere T, Van Soom A, Van Cruchten S, Coryn M, Gortz K, Maes D, de Kruif A (2004) Sperm distribution in the genital tract of the bitch following artificial insemination in relation to the time of ovulation. Reproduction 128:801–811
Riley JK, Heeley JM, Wyman AH, Schlichting EL, Moley KH (2004) TRAIL and KILLER are expressed and induce apoptosis in the murine preimplantation embryo. Biol Reprod 71:871–877
Robertson SA (2005) Seminal plasma and male factor signalling in the female reproductive tract. Cell Tissue Res 322:43–52
Robertson SA (2007) Seminal fluid signaling in the female reproductive tract: lessons from rodents and pigs. J Anim Sci 85:E36–44
Robertson SA (2010) Immune regulation of conception and embryo implantation-all about quality control? J Reprod Immunol 85:51–57
Robertson SA, Sharkey DJ (2001) The role of semen in induction of maternal immune tolerance to pregnancy. Semin Immunol 13:243–254
Robertson SA, Mayrhofer G, Seamark RF (1992) Uterine epithelial cells synthesize granulocyte-macrophage colony-stimulating factor and interleukin-6 in pregnant and nonpregnant mice. Biol Reprod 46:1069–1079
Robertson SA, Mau VJ, Tremellen KP, Seamark RF (1996) Role of high molecular weight seminal vesicle proteins in eliciting the uterine inflammatory response to semen in mice. J Reprod Fertil 107:265–277
Robertson SA, Allanson M, Mau VJ (1998) Molecular regulation of uterine leukocyte recruitment during early pregnancy in the mouse. Placenta 19:101–119
Robertson SA, Roberts CT, Farr KL, Dunn AR, Seamark RF (1999) Fertility impairment in granulocyte-macrophage colony-stimulating factor-deficient mice. Biol Reprod 60:251–261
Robertson SA, Sjoblom C, Jasper MJ, Norman RJ, Seamark RF (2001) Granulocyte-macrophage colony-stimulating factor promotes glucose transport and blastomere viability in murine preimplantation embryos. Biol Reprod 64:1206–1215
Robertson SA, Ingman WV, O'Leary S, Sharkey DJ, Tremellen KP (2002) Transforming growth factor beta-a mediator of immune deviation in seminal plasma. J Reprod Immunol 57:109–128
Robertson SA, Guerin LR, Moldenhauer LM, Hayball JD (2009a) Activating T regulatory cells for tolerance in early pregnancy—the contribution of seminal fluid. J Reprod Immunol 83:109–116
Robertson SA, Guerin LR, Bromfield JJ, Branson KM, Ahlstrom AC, Care AS (2009b) Seminal fluid drives expansion of the CD4+CD25+ T regulatory cell pool and induces tolerance to paternal alloantigens in mice. Biol Reprod 80:1036–1045
Robertson SA, Chin PY, Glynn DJ, Thompson JG (2011) Peri-conceptual cytokines–setting the trajectory for embryo implantation, pregnancy and beyond. Am J Reprod Immunol 66(Suppl 1):2–10
Rodriguez-Martinez H, Saravia F, Wallgren M, Martinez EA, Sanz L, Roca J, Vazquez JM, Calvete JJ (2010) Spermadhesin PSP-I/PSP-II heterodimer induces migration of polymorphonuclear neutrophils into the uterine cavity of the sow. J Reprod Immunol 84:57–65

Roldan ER, Gomendio M, Vitullo AD (1992) The evolution of eutherian spermatozoa and underlying selective forces: female selection and sperm competition. Biol Rev Camb Philos Soc 67:551–593

Rozeboom KJ, Troedsson MH, Hodson HH, Shurson GC, Crabo BG (2000) The importance of seminal plasma on the fertility of subsequent artificial inseminations in swine. J Anim Sci 78:443–448

Rubtsov YP, Rudensky AY (2007) TGFbeta signalling in control of T-cell-mediated self-reactivity. Nat Rev Immunol 7:443–453

Rudensky AY (2011) Regulatory T, cells and Foxp3. Immunol Rev 241:260–268

Sakaguchi S (2000) Regulatory T, cells: key controllers of immunologic self-tolerance. Cell 101:455–458

Sanford TR, De M, Wood GW (1992) Expression of colony-stimulating factors and inflammatory cytokines in the uterus of CD1 mice during days 1 to 3 of pregnancy. J Reprod Fertil 94:213–220

Sasaki Y, Sakai M, Miyazaki S, Higuma S, Shiozaki A, Saito S (2004) Decidual and peripheral blood CD4+CD25+ regulatory T cells in early pregnancy subjects and spontaneous abortion cases. Mol Hum Reprod 10:347–353

Sasaki Y, Darmochwal-Kolarz D, Suzuki D, Sakai M, Ito M, Shima T, Shiozaki A, Rolinski J, Saito S (2007) Proportion of peripheral blood and decidual CD4(+) CD25(bright) regulatory T cells in pre-eclampsia. Clin Exp Immunol 149:139–145

Schaefer TM, Desouza K, Fahey JV, Beagley KW, Wira CR (2004) Toll-like receptor (TLR) expression and TLR-mediated cytokine/chemokine production by human uterine epithelial cells. Immunology 112:428–436

Schieve LA (2002) Low and very low birth weight in infants conceived with use of assisted reproductive technology. N Engl J Med 346:731

Schjenken J, Robertson S (2014) Seminal fluid and immune adaptation for pregnancy—comparative biology in Mammalian species. Reprod Domest Anim 49(Suppl 3):27–36

Schjenken SE, Glynn DJ, Sharkey DJ, Robertson SA (2015) TLR4 signaling is a major mediator of the female tract response to seminal fluid in mice. Biol Reprod (in press)

Scott JL, Ketheesan N, Summers PM (2006) Presence of TGFbeta but not IL-8 or GM-CSF in ram seminal plasma. Society of Reproductive Biology, Gold Coast, QLD, p 49

Scott JL, Ketheesan N, Summers PM (2009) Spermatozoa and seminal plasma induce a greater inflammatory response in the ovine uterus at oestrus than dioestrus. Reprod Fertil Dev 21:817–826

Sharkey DJ, Macpherson AM, Tremellen KP, Robertson SA (2007) Seminal plasma differentially regulates inflammatory cytokine gene expression in human cervical and vaginal epithelial cells. Mol Hum Reprod 13:491–501

Sharkey DJ, Macpherson AM, Tremellen KP, Mottershead DG, Gilchrist RB, Robertson SA (2012a) TGF-beta mediates proinflammatory seminal fluid signaling in human cervical epithelial cells. J Immunol 189:1024–1035

Sharkey DJ, Tremellen KP, Jasper MJ, Gemzell-Danielsson K, Robertson SA (2012b) Seminal fluid induces leukocyte recruitment and cytokine and chemokine mRNA expression in the human cervix after coitus. J Immunol 188:2445–2454

Shen XH, Han YJ, Zhang DX, Cui XS, Kim NH (2009) A link between the interleukin-6/Stat3 anti-apoptotic pathway and microRNA-21 in preimplantation mouse embryos. Mol Reprod Dev 76:854–862

Shevach EM (2002) CD4+ CD25+ suppressor T cells: more questions than answers. Nat Rev Immunol 2:389–400

Shimada M, Yanai Y, Okazaki T, Noma N, Kawashima I, Mori T, Richards JS (2008) Hyaluronan fragments generated by sperm-secreted hyaluronidase stimulate cytokine/chemokine production via the TLR2 and TLR4 pathway in cumulus cells of ovulated COCs, which may enhance fertilization. Development 135:2001–2011

Short SM, Wolfner MF, Lazzaro BP (2012) Female Drosophila melanogaster suffer reduced defense against infection due to seminal fluid components. J Insect Physiol 58:1192–1201

Sjoblom C, Wikland M, Robertson SA (1999) Granulocyte-macrophage colony-stimulating factor promotes human blastocyst development in vitro. Hum Reprod 14:3069–3076
Sjoblom C, Roberts CT, Wikland M, Robertson SA (2005) Granulocyte-macrophage colony-stimulating factor alleviates adverse consequences of embryo culture on fetal growth trajectory and placental morphogenesis. Endocrinology 146:2142–2153
Skjervold H (1975) Comparison of litter size by use of natural and by artificial mating. Z Tierzucht Zucht Biol 92:252–259
Somerset DA, Zheng Y, Kilby MD, Sansom DM, Drayson MT (2004) Normal human pregnancy is associated with an elevation in the immune suppressive CD25(+) CD4(+) regulatory T-cell subset. Immunology 112:38–43
Suarez SS (2008) Regulation of sperm storage and movement in the mammalian oviduct. Int J Dev Biol 52:455–462
Suarez SS, Pacey AA (2006) Sperm transport in the female reproductive tract. Hum Reprod Update 12:23–37
Sutovsky P, Lovercamp K (2010) Molecular markers of sperm quality. Soc Reprod Fertil Suppl 67:247–256
Tanco VM, Van Steelandt MD, Ratto MH, Adams GP (2012) Effect of purified llama ovulation-inducing factor (OIF) on ovarian function in cattle. Theriogenology 78:1030–1039
Taylor U, Rath D, Zerbe H, Schuberth HJ (2008) Interaction of intact porcine spermatozoa with epithelial cells and neutrophilic granulocytes during uterine passage. Reprod Domest Anim 43:166–175
Templeton AA, Cooper I, Kelly RW (1978) Prostaglandin concentrations in the semen of fertile men. J Reprod Fertil 52:147–150
Thompson LA, Barratt CL, Bolton AE, Cooke ID (1992) The leukocytic reaction of the human uterine cervix. Am J Reprod Immunol 28:85–89
Thompson JG, Kind KL, Roberts CT, Robertson SA, Robinson JS (2002) Epigenetic risks related to assisted reproductive technologies: short- and long-term consequences for the health of children conceived through assisted reproduction technology: more reason for caution? Hum Reprod 17:2783–2786
Tomlinson MJ, White A, Barratt CL, Bolton AE, Cooke ID (1992) The removal of morphologically abnormal sperm forms by phagocytes: a positive role for seminal leukocytes? Hum Reprod 7:517–522
Townson DH, Warren JS, Flory CM, Naftalin DM, Keyes PL (1996) Expression of monocyte chemoattractant protein-1 in the corpus luteum of the rat. Biol Reprod 54:513–520
Tremellen KP, Seamark RF, Robertson SA (1998) Seminal transforming growth factor beta1 stimulates granulocyte-macrophage colony-stimulating factor production and inflammatory cell recruitment in the murine uterus. Biol Reprod 58:1217–1225
Tremellen KP, Valbuena D, Landeras J, Ballesteros A, Martinez J, Mendoza S, Norman RJ, Robertson SA, Simon C (2000) The effect of intercourse on pregnancy rates during assisted human reproduction. Hum Reprod 15:2653–2658
Troedsson MH, Loset K, Alghamdi AM, Dahms B, Crabo BG (2001) Interaction between equine semen and the endometrium: the inflammatory response to semen. Anim Reprod Sci 68:273–278
Trowsdale J, Betz AG (2006) Mother's little helpers: mechanisms of maternal-fetal tolerance. Nat Immunol 7:241–246
Turner EC, Hughes J, Wilson H, Clay M, Mylonas KJ, Kipari T, Duncan WC, Fraser HM (2011) Conditional ablation of macrophages disrupts ovarian vasculature. Reproduction 141:821–831
Van der Hoek KH, Maddocks S, Woodhouse CM, van Rooijen N, Robertson SA, Norman RJ (2000) Intrabursal injection of clodronate liposomes causes macrophage depletion and inhibits ovulation in the mouse ovary. Biol Reprod 62:1059–1066
Vazquez JM, Martinez EA, Roca J, Gil MA, Parrilla I, Cuello C, Carvajal G, Lucas X, Vazquez JL (2005) Improving the efficiency of sperm technologies in pigs: the value of deep intrauterine insemination. Theriogenology 63:536–547
Waberski D, Claassen R, Hahn T, Jungblut PW, Parvizi N, Kallweit E, Weitze KF (1997) LH profile and advancement of ovulation after transcervical infusion of seminal plasma at different stages of oestrus in gilts. J Reprod Fertil 109:29–34

Watson JG, Carroll J, Chaykin S (1983) Reproduction in mice: the fate of spermatozoa not involved in fertilization. Gamete Res 7:75–84
Wolfner MF (2002) The gifts that keep on giving: physiological functions and evolutionary dynamics of male seminal proteins in Drosophila. Heredity 88:85–93
Wong CL, Lee KH, Lo KM, Chan OC, Goggins W, Or WS, Chow PH (2007) Ablation of paternal accessory sex glands imparts physical and behavioural abnormalities to the progeny: an in vivo study in the golden hamster. Theriogenology 68:654–662
Wong CL, Chan OC, Lee KH, O WS, Chow PH (2008) Absence of paternal accessory sex glands dysregulates preimplantation embryo cell cycle and causes early oviductal-uterine transit in the golden hamster in vivo. Fertil Steril 89:1021–1024
Wu RJ, Van der Hoek KH, Ryan NK, Norman RJ, Robker RL (2004) Macrophage contributions to ovarian function. Hum Reprod Update 10:119–133
Ying Y, Chow PH (1998) O WS. Effects of male accessory sex glands on deoxyribonucleic acid synthesis in the first cell cycle of golden hamster embryos. Biol Reprod 58:659–663
Yoshida S, Ono M, Shono T, Izumi H, Ishibashi T, Suzuki H, Kuwano M (1997) Involvement of interleukin-8, vascular endothelial growth factor, and basic fibroblast growth factor in tumor necrosis factor alpha-dependent angiogenesis. Mol Cell Biol 17:4015–4023
Zheng WM, Nishibori M, Isobe N, Yoshimura Y (2001) An in situ hybridization study of the effects of artificial insemination on the localization of cells expressing MHC class II mRNA in the chicken oviduct. Reproduction 122:581–586
Ziebe S, Loft A, Povlsen BB, Erb K, Agerholm I, Aasted M, Gabrielsen A, Hnida C, Zobel DP, Munding B, Bendz SH, Robertson SA (2013) A randomized clinical trial to evaluate the effect of granulocyte-macrophage colony-stimulating factor (GM-CSF) in embryo culture medium for in vitro fertilization. Fertil Steril 99:1600–1609

Chapter 7
Role of Seminal Plasma in Human Female Reproductive Failure: Immunomodulation, Inflammation, and Infections

Deborah J. Anderson and Joseph A. Politch

Abstract Human seminal plasma contains factors that can regulate the female immune system and potentially promote reproductive fitness. Adverse effects on fertility and pregnancy may occur when seminal plasma provides insufficient, excessive, or altered signals or when the female partner is incapable of receiving these signals.

Keywords Seminal plasma • Prostate • Seminal vesicles • Vaginal epithelium • Endometrium • Cytokines • TGF-beta • PGE

Seminal plasma can theoretically affect the reproductive fitness of a female sexual partner in a number of ways: (1) in healthy couples, normal components of seminal plasma such as immunomodulatory, proinflammatory, and growth factors may positively impact the female reproductive tract by inducing tolerance and supporting early pregnancy; (2) the immunomodulatory/proinflammatory effects of seminal plasma thought to enhance fertility and early fetal survival may also predispose women to genital infections which could lead to reproductive failure; (3) alterations in the composition of semen in male partners with pathologic conditions such as genital infections could have detrimental effects on female reproductive function by providing insufficient, excessive, or altered signals. Much of the research on

D.J. Anderson, Ph.D. (✉)
Department of Obstetrics and Gynecology, Boston University School of Medicine, 670 Albany Street, Suite 516, Boston, MA 02118, USA

Department of Microbiology, Boston University School of Medicine, 670 Albany Street, Suite 516, Boston, MA 02118, USA

Department of Medicine, Boston University School of Medicine, 670 Albany Street, Suite 516, Boston, MA 02118, USA
e-mail: Deborah.Anderson@BMC.org

J.A. Politch, Ph.D.
Department of Obstetrics and Gynecology, Boston University School of Medicine, 670 Albany Street, Suite 516, Boston, MA 02118, USA

R. Bronson (ed.), *The Male Role in Pregnancy Loss and Embryo Implantation Failure*, Advances in Experimental Medicine and Biology 868,
DOI 10.1007/978-3-319-18881-2_7

beneficial effects of seminal plasma on fertility has been conducted in animal models and is reviewed elsewhere in this issue. This chapter will focus on human studies and will address potential effects of seminal plasma on human female reproduction. Specifically, we will review the origins and concentrations of immunomodulatory/proinflammatory factors in human seminal plasma in health and disease, their effects on immune defense in the female genital tract that may promote fertility and fetal well-being, and mechanisms whereby these factors may directly or indirectly contribute to female reproductive failure.

7.1 Tissue Origins of Seminal Plasma Components

The male genital tract is contained within an epithelial barrier system that normally limits the penetration of components from the peripheral blood and surrounding tissues (Anderson and Pudney 2005). Therefore, most components of the semen are derived from organs that comprise the male genital tract. Recent proteomics studies indicate that there are over 900 proteins in human seminal plasma (Rodriguez-Martinez et al. 2011; Cadavid et al. 2014). Most of the seminal plasma volume is produced in the seminal vesicles (65–75 %) and prostate (25–35 %) (Mann 1954). Several of the principal organs contribute specific immunomodulatory seminal plasma components which will be discussed in more detail below. The epididymis produces soluble HLA-G (Rajagopalan et al. 2006) and CD52g (Norton et al. 2002); the seminal vesicles are a principle source of prostaglandins (Kelley 1981) and TGF-β (Robertson et al. 2002); the prostate also produces TGF-β (Pannek et al. 1999) along with large amounts of exosomes, subcellular microvesicles enriched in bioactive components including cytokines and small RNAs (Burden et al. 2006; Vojtech et al. 2014). Local antibody production in the male genital tract occurs primarily in the penile urethra (Pudney and Anderson 2011). Other important components of semen, such as cytokines, chemokines, and mucins, are produced by epithelial, immune, and stromal cells throughout the genital tract (Anderson and Pudney 2005) (Fig. 7.1).

7.2 Beneficial Effects of Immunomodulatory and Proinflammatory Factors in Human Seminal Plasma

Seminal plasma contains very high concentrations of two potent immunomodulatory factors, TGF-β and PGE, which may promote female fertility by suppressing natural immune responses to sperm and the semi-allogeneic fetus. Humoral and cellular immunity to sperm and embryos have been associated with infertility and miscarriage (Kokcu et al. 2012; Hill 1995; Raghupathy 1997). The immunomodulatory roles of seminal TGF-β and PGE were recently reviewed in detail by Robertson (2013) and Doncel (2014). TGF-β is a ubiquitous pleiotropic immunomodulatory

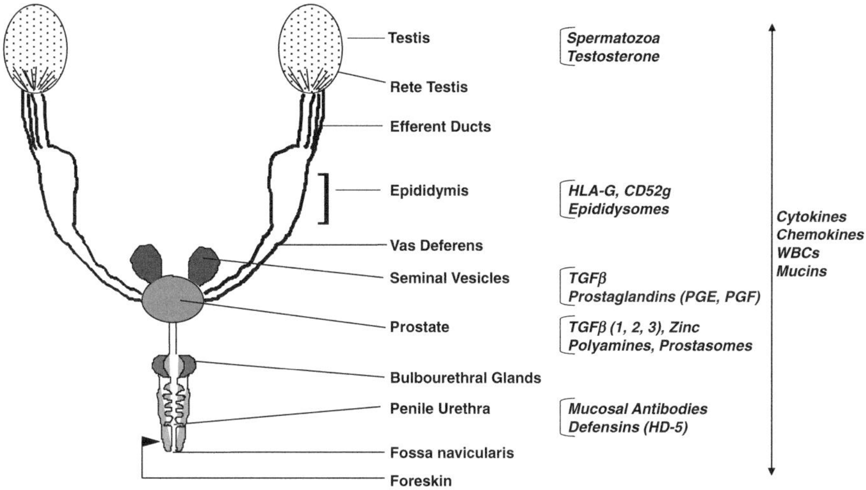

Fig. 7.1 Origin of seminal plasma components

factor that has been implicated as a key factor in the modulation of host defense (Wahl et al. 2006; Gorelik and Flavell 2002). Its potent immunoregulatory activity became evident from studies in Tgfb1 and Tgfbr1 knockout mice (mice without TGF-β or the TGF-β receptor), which develop lethal inflammatory disorders early in life (Kulkarni et al. 1993; Marie et al. 2006; Shull et al. 1992). Three isoforms of TGF-β are present in human seminal plasma (TGF-β1, TGF-β2, and TGF-β3); they exist primarily in latent form until activation in the female genital tract by proteases and acid pH (Robertson et al. 2002). TGF-β induces stimulatory or inhibitory effects in human T cells depending on the T cell differentiation status and the stimulatory conditions (Oh and Li 2013). The first evidence that TGF-β plays a critical role at mucosal sites was provided by early studies on TGF-β signaling in the intestine showing that induction of regulatory T cells (Tregs) by TGF-β promoted immunological tolerance to food antigens by active control of innate and adaptive immune responses (Harrison and Powrie 2013). TGF-β has been suggested to be one of the major factors inducing immune tolerance in the female genital tract by inducing differentiation of Treg cells and suppressing the activity of natural killer cells (Robertson et al. 2002), thereby suppressing immunity to antigens expressed on sperm and the implanting embryo. TGF-β also upregulates the expression of proinflammatory cytokines and chemokines in epithelial cells from the female reproductive tract which could stimulate cell growth and angiogenesis to support embryo implantation (Sharkey et al. 2012a) and thus be beneficial to early events in pregnancy. PGE is another potent immunomodulatory factor in semen, capable of modulating immune functions on multiple levels (Quayle et al. 1989). The concentration of PGE in seminal plasma is several orders of magnitude higher than that in blood plasma, although there is a high degree of interindividual variation (Templeton et al. 1978). PGE suppresses macrophage and neutrophil function and the cytotoxic activity of T lymphocytes and natural killer cells and upregulates the inflammatory mediator Cox 2 in vaginal epithelial cells (Templeton et al. 1978). Kelley et al. showed that PGE exposure results in upregulated IL-10 and downregulated IL-12 production, shifting the T cell response from a Th1 (cell-mediated immunity (CMI) dominant) to Th2 (humoral immunity dominant) immune response (Kelley 1981). Soluble HLA-G is also present in semen and has been proposed to suppress adverse NK cell activity directed against invading cytotrophoblast (Rajagopalan et al. 2006). CD52g, a sperm-coating glycoprotein derived from the epididymis, may also play an important role in preventing antisperm immunity and infertility, although it can itself be a target of antisperm antibodies in some infertility patients (Norton et al. 2002). Seminal plasma also contains high concentrations of proinflammatory cytokines and chemokines that could affect fertility by recruiting and activating immune cells in the reproductive tissues and stimulating the production of factors that stimulate cell growth and angiogenesis (Politch et al. 2007). Specifically, seminal plasma has very high concentrations of IL-7, a hematopoietic growth factor that promotes the proliferation of lymphoid progenitors, B cell maturation, and T and NK cell survival (Fry and Mackall 2005), and three chemokines, SDF-1, MCP-1, and IL-8, which may recruit leukocytes to the insemination site to participate in immune defense and scavenger functions. The concentrations and ranges of principal immunomodulatory and proinflammatory factors in semen are provided in Table 7.1. Unpublished data

Table 7.1 Concentrations of selected components in normal semen

Component	Concentration	References
Immunomodulatory factors		
TGF-β1		Sharkey et al. (2012a)
Total	219.3 ± 13.4 ng/ml[a]	
Bioactive (% total)	2.3 ± 0.4 ng/ml[a] (1.2 %)	
TGF-β2		
Total	5.3 ± 0.7 ng/ml[a]	
Bioactive (% total)	0.25 ± 0.04 ng/ml[a] (5.3 %)	
TGF-β3		
Total	172.2 ± 32.8 ng/ml[a]	
Bioactive (% total)	3.5 ± 1.2 ng/ml[a] (1.8 %)	
PGE-1	7.0 ± 6.0 μg/ml[a]	Gerozissis et al. (1982)
PGE-2	14.0 ± 11.0 μg/ml[a]	
PGF-1a	1.0 ± 0.7 μg/ml[a]	
PGF-2a	2.0 ± 2.0 μg/ml[a]	
IL-7	2,365.8 (1,109.5–3,985.5) pg/ml[b]	Politch et al. (2007)
HLA-G	82 (29–1,161) U/ml[c]	Dahl et al. (2014)
Exosomes	~1 Trillion/ejaculate	Vojtech et al. (2014)
Chemokines		
MCP-1	3.3 (0.3–81.5) ng/ml[b]	Politch et al. (2007)
IL-8	1.6 (0.4–14.7) ng/ml[b]	
SDF-1α	5.1 (ND–18.0) ng/ml[b]	
Proinflammatory cytokines		
TNF-α	1.5 (ND–40.3) pg/ml[b]	Politch et al. (2007)
GM-CSF	1.5 (ND–1,190.6) pg/ml[b]	

[a]Mean ± SE
[b]Geometric mean (range)
[c]Median (range)

from our laboratory indicate that TGF-β levels are decreased and IL-8 levels are significantly increased in semen from men with leukocytospermia (male genital inflammation) (Politch J, personal communication).

An exciting new area of research is the study of seminal exosomes. These highly abundant subcellular microvesicles, produced primarily by the prostate but also in the epididymis and seminal vesicles, are enriched in bioactive components including cytokines and small RNAs (miRNA, YRNA, and tRNA) and may play an important role in the fertilization and intercellular communication in the genital tract (Burden et al. 2006; Vojtech et al. 2014; Li et al. 2013). It is estimated that approximately one trillion exosomes are present in a human ejaculate. These small vesicles readily fuse with the plasma membrane of sperm and other cell types to deliver important signaling molecules. A number of immune-related mRNAs are targeted by miRNAs in seminal exosomes; whether miRNAs can be delivered by seminal exosomes in sufficient quantity to target genes and change cellular functions in the vaginal immune cell population is unknown (Vojtech et al. 2014). They have been shown to play a direct role in antiviral immune defense (Madison et al. 2014).

7.3 Evidence for Effects of Seminal Plasma on the Human Female Reproductive Tract

Recently, a meta-analysis was conducted on the role of seminal plasma for improved outcomes during in vitro fertilization. The outcome of IVF treatment in patients exposed to seminal plasma near the time of oocyte pickup or embryo transfer was compared to that of controls with no exposure to seminal plasma (a total of 2,204 patients in seven randomized control trials). They found a statistically significant improvement in the clinical pregnancy rate after seminal plasma exposure (RR 1.24, $p=0.003$), but no improvement in the ongoing pregnancy/live birth rate (Crawford et al. 2015). However, this topic is a matter of debate. Michael Bedford has pointed out that virgin animals are perfectly good embryo transfer recipients and that many human IVF programs do not use priming with seminal plasma in conjunction with IVF cycles and obtain good fertilization and pregnancy outcomes. He concludes that whereas a nuanced effect of seminal plasma on fertility outcome in humans cannot be dismissed, many experimental and clinical results demonstrate that the presence of seminal plasma is not essential for the transport and survival of spermatozoa in the female tract, for fertilization, or for implantation and embryonic development (Bedford 2015). We review below evidence from in vitro and clinical studies of the effect of seminal plasma on various regions of the human female genital tract.

7.3.1 *Vagina and External Genitalia*

In human reproduction, semen is deposited into the vagina, and it is this site and the external genitalia that have maximum exposure to semen components. Semen concentrations in vaginal fluid decline after intercourse to approximately 50 % after 1 h and reach baseline after 24 h (Macaluso et al. 1999; Graves et al. 1985). The external genitalia (labia majora and minora) are covered with keratinized skin, and the vagina and ectocervix are lined with specialized nonkeratinizing stratified squamous epithelia (Anderson 2007). These multilayered epithelia normally afford a barrier to external signaling by presenting a wall of cornified enucleated cells on the apical surface which lack most membrane receptors and signaling pathways (Anderson et al. 2014). Lipophilic molecules from semen could be absorbed through the vaginal epithelium to achieve local or systemic effects (Muranishi et al. 1993). In women with certain lower genital infections (e.g., HSV-2, HPV, GC) or epithelial lesions, living functional basal epithelial cells or leukocytic infiltrates in the vaginal epithelium could be exposed and react to seminal components. Vaginal and ectocervical cells grown as monolayers in vitro respond to semen challenge by producing GM-CSF, IL-6, IL-8, and MCP-1 (Sharkey et al. 2007); these cultures are not fully differentiated and represent the basal epithelial layer of the stratified epithelium that is exposed by vaginal lesions or infections. A clinical study that monitored the infiltration of lymphocytes and other WBC population into the ectocervical mucosa following intercourse reported increased numbers of T lymphocytes, macrophages, and dendritic cells (Sharkey et al. 2012b), but whether the signal was transmitted across the stratified squamous epithelium of the ectocervix or via the neighboring endocervix has yet to be determined.

7.3.2 Endocervix

The human endocervix is lined with a single layer of viable columnar epithelial cells that are highly responsive to external signals (Fichorova and Anderson 1999), including seminal plasma (Sharkey et al. 2012a). The opening of the endocervix (cervical os) is protected from bacteria in the vaginal compartment by a thick layer of secreted mucins (Gipson et al. 1997), but seminal components may diffuse through mucus (Cone 2009) or directly contact endocervical epithelial cells after intercourse due to disruption of the mucus barrier or other means of exposure (e.g., cervical ectopy). Studies have documented leukocytic exudates in the cervical canal following intercourse (Pandya and Cohen 1985; Thompson et al. 1992) and infiltrates of macrophages, dendritic cells, and T lymphocytes in the cervical epithelium and stroma (Sharkey et al. 2012b) indicating that the cervix is a region that commonly responds to seminal signaling. Leukocytic infiltrates could play a scavenger role in the clearance of sperm and other seminal factors after intercourse. Less well understood are the potential effects of endocervical epithelial factors produced in response to seminal plasma on other aspects of reproductive function such as sperm capacitation and effects registered in the upper tract pertaining to implantation.

7.3.3 Uterus/Endometrium

Seminal plasma effects on various sites in the female reproductive tract depend on the concentration of seminal factors reaching the tissue. In mice, semen rapidly enters the uterus after coitus (Zamboni 1972), and a number of studies have documented effects of semen on endometrial receptivity and implantation in mice (Robertson et al. 2013). However, there is debate on whether seminal plasma ascends beyond the cervix to enter the upper genital tract (uterus, fallopian tubes) in women. A series of magnetic resonance imaging studies using ^{99m}Tc-labeled human albumin microspheres showed radiolabel dispersion into the uterus and fallopian tubes following deposition into the vagina (Kunz et al. 1996; Venter and Iturralde 1979; Zervomanolakis et al. 2007). However, a number of other studies using semen surrogates or vaginal gels have failed to document this effect (Barnhart et al. 2004, 2005; Brown et al. 1997; Chatterton et al. 2004; Louissaint et al. 2012; Mauck et al. 2008). Even if appreciable amounts of soluble vaginal contents do not ascend into the uterus through the endocervix, seminal plasma components could signal cells in the upper reproductive tract if small amounts ascend into this region after intercourse through peristalsis as proposed by some studies or through absorption through the vaginal epithelium. A small percentage (1–15 %) of TGF-beta in semen is associated with the sperm fraction (Sharkey et al. 2012a), therefore making it possible for sperm to provide TGF-β signaling as they ascend into the upper genital tract.

7.4 Potential Adverse Effects of Seminal Plasma

7.4.1 Insufficient Seminal Plasma Immunoregulation Could Permit Immunity Against Sperm and the Conceptus

In some couples, the man's seminal plasma may be deficient in some of the immunoregulatory compounds described above, or the female partner could be nonresponsive to seminal plasma signaling (e.g., deficient in receptors or proteases that activate TGF-β) due to genetic abnormalities, genital infections, or other circumstances. Lack of immune tolerance/immunosuppression in the genital tract could promote the synthesis of antibodies associated with female reproductive failure including antisperm and antiphospholipid antibodies or the generation of a T cell response to the conceptus that can lead to miscarriage (Kokcu et al. 2012; Hill 1995; Raghupathy 1997).

7.4.2 Regulation of Cervicovaginal Immune Defense by Seminal Plasma: Increased Susceptibility to Infections?

Memory T cells, including antigen-specific cytotoxic CD8+ T cells, reside throughout the female genital tract and are thought to play a major role in host defense at this site. They are concentrated in the transformation zone of the cervix, where seminal plasma exerts its maximum effect (Pudney et al. 2005). Chemokines in semen attract T cells, along with PMNs and macrophages, to the endocervix after intercourse and may promote immune defense during this vulnerable window of infection. But TGF-β and PGE may also exert a tolerizing or immunosuppressive effect on these critical immune defense cells and impede their ability to clear an infection. Viral and bacterial genital infections have been associated with infertility and miscarriage in women (Novy et al. 2008).

7.4.3 Proinflammatory Factors in Seminal Plasma Could Promote HIV Infection

Genital inflammation is a risk factor for HIV transmission due to the recruitment of HIV target cells into the mucosal epithelium and their activation which makes cells more vulnerable to infection and enables cells to produce more HIV once infected (Mayer and Venkatesh 2011). TGF-beta and PGE act on cervical epithelial cells to upregulate the synthesis of proinflammatory cytokines and chemokines which recruit leukocytes and regulate their immune function (Sharkey et al. 2012a). Suppression of T cell immunoactivity and promotion of inflammation in the genital tract by seminal plasma could promote the sexual transmission of HIV-1. This topic has recently been reviewed by Doncel et al. (2014) and Rametse et al. (2014).

7.5 Conclusions

Seminal fluid improves the pregnancy rate in laboratory and farm animals, and a number of components in seminal plasma have been identified that have beneficial effects on reproduction. The extent to which seminal plasma affects fertility in humans is not fully understood, but there is considerable evidence that seminal plasma affects immune cell populations in the lower genital tract. Several studies have documented leukocytic infiltration into the endocervical canal following intercourse, possibly due to chemokines in semen, and these cells may play a scavenger role and/or secrete factors that can positively or negatively affect fertility. In vitro studies have demonstrated that seminal plasma components are immunosuppressive and induce the differentiation of Treg cells that can mediate tolerance in the genital tract to foreign antigens on sperm and trophoblast cells. This effect of seminal plasma must be tightly controlled so as not to weaken immune defense to sexually transmitted pathogens. If there is too little immunomodulation by semen, cellular and humoral immune responses could develop against antigens expressed on sperm or the conceptus and negatively affect female reproductive function. However, too much immunosuppression or the induction of inflammation by seminal plasma could predispose women to genital infections which also can adversely affect reproductive function.

Acknowledgment Supported by NIH grant U19AI096398 (DJA).

References

Anderson DJ (2007) Genitourinary immune defense. In: Holmes K, Sparling P, Stamm W, Piot P, Wasserheit J, Corey L, Cohen M (eds) Sexually transmitted diseases, 4th edn. McGraw-Hill, New York, NY, pp 271–288

Anderson DJ, Pudney J (2005) Mucosal immunology of the human male genital tract and experimental models. In: Mestecky J, Lamm M, Strober W, Bienenstock R, Mcghee JR, Mayer L (eds) Mucosal immunology, 3rd edn. Elsevier Academic, Boston, pp 1647–1660

Anderson DJ, Marathe J, Pudney J (2014) The structure of the human vaginal stratum corneum and its role in immune defense. Am J Reprod Immunol 71:618–623

Barnhart KT, Pretorius ES, Timbers K, Shera D, Shabbout M, Malamud D (2004) In vivo distribution of a vaginal gel: MRI evaluation of the effects of gel volume, time and simulated intercourse. Contraception 70:498–505

Barnhart KT, Pretorius ES, Shaunik A, Timbers K, Nasution M, Mauck C (2005) Vaginal distribution of two volumes of the novel microbicide gel cellulose sulfate (2.5 and 3.5mL). Contraception 72:65–70

Bedford JM (2015) The functions – or not – of seminal plasma? Biol Reprod 92:18

Brown J, Hooper G, Kenyon CJ, Haines S, Burt J, Humphries JM, Newman SP, Davis SS, Sparrow RA, Wilding IR (1997) Spreading and retention of vaginal formulations in post-menopausal women as assessed by gamma scintigraphy. Pharm Res 14:1073–1078

Burden HP, Holmes CH, Persad R, Whittington K (2006) Prostasomes—their effects on human male reproduction and fertility. Hum Reprod Update 12:283–292

Cadavid JA, Alvarez A, Markert UR, Cardona Maya W (2014) Differential protein expression in seminal plasma from fertile and infertile males. J Hum Reprod Sci 7:206–211

Chatterton BE, Penglis S, Kovacs JC, Presnell B, Hunt B (2004) Retention and distribution of two 99mTc-DTPA labelled vaginal dosage forms. Int J Pharm 271:137–143

Cone RA (2009) Barrier properties of mucus. Adv Drug Deliv Rev 61:75–85

Crawford G, Ray A, Gudi A, Shah A, Homburg R (2015) The role of seminal plasma for improved outcomes during in vitro fertilization treatment: review of the literature and meta-analysis. Hum Reprod Update 21:275–284

Dahl M, Perin TL, Djurisic S, Rasmussen M, Ohlsson J, Buus S, Lindhard A, Hviid TV (2014) Soluble human leukocyte antigen-G in seminal plasma is associated with HLA-G genotype: possible implications for fertility success. Am J Reprod Immunol 72:89–105

Doncel GF, Anderson S, Zalenskaya I (2014) Role of semen in modulating the female genital tract microenvironment—implications for HIV transmission. Am J Reprod Immunol 71:564–574

Fichorova RN, Anderson DJ (1999) Differential expression of immunobiological mediators by immortalized human cervical and vaginal epithelial cells. Biol Reprod 60:508–514

Fry TJ, Mackall CL (2005) The many faces of IL-7: from lymphopoiesis to peripheral T cell maintenance. J Immunol 174:6571–6576

Gerozissis K, Jouannet P, Soufir JC, Dray F (1982) Origin of prostaglandins in human semen. J Reprod Fertil 65:401–404

Gipson IK, Ho SB, Spurr-Michaud SJ, Tisdale AS, Zhan Q, Torlakovic E, Pudney J, Anderson DJ, Toribara NW, Hill JA 3rd (1997) Mucin genes expressed by human female reproductive tract epithelia. Biol Reprod 56:999–1011
Gorelik L, Flavell RA (2002) Transforming growth factor-beta in T-cell biology. Nat Rev Immunol 2:46–53
Graves HC, Sensabaugh GF, Blake ET (1985) Postcoital detection of a male-specific semen protein. Application to the investigation of rape. N Engl J Med 312:338–343
Harrison OJ, Powrie FM (2013) Regulatory T cells and immune tolerance in the intestine. Cold Spring Harb Perspect Biol 5:a018341. doi:10.1101/cshperspect.a018341
Hill JA (1995) T-helper 1-type immunity to trophoblast: evidence for a new immunological mechanism for recurrent abortion in women. Hum Reprod 10(Suppl 2):114–120
Kelley RW (1981) Prostaglandin synthesis in the male and female reproductive tract. J Reprod Fertil 62:293–304
Kokcu A, Yavuz E, Celik H, Bildircin D (2012) A panoramic view to relationships between reproductive failure and immunological factors. Arch Gynecol Obstet 286:1283–1289
Kulkarni AB, Huh CG, Becker D, Geiser A, Lyght M, Flanders KC, Roberts AB, Sporn MB, Ward JM, Karlsson S (1993) Transforming growth factor beta 1 null mutation in mice causes excessive inflammatory response and early death. Proc Natl Acad Sci USA 90:770–774
Kunz G, Beil D, Deininger H, Wildt L, Leyendecker G (1996) The dynamics of rapid sperm transport through the female genital tract: evidence from vaginal sonography of uterine peristalsis and hysterosalpingoscintigraphy. Hum Reprod 11:627–632
Li J, Liu K, Liu Y, Xu Y, Zhang F, Yang H, Liu J, Pan T, Chen J, Wu M, Zhou X, Yuan Z (2013) Exosomes mediate the cell-to-cell transmission of IFN-alpha-induced antiviral activity. Nat Immunol 14:793–803
Louissaint NA, Fuchs EJ, Bakshi RP, Nimmagadda S, Du Y, Macura KJ, King KE, Wahl R, Goldsmith AJ, Caffo B, Cao YJ, Anderson J, Hendrix CW (2012) Distribution of cell-free and cell-associated HIV surrogates in the female genital tract after simulated vaginal intercourse. J Infect Dis 205:725–732
Macaluso M, Lawson L, Akers R, Valappil T, Hammond K, Blackwell R, Hortin G (1999) Prostate-specific antigen in vaginal fluid as a biologic marker of condom failure. Contraception 59:195–201
Madison MN, Roller RJ, Okeoma CM (2014) Human semen contains exosomes with potent anti-HIV-1 activity. Retrovirology 11:102
Mann T (1954) The biochemistry of semen. Methuen, London
Marie JC, Liggitt D, Rudensky AY (2006) Cellular mechanisms of fatal early-onset autoimmunity in mice with the T cell-specific targeting of transforming growth factor-beta receptor. Immunity 25:441–454
Mauck CK, Katz D, Sandefer EP, Nasution MD, Henderson M, Digenis GA, Su I, Page R, Barnhart K (2008) Vaginal distribution of Replens and K-Y Jelly using three imaging techniques. Contraception 77:195–204
Mayer KH, Venkatesh KK (2011) Interactions of HIV, other sexually transmitted diseases, and genital tract inflammation facilitating local pathogen transmission and acquisition. Am J Reprod Immunol 65:308–316
Muranishi S, Yamamoto A, Okada H (1993) Rectal and vaginal absorption of peptides and proteins. Pharm Biotech 4:199–227
Norton EJ, Diekman AB, Westbrook VA, Mullins DW, Klotz KL, Gilmer LL, Thomas TS, Wright DC, Brisker J, Engelhard VH, Flickinger CJ, Herr JC (2002) A male genital tract-specific carbohydrate epitope on human CD52: implications for immunocontraception. Tissue Antigens 60:354–364
Novy MJ, Eschenbach DA, Witken SS (2008) Infections as a cause of infertility. Global Library of Women's Medicine. doi:10.3843/GLOWM.10328, ISSN 1756-2228
Oh SA, Li MO (2013) TGF-beta: guardian of T cell function. J Immunol 191:3973–3979
Pandya IJ, Cohen J (1985) The leukocytic reaction of the human uterine cervix to spermatozoa. Fertil Steril 43:417–421

Pannek J, Berges RR, Sauvageot J, Lecksell KL, Epstein JI, Partin AW (1999) Cell turnover in human seminal vesicles and the prostate: an immunohistochemical study. Prostate Cancer Prostatic Dis 2:200–203
Politch JA, Tucker L, Bowman FP, Anderson DJ (2007) Concentrations and significance of cytokines and other immunologic factors in semen of healthy fertile men. Hum Reprod 22:2928–2935
Pudney J, Anderson D (2011) Innate and acquired immunity in the human penile urethra. J Reprod Immunol 88:219–227
Pudney J, Quayle AJ, Anderson DJ (2005) Immunological microenvironments in the human vagina and cervix: mediators of cellular immunity are concentrated in the cervical transformation zone. Biol Reprod 73:1253–1263
Quayle AJ, Kelly RW, Hargreave TB, James K (1989) Immunosuppression by seminal prostaglandins. Clin Exp Immunol 75:387–391
Raghupathy R (1997) Th1-type immunity is incompatible with successful pregnancy. Immunol Today 18:478–482
Rajagopalan S, Bryceson YT, Kuppusamy SP, Geraghty DE, Van Der Meer A, Joosten I, Long EO (2006) Activation of NK cells by an endocytosed receptor for soluble HLA-G. PLoS Biol 4:e9
Rametse CL, Olivier AJ, Masson L, Barnabas S, Mckinnon LR, Ngcapu S, Liebenberg LJ, Jaumdally SZ, Gray CM, Jaspan HB, Passmore JA (2014) Role of semen in altering the balance between inflammation and tolerance in the female genital tract: does it contribute to HIV risk? Viral Immunol 27:200–206
Robertson SA, Ingman WV, O'leary S, Sharkey DJ, Tremellen KP (2002) Transforming growth factor beta–a mediator of immune deviation in seminal plasma. J Reprod Immunol 57:109–128
Robertson SA, Prins JR, Sharkey DJ, Moldenhauer LM (2013) Seminal fluid and the generation of regulatory T cells for embryo implantation. Am J Reprod Immunol 69:315–330
Rodriguez-Martinez H, Kvist U, Ernerudh J, Sanz L, Calvete JJ (2011) Seminal plasma proteins: what role do they play? Am J Reprod Immunol 66(Suppl 1):11–22
Sharkey DJ, Macpherson AM, Tremellen KP, Robertson SA (2007) Seminal plasma differentially regulates inflammatory cytokine gene expression in human cervical and vaginal epithelial cells. Mol Hum Reprod 13:491–501
Sharkey DJ, Macpherson AM, Tremellen KP, Mottershead DG, Gilchrist RB, Robertson SA (2012a) TGF-beta mediates proinflammatory seminal fluid signaling in human cervical epithelial cells. J Immunol 189:1024–1035
Sharkey DJ, Tremellen KP, Jasper MJ, Gemzell-Danielsson K, Robertson SA (2012b) Seminal fluid induces leukocyte recruitment and cytokine and chemokine mRNA expression in the human cervix after coitus. J Immunol 188:2445–2454
Shull MM, Ormsby I, Kier AB, Pawlowski S, Diebold RJ, Yin M, Allen R, Sidman C, Proetzel G, Calvin D et al (1992) Targeted disruption of the mouse transforming growth factor-beta 1 gene results in multifocal inflammatory disease. Nature 359:693–699
Templeton AA, Cooper I, Kelly RW (1978) Prostaglandin concentrations in the semen of fertile men. J Reprod Fertil 52:147–150
Thompson LA, Barratt CL, Bolton AE, Cooke ID (1992) The leukocytic reaction of the human uterine cervix. Am J Reprod Immunol 28:85–89
Venter PF, Iturralde M (1979) Migration of a particulate radioactive tracer from the vagina to the peritoneal cavity and ovaries. S Afr Med J 55:917–919
Vojtech L, Woo S, Hughes S, Levy C, Ballweber L, Sauteraud RP, Strobl J, Westerberg K, Gottardo R, Tewari M, Hladik F (2014) Exosomes in human semen carry a distinctive repertoire of small non-coding RNAs with potential regulatory functions. Nucleic Acids Res 42:7290–7304
Wahl SM, Wen J, Moutsopoulos N (2006) TGF-beta: a mobile purveyor of immune privilege. Immunol Rev 213:213–227
Zamboni L (1972) Fertilization in the mouse. In: Moghissi KS, Hafez ESE (eds) Biology of mammalian fertilization and Implantation. C. C. Thomas, Springfield, IL
Zervomanolakis I, Ott HW, Hadziomerovic D, Mattle V, Seeber BE, Virgolini I, Heute D, Kissler S, Leyendecker G, Wildt L (2007) Physiology of upward transport in the human female genital tract. Ann N Y Acad Sci 1101:1–20

Chapter 8
Human Leukocyte Antigen-G Within the Male Reproductive System: Implications for Reproduction

Thomas Vauvert F. Hviid

Abstract In sexual reproduction in humans, a man has a clear interest in ensuring that the immune system of his female partner accepts the semi-allogenic fetus. Increasing attention has been given to soluble immunomodulatory molecules in the seminal fluid as one mechanism of ensuring this, possibly by "priming" the woman's immune system before conception and at conception. Recent studies have demonstrated the presence of the immunoregulatory and tolerance-inducible human leukocyte antigen (HLA)-G in the male reproductive organs. The expression of HLA-G in the blastocyst and by extravillous trophoblast cells in the placenta during pregnancy has been well described. Highly variable amounts of soluble HLA-G (sHLA-G) in seminal plasma from different men have been reported, and the concentration of sHLA-G is associated with HLA-G genotype. A first pilot study indicates that the level of sHLA-G in seminal plasma may even be associated with the chance of pregnancy in couples, where the male partner has reduced semen quality. More studies are needed to verify these preliminary findings.

Keywords MHC • HLA class Ib • HLA-G • Male reproductive system • Human reproduction

8.1 Introduction

In reproduction, the focus has traditionally been on the oocyte and the spermatozoon and on the embryo that develops from these to initial germ cells. However, the sperm cells are bathed in the seminal fluid that contains a large number of active molecules. Not all possible physiological functions related to the seminal fluid have

---

T.V.F. Hviid, M.D., Ph.D., D.M.Sc. (✉)
Department of Clinical Biochemistry, Centre for Immune Regulation and Reproductive Immunology (CIRRI), Roskilde Hospital, Copenhagen University Hospital (Roskilde), University of Copenhagen, 7-13 Køgevej, 4000 Roskilde, Denmark
e-mail: tvh@regionsjaelland.dk; hviid@dadlnet.dk

R. Bronson (ed.), *The Male Role in Pregnancy Loss and Embryo Implantation Failure*, Advances in Experimental Medicine and Biology 868,
DOI 10.1007/978-3-319-18881-2_8

been given special attention. Only recently, it has been proposed—and evidence for this proposal has been presented—that soluble immunomodulatory factors in the seminal fluid, or seminal plasma, may influence the immune system of the female partner (Robertson et al. 2009, 2013; Sharkey et al. 2012; Schjenken and Robertson 2014). This may occur even before fertilization during a period of sexual cohabitation and at least at conception. An immunomodulation of the female immune response through tolerance-inducible immune factors in the seminal fluid may be important for the woman's acceptance of the semi-allogenic fetus. Thereby, it may also influence pregnancy success and certain pregnancy complications, such as preeclampsia, where immune maladaptation seems to be involved in the pathophysiology (Redman and Sargent 2005).

Semen and seminal fluid have been shown to contain several immune molecules, and one of the most extensively studied is transforming growth factor-β (Robertson et al. 2003). Recently, we screened the male reproductive system, including semen and seminal plasma, for the nonclassical human leukocyte antigen (HLA) class Ib molecule, HLA-G (Larsen et al. 2011; Dahl et al. 2014). This is an immunomodulatory and tolerance-inducible molecule with a restricted tissue distribution; however, it is particularly expressed in the placenta during pregnancy (Kovats et al. 1990; Ishitani et al. 2003; Hviid 2006). We detected soluble HLA-G (sHLA-G) in seminal plasma in widely varying amounts when samples from different men were tested and compared. Below, preliminary results will be summarized, and the potential importance of the immunoregulatory and immunosuppressive actions of sHLA-G in semen for possible "priming" of the immune system of the female partner before conception and pregnancy will be discussed in addition to the possible significance for pregnancy success and the development of preeclampsia.

8.2 A Rationale for Immunomodulatory Factors in Human Semen

When a man successfully impregnates his female partner, he contributes one important factor: genetic material from the sperm cell. This is essential and has drawn all the focus. One function of the seminal fluid is to protect against sperm damage. Another important issue for the male might be to induce mechanisms that will ensure that the fetal expression of potential allogeneic proteins derived from the paternal genome is tolerated by the immune system of his female partner. A mechanism for immunological "priming" of the female before and at conception could be through the sexual cohabitation with the female partner based on installation of immunomodulatory factors in the female reproductive tract at sexual intercourse. In this way, the female partner might be prepared for acceptance of the blastocyst at the time of implantation and for the development of the semi-allogenic placental tissue and fetus in the uterus.

Based on animal models and human studies, emerging evidence has shown that seminal fluid, or seminal plasma, actually contains immunomodulatory molecules and that these molecules induce changes in the female partner's immune system, at least locally in the female reproductive organs (Beer et al. 1975; Tremellen et al. 2000; Robertson et al. 2003, 2013; Robertson 2005; Sharkey et al. 2012).

8.3 The Immunoregulatory HLA-G Molecule

The HLA genes are part of the human major histocompatibility complex (MHC) located on the short arm of chromosome 6. It includes a substantial number of immune genes. The best described are the classical HLA class Ia and II genes (HLA-A, HLA-B, HLA-C, HLA-DR, HLA-DQ, and HLA-DP) (The MHC Sequencing Consortium 1999). Their physiological role is antigen-peptide presentation, and they are well known for their importance in organ transplantation and association with autoimmune diseases (Doherty and Zinkernagel 1975; Svejgaard et al. 1983). However, human MHC also includes another group of so-called nonclassical HLA class Ib genes, the HLA-E, HLA-F, and HLA-G gene loci (Redman et al. 1984; Geraghty et al. 1987; Ellis 1990; Schmidt and Orr 1991).

8.3.1 HLA-G Is Expressed in Immune-Privileged Sites

HLA-G has a restricted tissue distribution in non-pathological conditions. HLA expression has been detected in immune-privileged sites such as the uterus, placenta, eye, and testis (Kovats et al. 1990; Le Discorde et al. 2003; Larsen et al. 2011). Furthermore, HLA-G expression has been reported in the thymus, the matured cumulus–oocyte complex, and by certain immune cells as T cells, monocytes, and tolerogenic dendritic cells (Crisa et al. 1997; Rebmann et al. 2003; Feger et al. 2007). Soluble HLA-G protein has also been detected in the blood of pregnant and nonpregnant individuals, in follicular fluid, and in seminal fluid (Hviid et al. 2004a; Chen et al. 2008; Rizzo et al. 2009a; Larsen et al. 2011).

8.3.2 Functions of HLA-G

The HLA-G gene and protein are the most studied of the class Ib molecules. Table 8.1 lists some of the main functions of HLA-G. Some of these are inhibition of NK- and T-cell-mediated cell lysis through interaction with the immunoglobulin-like transcript (ILT) 2 receptor, the ILT4 receptor, and the killer Ig-like receptor 2 DL4 (KIR2DL4) (Ponte et al. 1999; Rajagopalan and Long 1999; Riteau et al. 2001; Menier et al. 2002). Furthermore, HLA-G may induce a shift from a proinflammatory T helper 1

Table 8.1 Important immunomodulatory and immunosuppressive functions of HLA-G reported in the literature based on different experimental settings

HLA-G function	HLA-G isoforms involved	Immune cells involved	References
Inhibition of NK- and T-cell-mediated cell lysis	Primarily HLA-G1 and HLA-G5 Also shown for the other alternatively spliced HLA-G isoforms (HLA-G2 to HLA-G4 and HLA-G6); however, an in vivo functionality is controversial	Decidual and peripheral NK cells $CD8^+$ cytotoxic T cells	Rouas-Freiss et al. (1997), Navarro et al. (1999), Ponte et al. (1999), Rajagopalan and Long (1999), Riteau et al. (2001), Le Discorde et al. (2005)
Inhibition of an allocytotoxic T lymphocyte response	HLA-G1 and soluble HLA-G (HLA-G5 and/or sHLA-G1)	Peripheral blood mononuclear cells (PBMCs)	Maejima et al. (1997), Kapasi et al. (2000)
Upregulation of inhibitory receptors	HLA-G1 and HLA-G5	NK cells and $CD4^+$ T cells	LeMaoult et al. (2005)
Shift from a proinflammatory Th1 response to a Th2 response	HLA-G1 and soluble HLA-G (HLA-G5 and/or sHLA-G1)	$CD4^+$ T cells	Maejima et al. (1997), Kapasi et al. (2000), Kanai et al. (2001), Rieger et al. (2002), van der Meer et al. (2004), McIntire et al. (2004)
Induction of $CD4^+$ T cell anergy/long-term unresponsiveness	HLA-G1	$CD4^+$ T cells	LeMaoult et al. (2004)
Possible induction of FoxP3-regulatory T cells Dendritic cell induction and immunosuppression	HLA-G5 (and HLA-G1?)	$CD4^+CD25^+FoxP3^+$ Tregs Dendritic cells $CD4^+$ and $CD8^+$ T cells	Ristich et al. (2005), Selmani et al. (2008), Castellaneta et al. (2011)

(Th1) cell-mediated response toward a Th2 response (Kapasi et al. 2000; McIntire et al. 2004). Finally, HLA-G can inhibit an allocytotoxic T lymphocyte (CTL) response, inhibit the proliferation of $CD4^+$ T cells, and induce $CD4^+$ T cell anergy, and this may contribute to long-term immune escape or tolerance (LeMaoult et al. 2004).

8.3.3 HLA-G Isoforms and Gene Polymorphism

HLA-G potentially exists in four membrane-bound isoforms, HLA-G1, HLA-G2, HLA-G3, and HLA-G4, and three soluble isoforms, HLA-G5, HLA-G6, and HLA-G7, all generated by alternative splicing of HLA-G mRNA (Ishitani and Geraghty 1992; Fujii et al. 1994; Hviid et al. 1998). The secreted soluble HLA-G isoforms are

Table 8.2 A comparison between the total number of DNA alleles and protein alleles of some of the classical HLA class Ia genes and the nonclassical HLA class Ib gene, HLA-G, according to the official WHO Nomenclature Committee for Factors of the HLA System and the International Immunogenetics Information System [IMGT/HLA Database (October 2014)]. HLA class Ia gene loci are among the most polymorphic genes in the human genome, whereas HLA-G at the protein level is nearly monomorphic

Gene locus	HLA-A	HLA-B	HLA-G
	Class Ia		Class Ib
Alleles (DNA)	2,946	3,693	50
Proteins	2,077	2,741	18
Null alleles	138	122	2

generated by the retention of intron 4 that includes a stop codon (Fujii et al. 1994; Hviid et al. 1998). The most important isoforms are the full-length membrane-bound isoform HLA-G1 and the full-length secreted isoform HLA-G5. Soluble HLA-G1 (sHLA-G1) is generated by the shedding of membrane-bound HLA-G1 molecules (Park et al. 2004).

In contrast to the extremely polymorphic HLA class Ia genes, the HLA-G gene is almost monomorphic. HLA-G has a low polymorphism in the coding regions (Hviid 2006; Dahl and Hviid 2012). According to the WHO Nomenclature Committee for Factors of the HLA System and the International Immunogenetics Information System (IMGT)/HLA Database, 18 HLA-G alleles have been described at the protein level; two of these are so-called null alleles coding for—at least some—nonfunctional protein isoforms. Table 8.2 shows a comparison between the numbers of reported alleles for some of the HLA class Ia genes and HLA-G. In short, only very few amino acid substitutions have been observed in the HLA-G protein. However, quite many single nucleotide polymorphisms (SNPs) have been described in the 5′-upstream regulatory region (5′URR) and in the 3′-untranslated region (3′UTR) of the HLA-G gene (Hviid et al. 1999; Ober et al. 2003; Castelli et al. 2011). In particular, a 14 bp insertion/deletion polymorphism in the 3′UTR has been widely studied in relation to HLA-G mRNA alternative splicing and HLA-G expression levels. This polymorphism has been associated with the risk of developing preeclampsia, recurrent spontaneous abortion, and the success of assisted reproduction, although not all studies show such associations (Harrison et al. 1993; Hviid et al. 2002, 2004b; Hylenius et al. 2004; Larsen et al. 2010; Iversen et al. 2008).

8.4 HLA Class Ib Molecules During Pregnancy

Follicular fluid has been shown to contain sHLA-G, as well as the matured cumulus–oocyte complex (Rizzo et al. 2009b). A range of studies have reported the expression of HLA-G mRNA and protein in blastocysts (Jurisicova et al. 1996; Yao et al. 2005; Verloes et al. 2011). Furthermore, consensus from a rather large number of studies is that the media from 2- to 3-day-old embryo cultures from in vitro fertilization treatments is positive for sHLA-G in approximately 30–40 % of the cases, although not supported by all studies (Fuzzi et al. 2002; Van Lierop et al. 2002; Sher et al. 2004; Noci et al. 2005; Yao et al. 2005; Sageshima et al. 2007; Vercammen et al. 2008; Kotze and Kruger 2013). Interestingly, a considerable number of independent studies have observed that the pregnancy rate in women who have embryos transferred from cultures, where sHLA-G is detected, is significantly higher than that in women, who have embryos transferred from sHLA-G-negative cultures (Fuzzi et al. 2002; Sher et al. 2004; Noci et al. 2005; Yie et al. 2005; Vercammen et al. 2008; Kotze and Kruger 2013). However, the source of the sHLA-G, or the detection assays, is controversial because it does not seem plausible that the two- to eight-cell blastocyst should be capable of producing the amounts of sHLA-G measured (Sargent et al. 2007). At least some of the sHLA-G might originate from follicular fluid adhering to the oocyte. One study has observed a significant association between an increased cleavage rate and detection of sHLA-G, and another study reported that downregulation of HLA-G attenuates the cleavage rate in human triploid embryos, which to some extent may explain the higher chance of pregnancy associated with positive detection of HLA-G (Yie et al. 2005; Sun et al. 2011). Additionally, the study by Yie et al. has also reported that although pregnancy and live births were observed in sHLA-G-negative IVF cycles, the rate of spontaneous abortions was higher in the HLA-G-negative group (25 %) versus the HLA-G-positive group (11 %) (Yie et al. 2005). Together, all these different studies support a role for HLA-G very early in pregnancy and even at the time of implantation.

Another interesting observation is a study of sHLA-G in maternal blood during early pregnancy after IVF treatment (Pfeiffer et al. 2000). The concentration of sHLA-G in serum samples from 20 women experiencing early spontaneous abortion was significantly reduced during the first 9 weeks of gestation, compared with those of 37 women with intact pregnancies. Even the mean preovulatory sHLA-G serum levels in the 20 women were significantly lower than the mean level in the women with successful pregnancies. These results are supported by another study by Sipak-Szmigiel et al. (2007); however, larger studies are needed to reproduce these initial findings. Soluble HLA-G concentrations in the maternal peripheral blood are two to five times higher than the levels in nonpregnant women and in men; it peaks at the end of the first trimester and the beginning of the second (Alegre et al. 2007; Rizzo et al. 2009a; Darmochwal-Kolarz et al. 2012). Almost all studies have reported significantly reduced sHLA-G levels in maternal blood in cases of preeclampsia in all three trimesters (Hackmon et al. 2007; Steinborn et al. 2007; Rizzo et al. 2009a; Bortolotti et al. 2012).

8.4.1 Expression of HLA Molecules in the Placenta

During pregnancy, HLA-G is expressed by the trophoblast cells in the placenta, especially the extravillous trophoblast cells that invade the uterine wall and the spiral arteries (Kovats et al. 1990; Le Bouteiller and Blaschitz 1999; Morales et al.

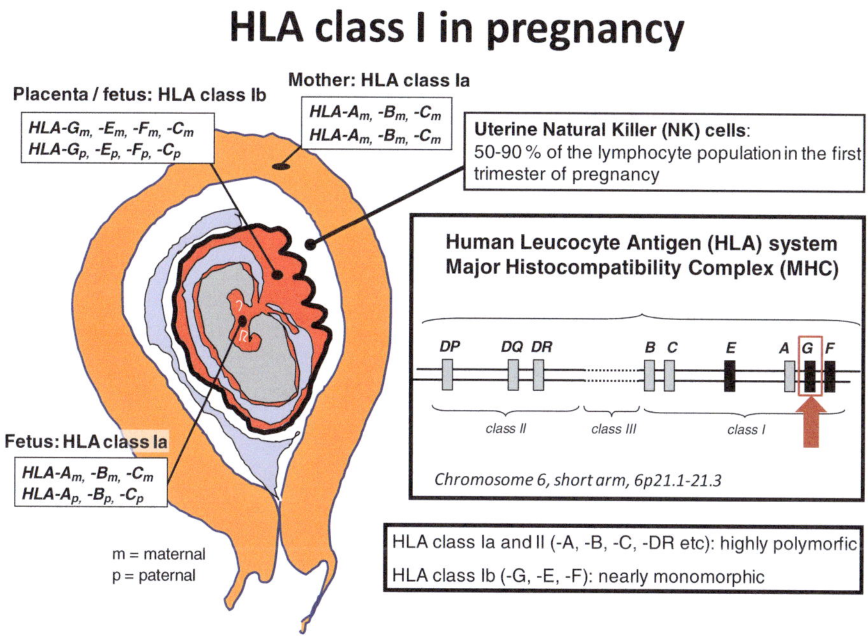

Fig. 8.1 Expression of human leukocyte antigen (*HLA*) molecules at the feto-maternal contact zone during pregnancy. The fetus inherits one HLA haplotype from the mother and one from the father; thereby, the fetus is semi-allogenic for the mother. The very polymorphic classical HLA class Ia and II molecules are not expressed by the trophoblast cells in the placenta, except for HLA-C. HLA class Ib proteins (HLA-E, HLA-F, and HLA-G) are expressed on the extravillous and invasive trophoblast cells, and they interact with specific receptors on uterine immune cells, especially natural killer (*NK*) cells. In this way, the trophoblast cells escape NK-cell-mediated lysis, and regulatory T cells may also be induced with the involvement of dendritic cells [Figure modified from (Hviid 2006)]

2003; Proll et al. 1999; Ishitani et al. 2003) (Fig. 8.1). It is in this feto-maternal contact zone that the HLA-G-expressing trophoblast cells, both as membrane-bound HLA-G and sHLA-G, are in intimate contact with the maternal immune cells. The leukocyte population in the decidua contains approximately 10 % T cells, 20 % macrophages, and 70 % NK cells (Loke et al. 1995). Therefore, the $CD16^{-/low}CD56^{high}$ NK cells represent the largest population of lymphocytes in the placenta, constituting 50–90 % of all resident leukocytes according to different studies (Bulmer et al. 1991; Koopman et al. 2003).

The extravillous trophoblast cells also express the other two HLA class Ib molecules, HLA-E and HLA-F, and may be the only cells in the body that do so (Ishitani et al. 2003) (Fig. 8.1). The trophoblast cells also express HLA-C at an apparently low level but not HLA-A or HLA-B or HLA class II molecules (King et al. 2000). Again, almost all studies have shown a significantly reduced expression of HLA-G mRNA and protein in the placenta in cases of preeclampsia compared to control pregnancies (Hara et al. 1996; Goldman-Wohl et al. 2000; Yie et al. 2004; Zhu et al. 2012).

8.5 HLA-G Expression in the Male Reproductive System

Most of the studies on HLA-G expression and function in reproduction have focused entirely on the female reproductive system, on pregnancy, and on certain pregnancy complications. However, two early studies of HLA class Ib gene expression in male gametogenic cells have been conducted (Guillaudeux et al. 1996; Fiszer et al. 1997). Guillaudeux et al. detected low levels of HLA class Ib mRNA in both spermatocytes and spermatids; three different alternatively spliced HLA-G mRNA isoforms were detected corresponding to HLA-G1, HLA-G2, and HLA-G3 (Guillaudeux et al. 1996). On the other hand, the same study did not observe any production of detectable HLA class I proteins in spermatogenic cells. These findings are partly in contrast to a study by Fiszer et al. that investigated HLA class Ib mRNA expression in male gametogenic cells from testicular tissue (Fiszer et al. 1997). Considerable levels of HLA-E mRNA were observed, very low levels of HLA-F, and no expression of HLA-G mRNA, even with RT-PCR techniques. HLA-E protein was observed on cells of the adluminal compartment within the seminiferous tubules. A new study by Yao et al. investigated HLA-G mRNA expression in testicular tissue with Johnson scores of 2–9 (Yao et al. 2014). The Johnson scoring system is a method of evaluating the quality of spermatogenesis in testicular biopsies. The HLA-G mRNA levels were significantly higher in testicular tissues with spermatocytes than those with only Sertoli cells and/or spermatogonia. Interestingly, the expression of HLA-G mRNA increased with higher Johnson score of the testicular tissue indicating an important role for HLA-G in spermatogenesis. In this study, HLA-G mRNA expression was also detected in ejaculated sperm. Investigation of HLA-G protein expression was not performed. Interestingly, by using siRNA techniques, it was found in the same study that silencing of the HLA-G gene impaired embryonic development indicating an important role for HLA-G in early pregnancy.

Langat et al. were the first to report the expression of HLA-G mRNA and protein in the normal human prostate (Langat et al. 2006). It was possible to detect mRNA for HLA-G1, HLA-G2, HLA-G5, and HLA-G6. However, only HLA-G5 protein was detectable. The HLA-G5 protein was prominent in the cytoplasm of tubuloglandular epithelia and in glandular secretions. In cases of prostatic adenocarcinomas, the HLA-G5 protein was detectable mainly in secretions.

Given this background—many studies of HLA-G in the female reproductive cycle and during pregnancy, and only a few published studies of specific issues regarding HLA-G in the male reproductive system—we decided to perform a systematic study of HLA-G protein expression in the male reproductive organs (Larsen et al. 2011). Immunohistochemical staining with the use of four different anti-HLA-G monoclonal antibodies (mAbs), two specific for all HLA-G isoforms and two specific for the soluble isoforms HLA-G5 and HLA-G6, was performed on paraffin-embedded tissue samples. Normal testis, testis with atrophy, prostate with hyperplasia, normal epididymis, normal ductus deferens, and normal seminal vesicle were studied. We detected HLA-G protein expression in normal testis in some of the Sertoli cells and in epididymal tissue. The ductuli efferentes stained very strongly for HLA-G. There was a weak expression of HLA-G in hyperplastic prostatic tissue.

Only mAbs against the soluble HLA-G isoforms stained positive, suggesting that soluble HLA-G5 is the predominantly expressed HLA-G protein isoform in the male reproductive organs. This is consistent with the findings of Langat et al. in the prostate (Langat et al. 2006). The seminal vesicle was negative for HLA-G protein expression (Larsen et al. 2011). Cells in seminal samples that were immobilized in a plasma–thrombin gel and paraffin embedded all stained negative for HLA-G indicating that leukocytes in the semen do not seem to contribute to sHLA-G in the seminal plasma.

However, based on Western blotting techniques and a sHLA-G ELISA, we detected sHLA-G in seminal plasma samples and in sperm samples. At least some of this sHLA-G was the HLA-G5 isoform (Larsen et al. 2011; Dahl et al. 2014). In a pilot study, we observed highly varied amounts of sHLA-G in seminal plasma samples from different men. This was also the case when the sHLA-G concentration was standardized to total protein concentration in the seminal plasma sample (Larsen et al. 2011). A very large variation in sHLA-G levels in seminal plasma samples was confirmed in a follow-up study of the male partners of 54 unselected couples attending a fertility clinic (Dahl et al. 2014).

It is possible that HLA-G in the testis might have a functional role serving as an immunosuppressive factor, thereby avoiding recognition of "self" sperm cells considered as autoantigens for the immune system. In this way, HLA-G might be a local factor among several that maintains the testes as an immune-privileged site. In support of this, the Sertoli cells seem to be immunoprotective, and they seem to locate HLA-G as described above (Mital et al. 2010; Larsen et al. 2011).

Interestingly, it has been reported that the rhesus monkey carries a nonclassical MHC class I gene named Mamu-AG (Ryan et al. 2002). The expression of Mamu-AG is very similar to HLA-G, and it is a putative homolog of HLA-G. Mamu-AG shares a number of features of HLA-G: generation of alternatively spliced mRNA isoforms, relatively low level of polymorphism, and a high level of expression at the feto-maternal interface (Ryan et al. 2002). However, Mamu-AG is also expressed as a soluble isoform, Mamu-AG5, in the rhesus monkey testis; it is generated by a premature stop codon in intron 4, just as in the case of HLA-G. The Sertoli cells were positive for Mamu-AG in immunostaining experiments. Semen or seminal plasma was not investigated, but late-stage primary and secondary spermatocytes and spermatids were positive for Mamu-AG5, while mature sperm was negative (Ryan et al. 2002). These similar observations across different species support a possible important role of MHC class Ib molecules in the male reproductive system and that they may serve a function in semen even before conception, at conception, and in very early stages of pregnancy.

8.6 HLA-G Genetics Influence HLA-G Protein Concentrations in Seminal Plasma

Several studies have shown significant associations between specific HLA-G genotypes, alleles and haplotypes, and different levels of soluble HLA-G in the blood from nonpregnant donors (Hviid et al. 2004a; Chen et al. 2008; Di Cristofaro et al. 2013; Martelli-Palomino et al. 2013). Therefore, we investigated whether soluble HLA-G levels in seminal plasma samples were associated with the HLA-G genotype of the men. We studied the concentration of sHLA-G in seminal plasma samples and the HLA-G 14 bp ins/del genotype in 40 men, half of them with reduced semen quality (Dahl et al. 2014). The concentration of sHLA-G in the seminal plasma samples was significantly associated with the HLA-G 14 bp ins/del genotype of the men. The del 14 bp/del 14 bp genotype showed the highest level of sHLA-G, and the ins 14 bp/ins 14 bp genotype showed the lowest level. These findings are exactly the same as reported for sHLA-G in blood plasma, or serum, in relation to the HLA-G 14 bp genotype (Hviid et al. 2004a; Chen et al. 2008). Measurements of total protein concentration in the seminal plasma samples were also performed to compensate for semen sample concentration. The same significant differences were observed, when the sHLA-G concentration in the seminal samples was corrected by the total protein concentration expressed as the ratio of sHLA-G to total protein. Furthermore, the same pattern was observed for the total amount of sHLA-G protein in the seminal sample obtained by multiplying the volume of semen with the sHLA-G concentration (Dahl et al. 2014). In conclusion, HLA-G genetics of the man clearly influences the amount of sHLA-G in his semen.

8.7 Immunomodulatory Factors in Seminal Fluid Influence the Female Immune Response

Several studies have indicated that repeated exposure to semen in animal models and in humans, respectively, improves reproductive success (Robertson et al. 2003; Robertson 2005). In mice, it seems that "uterine priming" with semen can promote implantation and fetal growth in subsequent pregnancies, even in a partner-specific manner (Beer et al. 1975). It seems that seminal fluid elicits an inflammation-like response in the female genital tract. Thereby, immune adaptations that can advance conception and pregnancy may be activated (Sharkey et al. 2012). In humans, live birthrates in couples undergoing IVF are significantly improved, when women are exposed to semen at the time of embryo transfer (Tremellen et al. 2000). In fertile women, immune cells and immune factors have been studied in cervix biopsies 12 h after unprotected vaginal coitus, vaginal coitus with the use of condom, or no coitus. After unprotected coitus, seminal fluid induced the recruitment of leukocytes and changes in cytokine and chemokine expression in the cervix and vagina (Sharkey et al. 2012).

8.7.1 HLA-G in Seminal Plasma Might Have Implications for Pregnancy Success

Several factors in seminal plasma may be involved in the modulation of the inflammatory response in the cervix and in the uterus. Two candidates for induction of tolerance to seminal antigens are transforming growth factor-β (TGF-β) and prostaglandin E_2, which can be detected at high concentrations within mammalian semen (Robertson et al. 2003; Robertson 2005). However, our studies also indicate sHLA-G as a possible tolerance-inducible and "priming" factor in human seminal fluid (Fig. 8.2). Regulatory T cells (Tregs) and tolerogenic dendritic cells of the woman are most likely to be important in this immunomodulation (Robertson et al. 2013). Interestingly, soluble HLA-G5 may be involved in the induction of $CD4^{+}CD25^{high}FoxP3^{+}$ Tregs (Selmani et al. 2008), and HLA-G seems to be implicated as a key regulator of tolerogenic dendritic cells (Ristich et al. 2005; Gregori et al. 2010; Amodio et al. 2013).

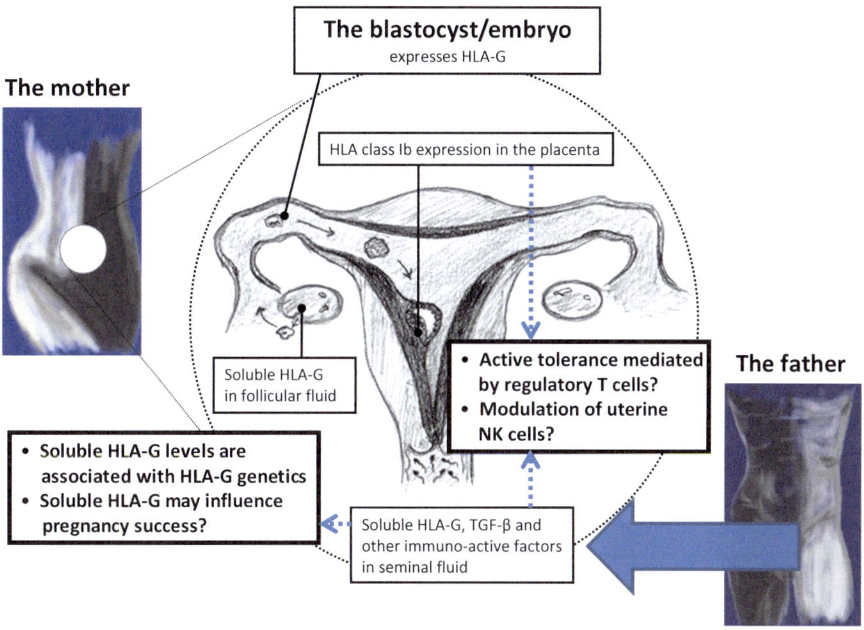

Fig. 8.2 Human leukocyte antigen-G is expressed in almost all of the phases of the reproductive cycle. Therefore, a central and important role for HLA-G in reproduction may be postulated. As shown, HLA-G is present in maternal blood, in follicular fluid, and in seminal plasma prior to implantation. After fertilization, membrane-bound HLA-G and secreted soluble HLA-G are expressed by the extravillous trophoblast cells in the placenta. The expression of HLA-G in the reproductive system during the reproductive cycle may modulate the local immune cells in the female reproductive organs toward immune tolerance of the semi-allogenic embryo. HLA-G gene polymorphisms influence HLA-G protein expression. Aberrant expression or reduced levels of HLA-G may influence pregnancy success and may modulate the risk of certain pregnancy complications, which seem to include immune maladaptation, such as preeclampsia [Figure modified from (Nilsson et al. 2014)]

In a pilot study of 54 unselected couples attending a fertility clinic, a trend for higher seminal plasma levels of sHLA-G per total protein and total sHLA-G in cases with reduced semen quality was observed when the female partner became pregnant after ART, compared with those couples, where no pregnancy was achieved (Dahl et al. 2014). Therefore, the amount of sHLA-G that the woman is exposed to before and at conception, especially in the genital tract, may influence the chance of obtaining a pregnancy. Most of the female partners to the males in the subgroup with reduced semen quality had normal fertility according to the results of the standard medical examination for female factors influencing fertility. It can be speculated that following successful ART procedures, these women might have been able to provide an optimal immunological response to high levels of sHLA-G in the semen of the partner (Dahl et al. 2014).

In conclusion, repeated female exposure to semen and paternal factors therein may be important for the success of pregnancy. One of these factors might be sHLA-G generating a state of local and maybe specific immunomodulation in the woman.

8.8 A Possible Importance of Seminal sHLA-G in Relation to the Risk of Developing Preeclampsia

Preeclampsia can be a very serious pregnancy complication, and it occurs in 2–8 % of all pregnancies. In the second half of pregnancy, the woman develops hypertension and proteinuria, which can be complicated by activation of the coagulation system and disseminated intravascular coagulation. Preeclampsia is a leading cause of maternal and fetal morbidity and mortality. The fetus and especially the placenta are central to the development of the syndrome, and in most cases, the symptoms disappear rapidly after delivery (Redman and Sargent 2005; Ahmed and Ramma 2014). Preeclampsia has been named the "disease of theories." However, a popular hypothesis for the etiology and pathogenesis of preeclampsia involves immune maladaptation in the early phases of pregnancy and placentation (Dekker and Sibai 1998; Saito et al. 2007). Experimental evidence exists for abnormal immunomodulation in the pregnant woman with preeclampsia when compared with uncomplicated pregnancies. This involves in cases of preeclampsia compared to controls: reduced fractions of regulatory FoxP3$^+$ T cells and CD4$^+$HLA-G$^+$ T cells in peripheral blood (Toldi et al. 2008; Santner-Nanan et al. 2009; Hsu et al. 2014), an apparently skewing of the immune response from T helper 2 (Th2) response toward a proinflammatory Th1 response (Darmochwal-Kolarz et al. 1999), reduced levels of sHLA-G in maternal blood (Hackmon et al. 2007; Steinborn et al. 2007; Rizzo et al. 2009a; Darmochwal-Kolarz et al. 2012), and aberrant expression of HLA-G in placentas (Goldman-Wohl et al. 2000; Yie et al. 2004; Zhu et al. 2012). In addition, several other immune parameters have been reported to be abnormal in cases of preeclampsia.

8.8.1 *Epidemiological Observations Support Immune Maladaptation as a Possibly Important Factor in Developing Preeclampsia*

It can be speculated that an abnormal exposure to immunomodulatory and tolerogenic factors in semen, either by reduced exposure or low concentration of these factors, might influence the fate of the pregnancy and especially the risk of developing preeclampsia. This has led to a theory of inadequate fetal, or paternal, tolerance induction in cases of preeclampsia, and this might already be of importance before conception involving a mechanism of immunological "priming" of the woman before or at conception. Several epidemiological observations support this and the hypothesis of immune maladaptation as an important factor in the development of preeclampsia: (1) preeclampsia is much more frequent in primipara/primigravida; (2) preeclampsia is more frequent in women with some of the autoimmune diseases, e.g., type 1 diabetes; and (3) there might be a higher risk of developing preeclampsia in a subsequent pregnancy for multipara, who changes partner. However, this may simply be attributed to a higher risk of preeclampsia as a consequence of longer duration to the next pregnancy according to one large study (Skjaerven et al. 2002). Furthermore, the use of donor sperm instead of partner (homologous) sperm in intrauterine insemination treatments seems to increase the risk of developing preeclampsia indicating a partner-specific dimension in a possible immunological "priming" of the woman (Gonzalez-Comadran et al. 2014). Finally, the sexual relationship with the father before preeclampsia seems to influence the risk of developing preeclampsia. A short sexual relationship with the father increases the risk of preeclampsia (Kho et al. 2009), and the use of barrier methods (condom and pessary) as contraception increases the risk of preeclampsia in a subsequent pregnancy (Einarsson et al. 2003). A recent study of 258 preeclampsia cases and 182 normotensive controls has confirmed that the risk of developing preeclampsia decreases significantly with increasing vaginal exposure to paternal semen (Saftlas et al. 2014). HLA typing for mother–offspring pairs, both cases and controls, was also performed for HLA-A, HLA-B, HLA-C, HLA-DRB1, and HLA-DQB1. The authors observed that HLA-A matching (or sharing), HLA class Ia matching, and combined HLA class Ia and II matching were associated with increased odds of preeclampsia (Triche et al. 2014). Very interestingly, the association with preeclampsia was influenced by prior vaginal exposure to paternal seminal fluid. For women with low semen exposure, the effects of HLA class Ia matching were amplified. With moderate to high semen exposure, HLA class II matching effects were predominant.

Therefore, there is accumulating evidence that seminal fluid exposure may induce immunological tolerance and "priming" in the woman to the semi-allogenic embryo and fetus in a subsequent pregnancy. Furthermore, reduced expression in seminal fluid may increase the risk of preeclampsia. As several studies have shown a link between reduced, or aberrant, HLA-G protein expression in the pregnant woman in cases of preeclampsia, it is possible to hypothesize that sHLA-G, as one of possibly several immunomodulatory factors in seminal fluid, may be involved in modifying the risk of preeclampsia through genetic association with the amount of sHLA-G in seminal fluid from a specific man and the degree of female exposure to sHLA-G influenced by the duration and the type of sexual relationship with the partner. However, a role for seminal sHLA-G in modifying the risk of preeclampsia is pure speculation at the moment, and experimental proof needs to be established.

8.9 Conclusions and Perspectives

New studies should clarify the associations between extended HLA-G gene haplotypes and the amount of sHLA-G in seminal fluid from individual men. From studies of soluble HLA-G concentrations in blood plasma, it has been shown that gene polymorphisms, especially in the 3′UTR but most likely also in the 5′URR, influence soluble HLA-G expression. Based on this, it might be possible to identify in a more specific way high and low producers of sHLA-G in seminal fluid, or seminal plasma, and the relevance to assisted reproduction treatments and to the risk of developing preeclampsia. Another interesting issue is whether a given man has fluctuating levels of sHLA-G in his seminal fluid over time or if the amount—adjusted for sperm volume variation—is fairly constant. In summary, the total amount of sHLA-G that a female partner is exposed to in semen in a given period of time, when the woman tries to conceive, is a combination of several factors: the frequency of sexual intercourse; the volume of semen per ejaculation, which may be related to the degree of sexual arousal by the male partner, although this is controversial; time of abstinence; and according to our studies, the HLA-G genotype of the male partner.

If our primary findings, which indicate a role of sHLA-G in seminal fluid for pregnancy success of the female partner, are verified in independent and larger studies, then it should be noted that the administration of purified, or recombinant, sHLA-G might actually be a possible co-treatment option in assisted reproduction.

Acknowledgments Support for this work was generously provided by grants from the Region Zealand Health Sciences Research Foundation.

References

Ahmed A, Ramma W (2014) Unraveling the theories of preeclampsia: are the protective pathways the new paradigm? Br J Pharmacol. doi:10.1111/bph.12977

Alegre E, Diaz-Lagares A, Lemaoult J, Lopez-Moratalla N, Carosella ED, Gonzalez A (2007) Maternal antigen presenting cells are a source of plasmatic HLA-G during pregnancy: longitudinal study during pregnancy. Hum Immunol 68(8):661–667. doi:10.1016/j.humimm.2007.04.007

Amodio G, Mugione A, Sanchez AM, Vigano P, Candiani M, Somigliana E, Roncarolo MG, Panina-Bordignon P, Gregori S (2013) HLA-G expressing DC-10 and CD4(+) T cells accumulate in human decidua during pregnancy. Hum Immunol 74(4):406–411. doi:10.1016/j.humimm.2012.11.031

Beer AE, Billingham RE, Scott JR (1975) Immunogenetic aspects of implantation, placentation and feto-placental growth rates. Biol Reprod 12(1):176–189

Bortolotti D, Gentili V, Melchiorri L, Rotola A, Rizzo R (2012) An accurate and reliable real time SNP genotyping assay for the HLA-G +3142 bp C>G polymorphism. Tissue Antigens 80(3):259–262. doi:10.1111/j.1399-0039.2012.01926.x

Bulmer JN, Morrison L, Longfellow M, Ritson A, Pace D (1991) Granulated lymphocytes in human endometrium: histochemical and immunohistochemical studies. Hum Reprod 6(6):791–798

Castellaneta A, Mazariegos GV, Nayyar N, Zeevi A, Thomson AW (2011) HLA-G level on monocytoid dendritic cells correlates with regulatory T-cell Foxp3 expression in liver transplant tolerance. Transplantation 91(10):1132–1140. doi:10.1097/TP.0b013e31821414c9

Castelli EC, Mendes-Junior CT, Veiga-Castelli LC, Roger M, Moreau P, Donadi EA (2011) A comprehensive study of polymorphic sites along the HLA-G gene: implication for gene regulation and evolution. Mol Biol Evol 28(11):3069–3086. doi:10.1093/molbev/msr138

Chen XY, Yan WH, Lin A, Xu HH, Zhang JG, Wang XX (2008) The 14 bp deletion polymorphisms in HLA-G gene play an important role in the expression of soluble HLA-G in plasma. Tissue Antigens 72(4):335–341. doi:10.1111/j.1399-0039.2008.01107.x

Crisa L, McMaster MT, Ishii JK, Fisher SJ, Salomon DR (1997) Identification of a thymic epithelial cell subset sharing expression of the class Ib HLA-G molecule with fetal trophoblasts. J Exp Med 186(2):289–298

Dahl M, Hviid TV (2012) Human leucocyte antigen class Ib molecules in pregnancy success and early pregnancy loss. Hum Reprod Update 18(1):92–109. doi:10.1093/humupd/dmr043

Dahl M, Perin TL, Djurisic S, Rasmussen M, Ohlsson J, Buus S, Lindhard A, Hviid TV (2014) Soluble human leukocyte antigen-G in seminal plasma is associated with HLA-G genotype: possible implications for fertility success. Am J Reprod Immunol 72(1):89–105. doi:10.1111/aji.12251

Darmochwal-Kolarz D, Leszczynska-Gorzelak B, Rolinski J, Oleszczuk J (1999) T helper 1- and T helper 2-type cytokine imbalance in pregnant women with pre-eclampsia. Eur J Obstet Gynecol Reprod Biol 86(2):165–170

Darmochwal-Kolarz D, Kolarz B, Rolinski J, Leszczynska-Gorzelak B, Oleszczuk J (2012) The concentrations of soluble HLA-G protein are elevated during mid-gestation and decreased in pre-eclampsia. Folia Histochem Cytobiol 50(2):286–291

Dekker GA, Sibai BM (1998) Etiology and pathogenesis of preeclampsia: current concepts. Am J Obstet Gynecol 179(5):1359–1375

Di Cristofaro J, El Moujally D, Agnel A, Mazieres S, Cortey M, Basire A, Chiaroni J, Picard C (2013) HLA-G haplotype structure shows good conservation between different populations and good correlation with high, normal and low soluble HLA-G expression. Hum Immunol 74(2):203–206. doi:10.1016/j.humimm.2012.10.027

Doherty PC, Zinkernagel RM (1975) A biological role for the major histocompatibility antigens. Lancet 1(7922):1406–1409

Einarsson JI, Sangi-Haghpeykar H, Gardner MO (2003) Sperm exposure and development of preeclampsia. Am J Obstet Gynecol 188(5):1241–1243

Ellis S (1990) HLA G: at the interface. Am J Reprod Immunol 23(3):84–86

Feger U, Tolosa E, Huang YH, Waschbisch A, Biedermann T, Melms A, Wiendl H (2007) HLA-G expression defines a novel regulatory T-cell subset present in human peripheral blood and sites of inflammation. Blood 110(2):568–577. doi:10.1182/blood-2006-11-057125

Fiszer D, Ulbrecht M, Fernandez N, Johnson JP, Weiss EH, Kurpisz M (1997) Analysis of HLA class Ib gene expression in male gametogenic cells. Eur J Immunol 27(7):1691–1695. doi:10.1002/eji.1830270715

Fujii T, Ishitani A, Geraghty DE (1994) A soluble form of the HLA-G antigen is encoded by a messenger ribonucleic acid containing intron 4. J Immunol 153(12):5516–5524

Fuzzi B, Rizzo R, Criscuoli L, Noci I, Melchiorri L, Scarselli B, Bencini E, Menicucci A, Baricordi OR (2002) HLA-G expression in early embryos is a fundamental prerequisite for the obtainment of pregnancy. Eur J Immunol 32(2):311–315. doi:10.1002/1521-4141(200202)32:2<311::AID-IMMU311>3.0.CO;2-8

Geraghty DE, Koller BH, Orr HT (1987) A human major histocompatibility complex class I gene that encodes a protein with a shortened cytoplasmic segment. Proc Natl Acad Sci USA 84(24):9145–9149
Goldman-Wohl DS, Ariel I, Greenfield C, Hanoch J, Yagel S (2000) HLA-G expression in extravillous trophoblasts is an intrinsic property of cell differentiation: a lesson learned from ectopic pregnancies. Mol Hum Reprod 6(6):535–540
Gonzalez-Comadran M, Avila JU, Tascon AS, Jimenez R, Sola I, Brassesco M, Carreras R, Checa MA (2014) The impact of donor insemination on the risk of preeclampsia: a systematic review and meta-analysis. Eur J Obstet Gynecol Reprod Biol 182C:160–166. doi:10.1016/j.ejogrb.2014.09.022
Gregori S, Tomasoni D, Pacciani V, Scirpoli M, Battaglia M, Magnani CF, Hauben E, Roncarolo MG (2010) Differentiation of type 1 T regulatory cells (Tr1) by tolerogenic DC-10 requires the IL-10-dependent ILT4/HLA-G pathway. Blood 116(6):935–944. doi:10.1182/blood-2009-07-234872
Guillaudeux T, Gomez E, Onno M, Drenou B, Segretain D, Alberti S, Lejeune H, Fauchet R, Jegou B, Le Bouteiller P (1996) Expression of HLA class I genes in meiotic and post-meiotic human spermatogenic cells. Biol Reprod 55(1):99–110
Hackmon R, Koifman A, Hyodo H, Glickman H, Sheiner E, Geraghty DE (2007) Reduced third-trimester levels of soluble human leukocyte antigen G protein in severe preeclampsia. Am J Obstet Gynecol 197(3):251.e1–251.e5. doi:10.1016/j.ajog.2007.06.033
Hara N, Fujii T, Yamashita T, Kozuma S, Okai T, Taketani Y (1996) Altered expression of human leukocyte antigen G (HLA-G) on extravillous trophoblasts in preeclampsia: immunohistological demonstration with anti-HLA-G specific antibody "87G" and anti-cytokeratin antibody "CAM5.2". Am J Reprod Immunol 36(6):349–358
Harrison GA, Humphrey KE, Jakobsen IB, Cooper DW (1993) A 14 bp deletion polymorphism in the HLA-G gene. Hum Mol Genet 2(12):2200
Hsu P, Santner-Nanan B, Joung S, Peek MJ, Nanan R (2014) Expansion of CD4(+) HLA-G(+) T Cell in human pregnancy is impaired in pre-eclampsia. Am J Reprod Immunol 71(3):217–228. doi:10.1111/aji.12195
Hviid TV (2006) HLA-G in human reproduction: aspects of genetics, function and pregnancy complications. Hum Reprod Update 12(3):209–232. doi:10.1093/humupd/dmi048
Hviid TV, Moller C, Sorensen S, Morling N (1998) Co-dominant expression of the HLA-G gene and various forms of alternatively spliced HLA-G mRNA in human first trimester trophoblast. Hum Immunol 59(2):87–98
Hviid TV, Sorensen S, Morling N (1999) Polymorphism in the regulatory region located more than 1.1 kilobases 5′ to the start site of transcription, the promoter region, and exon 1 of the HLA-G gene. Hum Immunol 60(12):1237–1244
Hviid TV, Hylenius S, Hoegh AM, Kruse C, Christiansen OB (2002) HLA-G polymorphisms in couples with recurrent spontaneous abortions. Tissue Antigens 60(2):122–132
Hviid TV, Rizzo R, Christiansen OB, Melchiorri L, Lindhard A, Baricordi OR (2004a) HLA-G and IL-10 in serum in relation to HLA-G genotype and polymorphisms. Immunogenetics 56(3):135–141. doi:10.1007/s00251-004-0673-2
Hviid TV, Hylenius S, Lindhard A, Christiansen OB (2004b) Association between human leukocyte antigen-G genotype and success of in vitro fertilization and pregnancy outcome. Tissue Antigens 64:66–69
Hylenius S, Andersen AM, Melbye M, Hviid TV (2004) Association between HLA-G genotype and risk of pre-eclampsia: a case-control study using family triads. Mol Hum Reprod 10(4):237–246. doi:10.1093/molehr/gah035
Ishitani A, Geraghty DE (1992) Alternative splicing of HLA-G transcripts yields proteins with primary structures resembling both class I and class II antigens. Proc Natl Acad Sci USA 89(9):3947–3951
Ishitani A, Sageshima N, Lee N, Dorofeeva N, Hatake K, Marquardt H, Geraghty DE (2003) Protein expression and peptide binding suggest unique and interacting functional roles for HLA-E, F, and G in maternal-placental immune recognition. J Immunol 171(3):1376–1384

Iversen AC, Nguyen OT, Tommerdal LF, Eide IP, Landsem VM, Acar N, Myhre R, Klungland H, Austgulen R (2008) The HLA-G 14 bp gene polymorphism and decidual HLA-G 14 bp gene expression in pre-eclamptic and normal pregnancies. J Reprod Immunol 78(2):158–165. doi:10.1016/j.jri.2008.03.001

Jurisicova A, Casper RF, MacLusky NJ, Mills GB, Librach CL (1996) HLA-G expression during preimplantation human embryo development. Proc Natl Acad Sci USA 93(1):161–165

Kanai T, Fujii T, Kozuma S, Yamashita T, Miki A, Kikuchi A, Taketani Y (2001) Soluble HLA-G influences the release of cytokines from allogeneic peripheral blood mononuclear cells in culture. Mol Hum Reprod 7(2):195–200

Kapasi K, Albert SE, Yie S, Zavazava N, Librach CL (2000) HLA-G has a concentration-dependent effect on the generation of an allo-CTL response. Immunology 101(2):191–200

Kho EM, Mccowan LM, North RA, Roberts CT, Chan E, Black MA, Taylor RS, Dekker GA, Consortium S (2009) Duration of sexual relationship and its effect on preeclampsia and small for gestational age perinatal outcome. J Reprod Immunol 82(1):66–73. doi:10.1016/j.jri.2009.04.011

King A, Burrows TD, Hiby SE, Bowen JM, Joseph S, Verma S, Lim PB, Gardner L, Le Bouteiller P, Ziegler A, Uchanska-Ziegler B, Loke YW (2000) Surface expression of HLA-C antigen by human extravillous trophoblast. Placenta 21(4):376–387. doi:10.1053/plac.1999.0496

Koopman LA, Kopcow HD, Rybalov B, Boyson JE, Orange JS, Schatz F, Masch R, Lockwood CJ, Schachter AD, Park PJ, Strominger JL (2003) Human decidual natural killer cells are a unique NK cell subset with immunomodulatory potential. J Exp Med 198(8):1201–1212. doi:10.1084/jem.20030305

Kotze D, Kruger TF (2013) HLA-G as a marker for embryo selection in assisted reproductive technology. Fertil Steril 100(6), e44. doi:10.1016/j.fertnstert.2013.10.014

Kovats S, Main EK, Librach C, Stubblebine M, Fisher SJ, DeMars R (1990) A class I antigen, HLA-G, expressed in human trophoblasts. Science 248(4952):220–223

Langat DK, Sue Platt J, Tawfik O, Fazleabas AT, Hunt JS (2006) Differential expression of human leukocyte antigen-G (HLA-G) messenger RNAs and proteins in normal human prostate and prostatic adenocarcinoma. J Reprod Immunol 71(1):75–86. doi:10.1016/j.jri.2006.01.006

Larsen MH, Hylenius S, Andersen AM, Hviid TV (2010) The 3′-untranslated region of the HLA-G gene in relation to pre-eclampsia: revisited. Tissue Antigens 75(3):253–261. doi:10.1111/j.1399-0039.2009.01435.x

Larsen MH, Bzorek M, Pass MB, Larsen LG, Nielsen MW, Svendsen SG, Lindhard A, Hviid TV (2011) Human leukocyte antigen-G in the male reproductive system and in seminal plasma. Mol Hum Reprod 17(12):727–738. doi:10.1093/molehr/gar052

Le Bouteiller P, Blaschitz A (1999) The functionality of HLA-G is emerging. Immunol Rev 167:233–244

Le Discorde M, Moreau P, Sabatier P, Legeais JM, Carosella ED (2003) Expression of HLA-G in human cornea, an immune-privileged tissue. Hum Immunol 64(11):1039–1044

Le Discorde M, Le Danff C, Moreau P, Rouas-Freiss N, Carosella ED (2005) HLA-G*0105 N null allele encodes functional HLA-G isoforms. Biol Reprod 73(2):280–288. doi:10.1095/biolreprod.104.037986

LeMaoult J, Krawice-Radanne I, Dausset J, Carosella ED (2004) HLA-G1-expressing antigen-presenting cells induce immunosuppressive CD4+ T cells. Proc Natl Acad Sci USA 101(18):7064–7069. doi:10.1073/pnas.0401922101

LeMaoult J, Zafaranloo K, Le Danff C, Carosella ED (2005) HLA-G up-regulates ILT2, ILT3, ILT4, and KIR2DL4 in antigen presenting cells, NK cells, and T cells. FASEB J 19(6):662–664. doi:10.1096/fj.04-1617fje

Loke YW, King A, Burrows TD (1995) Decidua in human implantation. Hum Reprod 10(Suppl 2):14–21

Maejima M, Fujii T, Kozuma S, Okai T, Shibata Y, Taketani Y (1997) Presence of HLA-G-expressing cells modulates the ability of peripheral blood mononuclear cells to release cytokines. Am J Reprod Immunol 38(2):79–82

Martelli-Palomino G, Pancotto JA, Muniz YC, Mendes-Junior CT, Castelli EC, Massaro JD, Krawice-Radanne I, Poras I, Rebmann V, Carosella ED, Rouas-Freiss N, Moreau P, Donadi EA

(2013) Polymorphic Sites at the 3′ Untranslated Region of the HLA-G Gene Are Associated with Differential hla-g Soluble Levels in the Brazilian and French Population. PLoS One 8(10), e71742. doi:10.1371/journal.pone.0071742
McIntire RH, Morales PJ, Petroff MG, Colonna M, Hunt JS (2004) Recombinant HLA-G5 and -G6 drive U937 myelomonocytic cell production of TGF-beta1. J Leukoc Biol 76(6):1220–1228. doi:10.1189/jlb.0604337
Menier C, Riteau B, Carosella ED, Rouas-Freiss N (2002) MICA triggering signal for NK cell tumor lysis is counteracted by HLA-G1-mediated inhibitory signal. Int J Cancer 100(1):63–70. doi:10.1002/ijc.10460
Mital P, Kaur G, Dufour JM (2010) Immunoprotective sertoli cells: making allogeneic and xenogeneic transplantation feasible. Reproduction 139(3):495–504. doi:10.1530/REP-09-0384
Morales PJ, Pace JL, Platt JS, Phillips TA, Morgan K, Fazleabas AT, Hunt JS (2003) Placental cell expression of HLA-G2 isoforms is limited to the invasive trophoblast phenotype. J Immunol 171(11):6215–6224
Navarro F, Llano M, Bellon T, Colonna M, Geraghty DE, Lopez-Botet M (1999) The ILT2(LIR1) and CD94/NKG2A NK cell receptors respectively recognize HLA-G1 and HLA-E molecules co-expressed on target cells. Eur J Immunol 29(1):277–283. doi:10.1002/(SICI)1521-4141(199901)29:01<277::AID-IMMU277>3.0.CO;2-4
Nilsson LL, Djurisic S, Hviid TV (2014) Controlling the immunological crosstalk during conception and pregnancy: HLA-G in reproduction. Front Immunol 5:198. doi:10.3389/fimmu.2014.00198
Noci I, Fuzzi B, Rizzo R, Melchiorri L, Criscuoli L, Dabizzi S, Biagiotti R, Pellegrini S, Menicucci A, Baricordi OR (2005) Embryonic soluble HLA-G as a marker of developmental potential in embryos. Hum Reprod 20(1):138–146. doi:10.1093/humrep/deh572
Ober C, Aldrich CL, Chervoneva I, Billstrand C, Rahimov F, Gray HL, Hyslop T (2003) Variation in the HLA-G promoter region influences miscarriage rates. Am J Hum Genet 72(6): 1425–1435. doi:10.1086/375501
Park GM, Lee S, Park B, Kim E, Shin J, Cho K, Ahn K (2004) Soluble HLA-G generated by proteolytic shedding inhibits NK-mediated cell lysis. Biochem Biophys Res Commun 313(3):606–611
Pfeiffer KA, Rebmann V, van der Ven K et al (2000) Soluble histocompatibility antigen levels in early pregnancy after IVF. Hum Immunol 61:559–564
Ponte M, Cantoni C, Biassoni R, Tradori-Cappai A, Bentivoglio G, Vitale C, Bertone S, Moretta A, Moretta L, Mingari MC (1999) Inhibitory receptors sensing HLA-G1 molecules in pregnancy: decidua-associated natural killer cells express LIR-1 and CD94/NKG2A and acquire p49, an HLA-G1-specific receptor. Proc Natl Acad Sci USA 96(10):5674–5679
Proll J, Blaschitz A, Hutter H, Dohr G (1999) First trimester human endovascular trophoblast cells express both HLA-C and HLA-G. Am J Reprod Immunol 42(1):30–36
Rajagopalan S, Long EO (1999) A human histocompatibility leukocyte antigen (HLA)-G-specific receptor expressed on all natural killer cells. J Exp Med 189(7):1093–1100
Rebmann V, Busemann A, Lindemann M, Grosse-Wilde H (2003) Detection of HLA-G5 secreting cells. Hum Immunol 64(11):1017–1024
Redman CW, Sargent IL (2005) Latest advances in understanding preeclampsia. Science 308(5728):1592–1594. doi:10.1126/science.1111726
Redman CW, McMichael AJ, Stirrat GM, Sunderland CA, Ting A (1984) Class 1 major histocompatibility complex antigens on human extra-villous trophoblast. Immunology 52(3):457–468
Rieger L, Hofmeister V, Probe C, Dietl J, Weiss EH, Steck T, Kammerer U (2002) Th1- and Th2-like cytokine production by first trimester decidual large granular lymphocytes is influenced by HLA-G and HLA-E. Mol Hum Reprod 8(3):255–261
Ristich V, Liang S, Zhang W, Wu J, Horuzsko A (2005) Tolerization of dendritic cells by HLA-G. Eur J Immunol 35(4):1133–1142. doi:10.1002/eji.200425741
Riteau B, Menier C, Khalil-Daher I, Martinozzi S, Pla M, Dausset J, Carosella ED, Rouas-Freiss N (2001) HLA-G1 co-expression boosts the HLA class I-mediated NK lysis inhibition. Int Immunol 13(2):193–201
Rizzo R, Andersen AS, Lassen MR, Sorensen HC, Bergholt T, Larsen MH, Melchiorri L, Stignani M, Baricordi OR, Hviid TV (2009a) Soluble human leukocyte antigen-G isoforms in mater-

nal plasma in early and late pregnancy. Am J Reprod Immunol 62(5):320–338. doi:10.1111/j.1600-0897.2009.00742.x

Rizzo R, Stignani M, Melchiorri L, Baricordi OR (2009b) Possible role of human leukocyte antigen-G molecules in human oocyte/embryo secretome. Hum Immunol 70:970–975

Robertson SA (2005) Seminal plasma and male factor signalling in the female reproductive tract. Cell Tissue Res 322:43–52

Robertson SA, Bromfield JJ, Tremellen KP (2003) Seminal 'priming' for protection from pre-eclampsia-a unifying hypothesis. J Reprod Immunol 59(2):253–265

Robertson SA, Guerin LR, Bromfield JJ, Branson KM, Ahlstrom AC, Care AS (2009) Seminal fluid drives expansion of the CD4+CD25+ T regulatory cell pool and induces tolerance to paternal alloantigens in mice. Biol Reprod 80(5):1036–1045. doi:10.1095/biolreprod.108.074658

Robertson SA, Prins JR, Sharkey DJ, Moldenhauer LM (2013) Seminal fluid and the generation of regulatory T cells for embryo implantation. Am J Reprod Immunol 69(4):315–330. doi:10.1111/aji.12107

Rouas-Freiss N, Goncalves RM, Menier C, Dausset J, Carosella ED (1997) Direct evidence to support the role of HLA-G in protecting the fetus from maternal uterine natural killer cytolysis. Proc Natl Acad Sci USA 94(21):11520–11525

Ryan AF, Grendell RL, Geraghty DE, Golos TG (2002) A soluble isoform of the rhesus monkey nonclassical MHC class I molecule Mamu-AG is expressed in the placenta and the testis. J Immunol 169(2):673–683

Saftlas AF, Rubenstein L, Prater K, Harland KK, Field E, Triche EW (2014) Cumulative exposure to paternal seminal fluid prior to conception and subsequent risk of preeclampsia. J Reprod Immunol 101–102:104–110. doi:10.1016/j.jri.2013.07.006

Sageshima N, Shobu T, Awai K, Hashimoto H, Yamashita M, Takeda N, Odawara Y, Nakanishi M, Hatake K, Ishitani A (2007) Soluble HLA-G is absent from human embryo cultures: a reassessment of sHLA-G detection methods. J Reprod Immunol 75(1):11–22. doi:10.1016/j.jri.2007.02.010

Saito S, Shiozaki A, Nakashima A, Sakai M, Sasaki Y (2007) The role of the immune system in preeclampsia. Mol Aspects Med 28(2):192–209. doi:10.1016/j.mam.2007.02.006

Santner-Nanan B, Peek MJ, Khanam R, Richarts L, Zhu E, de St F, Groth B, Nanan R (2009) Systemic increase in the ratio between Foxp3+ and IL-17-producing CD4+ T cells in healthy pregnancy but not in preeclampsia. J Immunol 183(11):7023–7030. doi:10.4049/jimmunol.0901154

Sargent I, Swales A, Ledee N, Kozma N, Tabiasco J, Le Bouteiller P (2007) sHLA-G production by human IVF embryos: can it be measured reliably? J Reprod Immunol 75(2):128–132. doi:10.1016/j.jri.2007.03.005

Schjenken JE, Robertson SA (2014) Seminal fluid and immune adaptation for pregnancy–comparative biology in mammalian species. Reprod Domest Anim 49(Suppl 3):27–36. doi:10.1111/rda.12383

Schmidt CM, Orr HT (1991) A physical linkage map of HLA-A, -G, -7.5p, and -F. Hum Immunol 31(3):180–185

Selmani Z, Naji A, Zidi I, Favier B, Gaiffe E, Obert L, Borg C, Saas P, Tiberghien P, Rouas-Freiss N, Carosella ED, Deschaseaux F (2008) Human leukocyte antigen-G5 secretion by human mesenchymal stem cells is required to suppress T lymphocyte and natural killer function and to induce CD4+CD25highFOXP3+ regulatory T cells. Stem Cells 26(1):212–222. doi:10.1634/stemcells.2007-0554

Sharkey DJ, Tremellen KP, Jasper MJ, Gemzell-Danielsson K, Robertson SA (2012) Seminal fluid induces leukocyte recruitment and cytokine and chemokine mRNA expression in the human cervix after coitus. J Immunol 188(5):2445–2454. doi:10.4049/jimmunol.1102736

Sher G, Keskintepe L, Nouriani M, Roussev R, Batzofin J (2004) Expression of sHLA-G in supernatants of individually cultured 46-h embryos: a potentially valuable indicator of 'embryo competency' and IVF outcome. Reprod Biomed Online 9(1):74–78

Sipak-Szmigiel O, Ronin-Walknowska E, Cybulski C, Plonka T, Lubinski J (2007) Antigens HLA-G, sHLA- G and sHLA- class I in reproductive failure. Folia Histochem Cytobiol 45(Suppl 1): S137–S141

Skjaerven R, Wilcox AJ, Lie RT (2002) The interval between pregnancies and the risk of preeclampsia. N Engl J Med 346(1):33–38. doi:10.1056/NEJMoa011379
Steinborn A, Varkonyi T, Scharf A, Bahlmann F, Klee A, Sohn C (2007) Early detection of decreased soluble HLA-G levels in the maternal circulation predicts the occurrence of preeclampsia and intrauterine growth retardation during further course of pregnancy. Am J Reprod Immunol 57(4):277–286. doi:10.1111/j.1600-0897.2007.00475.x
Sun LL, Wang AM, Haines CJ, Han Y, Yao YQ (2011) Down-regulation of HLA-G attenuates cleavage rate in human triploid embryos. J Reprod Infertil 12(3):215–220
Svejgaard A, Platz P, Ryder LP (1983) HLA and disease 1982–a survey. Immunol Rev 70: 193–218
The MHC Sequencing Consortium (1999) Complete sequence and gene map of a human major histocompatibility complex. The MHC sequencing consortium. Nature 401(6756):921–923. doi:10.1038/44853
Toldi G, Svec P, Vasarhelyi B, Meszaros G, Rigo J, Tulassay T, Treszl A (2008) Decreased number of FoxP3+ regulatory T cells in preeclampsia. Acta Obstet Gynecol Scand 87(11):1229–1233. doi:10.1080/00016340802389470
Tremellen KP, Valbuena D, Landeras J, Ballesteros A, Martinez J, Mendoza S, Norman RJ, Robertson SA, Simon C (2000) The effect of intercourse on pregnancy rates during assisted human reproduction. Hum Reprod 15(12):2653–2658
Triche EW, Harland KK, Field EH, Rubenstein LM, Saftlas AF (2014) Maternal-fetal HLA sharing and preeclampsia: variation in effects by seminal fluid exposure in a case-control study of nulliparous women in Iowa. J Reprod Immunol 101–102:111–119. doi:10.1016/j.jri.2013.06.004
van der Meer A, Lukassen HG, van Lierop MJ, Wijnands F, Mosselman S, Braat DD, Joosten I (2004) Membrane-bound HLA-G activates proliferation and interferon-gamma production by uterine natural killer cells. Mol Hum Reprod 10(3):189–195. doi:10.1093/molehr/gah032
Van Lierop MJ, Wijnands F, Loke YW, Emmer PM, Lukassen HG, Braat DD, van der Meer A, Mosselman S, Joosten I (2002) Detection of HLA-G by a specific sandwich ELISA using monoclonal antibodies G233 and 56B. Mol Hum Reprod 8(8):776–784
Vercammen MJ, Verloes A, Van de Velde H, Haentjens P (2008) Accuracy of soluble human leukocyte antigen-G for predicting pregnancy among women undergoing infertility treatment: meta-analysis. Hum Reprod Update 14(3):209–218. doi:10.1093/humupd/dmn007
Verloes A, Van de Velde H, LeMaoult J, Mateizel I, Cauffman G, Horn PA, Carosella ED, Devroey P, De Waele M, Rebmann V, Vercammen M (2011) HLA-G expression in human embryonic stem cells and preimplantation embryos. J Immunol 186(4):2663–2671. doi:10.4049/jimmunol.1001081
Yao YQ, Barlow DH, Sargent IL (2005) Differential expression of alternatively spliced transcripts of HLA-G in human preimplantation embryos and inner cell masses. J Immunol 175(12): 8379–8385
Yao GD, Shu YM, Shi SL, Peng ZF, Song WY, Jin HX, Sun YP (2014) Expression and potential roles of HLA-G in human spermatogenesis and early embryonic development. PLoS One 9(3), e92889. doi:10.1371/journal.pone.0092889
Yie SM, Li LH, Li YM, Librach C (2004) HLA-G protein concentrations in maternal serum and placental tissue are decreased in preeclampsia. Am J Obstet Gynecol 191(2):525–529. doi:10.1016/j.ajog.2004.01.033
Yie SM, Balakier H, Motamedi G, Librach CL (2005) Secretion of human leukocyte antigen-G by human embryos is associated with a higher in vitro fertilization pregnancy rate. Fertil Steril 83(1):30–36. doi:10.1016/j.fertnstert.2004.06.059
Zhu X, Han T, Yin G, Wang X, Yao Y (2012) Expression of human leukocyte antigen-G during normal placentation and in preeclamptic pregnancies. Hypertens Pregnancy 31(2):252–260. doi:10.3109/10641955.2011.638955

Chapter 9
Prostasomes: Their Characterisation: Implications for Human Reproduction

Prostasomes and Human Reproduction

Gunnar Ronquist

Abstract The prostate is a principal accessory genital gland that is vital for normal fertility. Epithelial cells lining the prostate acini release in a defined fashion (exocytosis) organellar nanosized structures named prostasomes. They are involved in the protection of sperm cells against immune response in the female reproductive tract by modulating the complement system and by inhibiting monocyte and neutrophil phagocytosis and lymphocyte proliferation. The immunomodulatory function most probably involves small non-coding RNAs present in prostasomes. Prostasomes have also been proposed to regulate the timing of sperm cell capacitation and induction of the acrosome reaction, since they are rich in various transferable bioactive molecules (e.g. receptors and enzymes) that promote the fertilising ability of sperm cells. Antigenicity of sperm cells has been well documented and implicated in involuntary immunological infertility of human couples, and antisperm antibodies (ASA) occur in several body fluids. The propensity of sperm cells to carry attached prostasomes suggests that they are a new category of sperm antigens. Circulating human ASA recognise prostasomes, and among 12 identified prostasomal antigens, prolactin- inducible protein (95 %) and clusterin (85 %) were immunodominant at the expense of the other 10 that were sporadically occurring.

Keywords Prostasomes • Non-coding RNA • Seminal fluid • Spermatozoa • Antioxidants • Immunosuppression

---

G. Ronquist, M.D. (✉)
Department of Medical Sciences, Clinical Chemistry, The University of Uppsala, 751 85 Uppsala, Sweden
e-mail: gunnar.ronquist@akademiska.se

R. Bronson (ed.), *The Male Role in Pregnancy Loss and Embryo Implantation Failure*, Advances in Experimental Medicine and Biology 868,
DOI 10.1007/978-3-319-18881-2_9

9.1 Introduction

The acinar, epithelial cells of the prostate gland have the capacity for secretion. These cells are not alone (although being in majority) in the epithelial lining of prostatic ducts, since there are also minor elements of basal and endocrine–paracrine (APUD) cells. The secretion is not only the result of the synthesising activity of the epithelial cells but depends also on transudation from blood plasma. Besides soluble components, the human prostate gland secretes a particular fraction organised in well-defined organelles (extracellular vesicles, EVs) termed prostasomes (Ronquist and Brody 1985) that were first reported in the 1970s (Ronquist and Hedstrom 1977). They are surrounded by a bilayered membrane and have a globular appearance (Fig. 9.1) and can be recovered in prostatic and seminal fluids with similar ultrastructure. Subsequent electron microscopy examinations revealed that the EVs corresponded to intracellular vesicles inside another larger vesicle, a so-called storage vesicle, equivalent to a multivesicular body (MVB) of late endosomal origin (Ronquist and Brody 1985). Prostasomes exhibited similar mean diameters of about 150 nm regardless of being intracellular or extracellular (after isolation) (Ronquist and Brody 1985), but the size varied and the diameter of a vast majority of prostasomes was within the range of 30–200 nm (Ronquist 2012). Ultrastructural studies

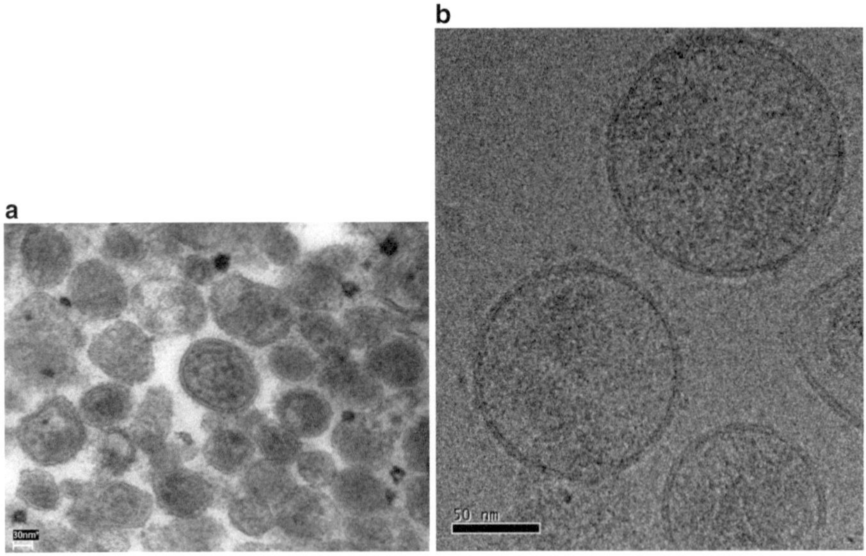

Fig. 9.1 Ultrastructure of prostasomes. (**a**) Thin-section transmission electron microscopy (*EM*) image of human prostasomes isolated from seminal plasma. Prostasomes display rounded structures with varied sizes more or less filled with electron-dense material. (**b**) Ultrastructural appearance of prostasomes by cryo-EM. The samples have been vitrified in liquid ethane to prevent the formation of perturbating ice crystals. The rounded prostasomes are surrounded by classical biological membranes (Brisson A & Ronquist G, unpublished 2013)

revealed that prostasomes were delivered into the glandular duct by so-called exocytosis (Ronquist and Brody 1985; Sahlen et al. 2002). Functional effects of prostasomes on sperm cells were early registered (Stegmayr and Ronquist 1982).

9.2 Biochemical Characteristics of Prostasomes

9.2.1 Proteins

The protein composition of human prostasomes is varied and has been comprehensively examined (Ronquist et al. 2013). There are almost 1,000 different proteins in prostasomes and many proteins are enzymes. In order to investigate possible contamination from testes and epididymides in our prostasome preparation, we followed the Mg^{2+} and/or Ca^{2+} ATPase activity (a marker enzyme of prostasomes) in seminal plasma of 13 men before and after vasectomy. Interestingly, there was no change in ATPase activity meaning that the contribution of sperm cells with regard to ATPase activity and therewith prostasomes was nil. Similar precautions were carried out by Aalberts et al. (2012), and they isolated from vasectomised men two distinct seminal prostasome populations. Both types of prostasomes resembled exosomes in terms of their buoyant density, size and presence of characteristic exosome markers.

Sodium dodecyl sulphate-polyacrylamide gel electrophoresis (SDS-PAGE) separation of prostasomal proteins gave rise to a characteristic and consistent banding pattern (Ronquist et al. 2011) (Fig. 9.2) that was distinctly different from that obtained after SDS-PAGE separation of pellets from seminal vesicle secretion subjected to ultracentrifugation, again emphasising the prostate gland as solitary origin of prostasomes. The SDS-PAGE characteristics of prostasomes were three distinct bands identified as aminopeptidase N (CD13; 110 kDa), dipeptidyl peptidase 4 (CD26; 88 kDa) and enkephalinase or neutral endopeptidase, NEP (CD10; 86 kDa) (Fig. 9.2). Mammalian aminopeptidase N (APN) is an important player in many physiological processes, such as sperm motility, cell–cell adhesion, immune cell chemotaxis, tumour angiogenesis and metastasis and coronavirus entry.

We identified dipeptidyl peptidase 4 (DPP-4) as the antigen of a monoclonal antiprostasome antibody on human prostasomes, and the specific activity of DPP-4 in its prostasomal context is unprecedentedly high (Vanhoof et al. 1992). The DPP-4 antigen (CD26) and enzymatic activity were present in human prostatic fluid but absent from that of the seminal vesicles (Wilson et al. 1998) in harmony with the idea of a solitary prostate gland origin of prostasomes. On the T-cell surface, DPP-4 functions as adenosine deaminase binding protein. A transfer of CD26 from prostasomes to sperm cells was possible, followed by an interaction of prostasomal adenosine deaminase and the transferred CD26 on sperm cells, ultimately leading to fusion between prostasomes and sperm cells (Minelli et al. 1999).

Prostasome-associated enkephalinase/neutral endopeptidase, NEP (CD10), was characterised by Aumüller's group (Renneberg et al. 2001). Endogenous opioid

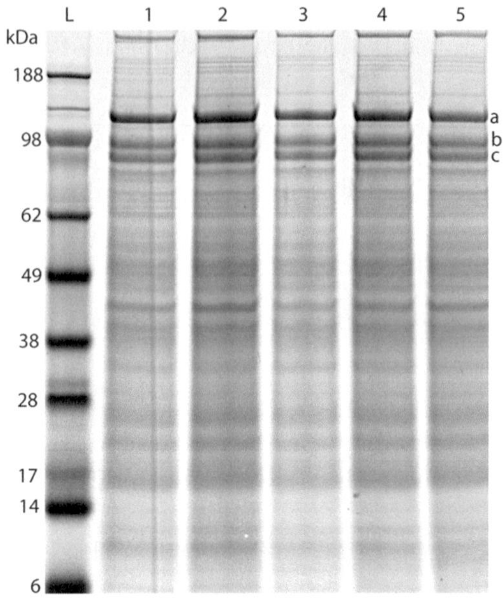

Fig. 9.2 Highly reproducible SDS-PAGE pattern of preparations from different pools of seminal prostasomes (1–5) over a 2-year period. The three marked bands, *a–c*, in the high molecular weight range have been identified as follows: (*a*) CD13—aminopeptidase N (approximately 110 kDa), (*b*) CD26—dipeptidylpeptidase 4 (approximately 88 kDa), (*c*) CD10—neprilysin (approximately 86 kDa) (Reprinted from Ronquist et al. (2011) with permission of the publisher. Copyright 2014 John Wiley & Sons A/S)

peptides, among which the naturally occurring opioid pentapeptides (enkephalins) are to be found, participate in the regulation of reproductive physiology at multiple sites.

Three of the glycolytic enzymes detected in prostasomes on proteomic examinations (Ronquist et al. 2013) belong to the top ten proteins found in most exosomes. Moreover, not only human prostasomes but also bovine, canine and equine prostasomes demonstrated a capacity for adenosine triphosphate (ATP) production (Ronquist et al. 2013). The effect of extracellular ATP on the activation of sperm cells has revealed a signal transduction mechanism for ATP involving the purinergic receptor-mediated release of second messengers culminating in acrosomal exocytosis (Luria et al. 2002).

9.2.2 Lipids

Prostasomes exhibit a peculiar lipid composition with an exceptionally high cholesterol-to-phospholipid molar ratio around 2.0. Sphingomyelin (at the expense of phosphatidylcholine) was the predominant phospholipid class representing nearly half of the total phospholipid measured. One third of the fatty acids in sphingomyelin were palmitic acid. Remaining fatty acids consisted largely of saturated and monounsaturated fatty acids (Arvidson et al. 1989). This peculiar pattern suggested the lipids in the prostasome membrane to be highly ordered. This conclusion was corroborated by our electron spin-labelling experiments showing that the order parameters for prostasomes and extracted prostasome lipids were very high (Arvidson et al. 1989) rendering the prostasomal membrane robust features withstanding physical violence, e.g. freezing and thawing. The two types of "true" prostasomes mentioned above and described by Aalberts et al. (2012) also harboured the characteristic lipid pattern. In addition, these authors reported the presence of monohexosylceramides. Prostasome/exosome formation may be dependent on hydrolysis of sphingomyelin by neutral sphingomyelinase in the lipid raft membrane domain of endosomes (Trajkovic et al. 2008). The hydrolysis product, ceramide, may serve as a trigger of endocytosis. The lipid raft hypothesis was launched in 1997 alleging that lipids play a regulatory role in which they mediate protein clustering and protein diffusion within the lipid bilayer of the membrane (Simons and Ikonen 1997). In this model, the membrane can undergo phase separation into coexisting liquid-disordered and liquid-ordered phases. The liquid-ordered phase, termed lipid raft (Simons and Ikonen 1997), was envisaged as enriched in cholesterol and saturated sphingolipids and phospholipids and characterised by tight lipid packing and reduced molecular diffusion, as we previously noticed for prostasomes (Arvidson et al. 1989).

9.2.3 Nucleic Acids

Different types of RNA have been detected in EVs from various sources (Raposo and Stoorvogel 2013). Technological developments have allowed for the deep sequencing of RNA involving also prostasomes isolated from the ejaculate of vasectomised men, revealing that the majority is neither mRNA nor miRNA (Aalberts et al. 2014). However, recent data provide evidence that prostasomes carry several small RNA biotypes with immunomodulatory functions when delivered to target cells (Vojtech et al. 2014). Fragments of human chromosomal DNA were identified inside purified prostasomes (Ronquist et al. 2009). A genome-wide DNA copy number analysis revealed that they contained fragments of DNA randomly selected from the entire genome (Ronquist et al. 2011). It has been argued that it cannot be excluded that small apoptotic vesicles, which are known to contain fragmented DNA, were co-isolated with prostasomes (Aalberts et al. 2014). Round bodies agreeing with apoptotic vesicles have indeed been identified in human semen, but their dimensions ranged from those similar to sperm head to much larger (Marchiani et al. 2007) meaning a size range considerably larger than that of prostasomes from which DNA was isolated. We prepared human seminal prostasomes in accordance with the prevailing protocol for exosome preparation including passage through a 0.2 μm filter and centrifugation in a sucrose gradient (Ronquist et al. 2012). Filterable prostasomes contained about half the amount of DNA when compared with nonfilterable prostasomes (Table 9.1). The DNA pattern in both types of prostasomes ranged from 1–2 kbp (kilobase pairs) to 10–14 kbp which was similar to what was found when examining prostasomes not subjected to filtration and sucrose gradient separation (Ronquist et al. 2009, 2011).

Table 9.1 DNA content in filterable and nonfilterable prostasomes (duplicate assays) obtained from the main band after sucrose gradient centrifugation

Prostasome type	Prostasome concentration (mg protein)/mL	Volume of prostasomes (mL)	Total prostasome content (mg)	DNA concentration (ng/μL)	Elution volume (μL)	Total DNA (μg)
Filterable	2	1.2	2.4	57	50	2.85
Filterable	2	1.2	2.4	96	50	4.8
Nonfilterable	2	1.2	2.4	160	50	8
Nonfilterable	2	1.2	2.4	160	50	8

From Ronquist et al. (2012)

9.2.4 *Neuroendocrine Components*

Prostasomes contain the neuroendocrine markers chromogranin B, neuropeptide Y and vasoactive intestinal polypeptide in about equimolar amount. Synaptophysin and chromogranin A were found in about 10 % and 2 %, respectively, of that amount (Stridsberg et al. 1996). It was reported that prostasomes express a common secretory granule protein, viz. granulophysin (Skibinski et al. 1994). This molecule has a similar structure as the neuroprotein synaptophysin mentioned above.

9.3 Human Reproduction and Role of Prostasomes

9.3.1 *Prostasome Regulation of Sperm Cell Function*

The physiological relevance of prostasomes was brought out by the finding that prostasomes are able to interact with sperm cells, albeit that both prostasomes and sperm cells display a net negative surface charge favouring repulsive forces (Ronquist et al. 1990). This important extracellular reaction between a cell and an organelle was subsequently confirmed in different ways. Accordingly, prostasomes can carry information from prostate cells to sperm cells. Transfer of a message to target cells could occur by three possible mechanisms: (1) by direct contact between the prostasomal membrane and the sperm cell plasma membrane, (2) by fusion of the two membranes and (3) by sperm cell internalisation of the prostasome. The female reproductive tract is equipped with a well-balanced immune system, and prostasomes are able to mediate different abilities to sperm cells which are important for their survival in a hostile environment to reach and penetrate the zona pellucida for fertilisation of the ovum (Park et al. 2011). Achievement of zona penetration by sperm cells means an ability of both lysis (hydrolytic enzymes) and thrust (hyperactivated motility). It seems reasonable that at least a part of the prostasomes (that are in great excess over sperm cells in an ejaculate) are able to deliver their cargo to sperm cells. This is well-founded, since there might be a conflict within the sperm cell between the critical demand of functional abilities and the silence of protein translation. Transcription ceases several days before the end of spermiogenesis, and the time between when expression is shut down and when acquisition of a distinct pattern of motility known as hyperactivation is needed may be weeks.

9.3.2 Sperm Motility, Capacitation and the Acrosome Reaction

Sperm motility is a critical factor in judging semen quality, and the motility pattern influences the fertilising capability of sperm cells. In the lower female reproductive tract, motility is important to penetrate the cervical mucus, while vigorous beating of the sperm tail is necessary for penetration of zona pellucida in the upper tract (Aitken et al. 1985). The motility pattern of sperm cells evoked by prostasomes (Fabiani et al. 1994; Arienti et al. 2004) helps oocyte fertilisation. We and others have suggested that prostasomes may be able to regulate the divalent cation concentrations in the microenvironment of sperm cells to promote motility (Arienti et al. 2004; Ronquist 1987). On the other hand, it has been claimed that the sperm plasma membrane is extremely impermeable to direct calcium entry into the cytoplasm (Vijayaraghavan and Hoskins 1989). In comparison with the surrounding seminal plasma, an unambiguous enrichment of calcium was observed in prostasomes (Stegmayr et al. 1982). Prostasomes may deliver calcium to sperm cells after fusion, and a detectable increase was indeed noted after 2 min of fusion (Palmerini et al. 1999). Progesterone was influential on the process, and the increased calcium accumulation in sperm cells produced by the fusion with prostasomes and by stimulation of progesterone was independent and additive phenomenon (Arienti et al. 2001). New evidence indicates that human sperm cells have a clever way to solve the conflict between the critical demand for Ca^{2+} signalling tools and the silence of protein translation (Park et al. 2011).

Natural fertilisation to occur implies interrelationships between the female and the sperm cells, and fertilising ability is only acquired in the female reproductive tract through a functional maturation process that is capacitation. It means a complex of structural and functional changes in sperm cells throughout their transit through the female reproductive tract and is considered to be complete when the sperm cells are able to respond to ligands in the zona pellucida by undergoing the acrosome reaction. The capacitation concept involves sperm alterations such as loss of cholesterol from the membrane, increased protein phosphorylation, increased intracellular concentrations of Ca^{2+} and cyclic nucleotides and hyperpolarisation of membrane potential (Visconti 2009). Capacitated sperm cells change their motility characteristics probably in order to facilitate their passage through the latter parts of the female reproductive tract (Ho et al. 2009). Herewith, they are primed to undergo the acrosome reaction in case they should come in contact with the zona pellucida and/or cumulus cells surrounding the ovum (Publicover et al. 2007), and sperm cells that acrosome react before a contact with these structures are invalid to fertilise. Hence, the acrosome reaction must take place after binding to a homologous zona.

Prostasomes bind primarily to the head of live sperm cells, and in vivo, prostasomes may bind to sperm cells in the uterus, to be carried (as "rucksacks") into oviduct and to fuse with the sperm cell only during the final approach of the ovum (Aalberts et al. 2013). This is in line with the observation that prostasome fusion with the sperm cell was an obligatory prerequisite for the transfer from prostasomes of a range of calcium ion signalling tools (including receptors and enzymes) for regulating sperm flagella (Park et al. 2011) and guaranteeing hyperactivated motility necessary for zona pellucida penetration. Apparently, this might be interpreted as a logical consequence of the limited outfit of the sperm cell. Progesterone released by cumulus cells surrounding the ovum is a potent stimulator of the acrosome reaction (Lishko et al. 2011). Human sperm cells are extremely sensitive to progesterone, demonstrating a chemotactic response to picomolar concentration of the hormone (Teves et al. 2006). Park et al. (2011) found that picomolar concentrations of progesterone induced a well-adapted, high-amplitude, calcium ion signal in sperm cells, provided they had fused with prostasomes. This is concordant with other data corroborating the view that prostasome–sperm cell fusion can stimulate the acrosome reaction making sperm cells more sensitive to the effect of progesterone (Palmerini et al. 2003).

9.3.3 Protective Ability Against Oxidative Damage

Reactive oxygen species (ROS) are a major cause of idiopathic male infertility. An abnormally high production of ROS has been demonstrated in 40 % of semen samples from infertile individuals (Iwasaki and Gagnon 1992). Leukocytes infiltrating the semen, particularly the polymorphonuclear neutrophils, seem to be the major source of ROS generation (Saez et al. 1998). Prostasomes have the ability to interact with neutrophils and reduce their capacity to produce superoxide anion after stimulation (Skibinski et al. 1992). Hence, prostasomes could play a role as an antioxidant factor, and it was demonstrated that prostasomes indeed reduced ROS production by sperm preparations containing polymorphonuclear neutrophils (Saez et al. 1998). Subsequent work suggested that prostasomes inhibit the NADPH (nicotinamide adenine dinucleotide phosphate, reduced form) oxidase activity of polymorphonuclear neutrophils by lipid transfer from prostasomes to the plasma membrane of these cells (Saez et al. 2000), therewith inhibiting the ongoing machinery of ROS production.

9.3.4 Protective Ability Against Bacteria

Prostasomal chromogranin B is in abundance over chromogranin A, which is unusual (Stridsberg et al. 1996). What is more, a C-terminal fragment of chromogranin B possesses a potent antibacterial ability. This peptide (secretolytin) forms a three-dimensional structure similar to the insect-derived proteins, cecropins, that are antibacterial as well. The biological activity of these peptides results from their ability to form channels through the bacterial wall leading to bacterial cell death. Also, other parts of both chromogranin B and chromogranin A were antibacterial (Metz-Boutigue et al. 1998). Prostasomes at low concentrations displayed distinct antibacterial effects associated with membrane cavities and bacterial cell death (Carlsson et al. 2000).

A group of peptide antibiotics is the cathelicidin family. The human cationic antimicrobial protein (hCAP-18) is the only known member of this family of proteins in humans (Larrick et al. 1994). The antimicrobial peptide LL-37 becomes activated when released from the C-terminal end of the hCAP-18 holoprotein. In addition to its antimicrobial activity, LL-37 has chemotactic activity for neutrophils, monocytes and T-cells (Agerberth et al. 2000). Hence, LL-37 may contribute to both innate and adaptive immunity, the latter by recruiting immunocompetent cells to sites of microbial invasion. hCAP-18 is expressed in the male reproductive system and high levels were found in seminal plasma (Malm et al. 2000). It appeared in two distinct fractions after gel chromatography of seminal plasma, and the high molecular fraction (the major part) represented hCAP-18 bound to prostasomes (Andersson et al. 2002). In other words, prostasomes can well serve as a reservoir of this precursor of the antibiotic peptide LL-37. It seems like prostasomes may exert antibacterial activities by more than one route.

9.4 Involvement of Prostasomes in Immune Responses

9.4.1 Seminal Plasma Immunosuppressive Activity

The prostate gland is equipped with an active immunological armamentarium and is able to respond to several nonself antigens. It has CD8+ suppressor/cytotoxic T lymphocytes, macrophages and B cells. Prostate epithelium and stromal cells express toll-like receptors and respond to various antigens (Hoover and Naz 2012). This is somewhat contrary to the general belief that the blood–testis barrier forms an immunological barrier excluding the entry of immunoglobulins and lymphocytes into the luminal compartment and preventing germ cell components crossing the barrier to elicit an immunological response in the body. Still, immunosuppressive factors exist and play a pivotal role in successful fertilisation, implantation and fetal growth by protecting female and male reproductive cells from the immunosurveillance system. Since the female genital tract is not an immunologically privileged site, the presence within seminal fluid of powerful immunosuppressive agents is called for. The activity of seminal plasma in this regard has been measured by suppression of the proliferation of lymphocytes that have been activated by mitogen and also by suppression of natural killer (NK) cell activity. The studies of inhibition of NK cell function have all concluded that the prostaglandins in semen are responsible for the inhibition. However, in studies that used the mitogen induced lymphoproliferation assay, activity was detected in high molecular weight fractions (Lee and Ha 1989), and this led to some confusion about the active substances in human semen. Kelly et al. (1991) solved this issue by identifying the high molecular weight immunosuppressive factor as prostasomes. The prostasomes not only inhibited lymphoproliferation but also the ability of macrophages to phagocytose latex beads. As a matter of fact, prostasomes bound rapidly to the leukocyte plasma membrane followed by endocytosis that was mediated by an undefined plasma membrane receptor. Interactions of prostasomes with neutrophils and monocytes inhibited their ability to phagocytose latex particles (Skibinski et al. 1992).

NK cells may represent an important component of innate immunity in the female reproductive tract, and the role of prostasomes in the regulation of NK cell activity showed that prostasomes expressed high levels of CD48, the ligand for the activating receptor CD244 (Tarazona et al. 2011). The interactions between NK cells and prostasomes resulted in a decrease of CD244 expression. Furthermore, the decreased NK cell activity observed in NK cells cultured in the presence of prostasomes suggested that prostasomes may immunomodulate the local environment within the female reproductive tract preventing immune-mediated sperm destruction and prolonging their survival rate (Tarazona et al. 2011). Hence, prostasomes play a significant role in neutralising immune defences against sperm cells in the hostile environment that the female reproductive tract constitutes.

The complement system consists of about 30 plasma and cellular proteins (receptors and regulators) with a primary function in host defence, to differentiate between self and nonself, and as a purging system of the body. The complement system destroys invading foreign cells and organisms, either by direct lysis or by recruitment of leukocytes. The main event in the activation of complement is the activation of C3 into C3b and C3a. This is achieved by two enzyme complexes called convertases, and the classic pathway is triggered by the formation of antigen–antibody complexes (Doekes et al. 1984). The complement system is controlled by several soluble and membrane-bound regulators. Most of the regulators are members of the "regulators of complement activation" (RCA) superfamily encoded by a gene cluster on chromosome 1 that mainly regulates the two types of convertases. Decay accelerating factor (DAF, CD55) and membrane cofactor protein (MCP, CD46) belong to this family. CD59 antigen is an 18–20 kDa membrane protein that is a regulator of membrane attack complex (MAC). Seminal plasma is known to have complement-inhibiting activity, (Tarter and Alexander 1984) and the trophoblast/leukocyte common (TLX) antigen occurring on sperm cells (Anderson et al. 1989) is identical with the above-mentioned MCP, an inhibitor of complement activation (Purcell et al. 1990). Rooney et al. described that all detectable CD59 in seminal plasma was associated with prostasomes representing a protection to sperm cells by being a pool of CD59 from which protein lost from sperm cells, perhaps as a result of low level complement attack or of normal membrane turnover, can be replenished (Rooney et al. 1993). CD46 and CD55 were subsequently found to be prostasome-associated as well (Rooney et al. 1996). These findings should be considered in the light of findings suggesting a functionally active complement system in the female reproductive tract (Jin et al. 1991). Therefore, it is reasonable to conclude that sperm cells are at risk within the female reproductive tract. The interaction of prostasomes with the local female immune system including immunomodulatory skills exerted by prostasomal small RNA biotypes (Vojtech et al. 2014) may prevent sperm cells from being phagocytosed, damaged or killed and therewith prolongs their lifespan in an otherwise hostile environment.

9.5 Sperm Cell Immunogenicity

9.5.1 Antisperm Antibodies

The addition of carbohydrates (glycosylation) is the most common form of covalent posttranslational modification of newly synthesised proteins. A variety of functions can be ascribed to the carbohydrate modifications of proteins. The common feature of the varied functions is that they either mediate specific recognition events (e.g. cell–cell, cell–matrix, immune responses) or that they modulate protein function (e.g. ligand binding; intra- and intercellular trafficking) (Varki 1993). Important constituents of a biological membrane (including that of the sperm cell) are peripherally located and integral proteins. Therefore, surface coat antigens that are peripherally associated with the sperm plasma membrane should be distinguished from those linked to integral plasma membrane proteins. A salient feature of this reasoning is the maintenance of the strict distinction between on the one hand immunity to sperm surface antigens that could theoretically play a role in gamete interactions leading to fertilisation and, on the other hand, due to, e.g. steric hindrance, hidden antigens of sperm cells that would not (Bronson 1999). The other variable in this complex of problems is that the immunoglobulin class responsible for the reactivity with the sperm cell antigens is not a single one but rather three (IgG, IgA and IgM) that appear in different concentrations, among which antisperm IgM in blood plasma hardly enters the male genital tract secretions due to its ungainly molecular structure (Bronson 1999). Despite the antigenic nature of sperm cells, the vast majority of males do not produce antisperm antibodies (ASA). This would in turn presuppose mechanisms that suppress this reaction, involving genital tract structures and sperm cells themselves. Freshly ejaculated and capacitated human sperm cells have been claimed to react with different types of ASA (Fusi and Bronson 1990). The alterations in antigenicity associated with capacitation may reflect a surface redistribution associated with changes in the functional state of sperm cells. ASA, which are more frequent than oocyte antibodies, may fulfil the criteria of an autoimmune disease in the male (Omu et al. 1999). ASA can inhibit fertility by interfering with motility, penetration of the cervical mucus, capacitation/acrosomal exocytosis, zona pellucida binding and sperm–oocyte fusion (Bohring and Krause 2003; Naz and Butler 2013). ASA are prevalent in the infertile male population, and the presence of ASA in males can reduce fecundity, but the causality is not strictly clear. It should be mentioned that ASA occur in several body fluids like seminal plasma and bound to the sperm cell surface and blood plasma of men and women but also in oviduct fluid, cervical mucus and follicular fluid of women. The presence of ASA has been described in 1.2–19 % of fertile couples, suggesting that not all ASA cause infertility. ASA from infertile patients may be directed to dissimilar sperm antigens and/or clusters of antigens or possess different antigen-binding characteristics differing from those of fertile individuals (Chamley and Clarke 2007). Bohring et al. (2001) investigated seminal plasma samples from 20 infertile patients, and 18 proteins associated with sperm membranes were detected as antigens and 6 proteins were biochemically identified.

9.5.2 Circulating ASA Recognise Prostasomes

A prostasome coat on swim-up sperm cells was found when we immunostained the samples with seven different monoclonal antibodies directed against purified prostasomes and all monoclonal antibodies tested demonstrated a similar staining pattern (Allegrucci et al. 2001). Therefore, the probability of a possible cross-reactivity with some other seminal plasma component was weak and prompted us to suggest that prostasomes may be a candidate antigen for ASA and we raised polyclonal chicken antibodies against purified seminal prostasomes (Allegrucci et al. 2001). Chicken antibodies have the advantage that they eliminate interference caused by the human complement system, rheumatoid factor and cellular Fc receptors and they resemble human autoantibodies in their reactivity. Human sperm cells incubated with increasing concentrations of chicken polyclonal antiprostasome antibodies caused approximately 80 % of the sperm cells to agglutinate. The agglutination displayed several types of sperm formation, mostly tail to tail contacts, but the design of interaction was dependent upon the concentration of the antiprostasome antibody. It should be pointed out that the agglutination of sperm cells by the chicken prostasome antibody was similar to the agglutination caused by sera from patients having ASA. When antiprostasome antibodies were preincubated with high concentrations of prostasomes, no agglutination was observed during the subsequent contact with sperm cells (Allegrucci et al. 2001) herewith underlining the specific role of prostasomes as antigens in this context. IgG antibodies against sperm cells were detected in the 20 sera of ASA-positive patients investigated for infertility (15 men and 5 women). In the majority of cases (90 %), the sera of the patients elicited complement activation, measured by the deposition of C3 on the sperm cells. A significant positive correlation was found between sperm cell-bound C3 and ASA titer and especially between deposition of C3 and IgG (Allegrucci et al.

2001). ASA of the IgG type in serum of infertile men and women do recognise prostasomes as antigens, and prostasomes are strongly immunogenic and they should not be overlooked in immunological infertility assessment.

As already mentioned, prostasomes have the ability to adhere to sperm cells albeit that both prostasomes and sperm cells have a net negative surface charge of their membranes favouring repulsive forces. This propensity of sperm cells to carry attached prostasomes, allowed us to regard prostasomes as a new category of sperm antigens (Carlsson et al. 2004a). In a study of the reactivity of ASA-positive sera from 116 suspected immunoinfertile patients, a binding of IgG antibodies to prostasomes was clearly visible in 113 patient sera (97 %) (Carlsson et al. 2004b). Those sera with well-expressed antibody reaction showed 3–10 different prostasomal bands in one-dimensional electrophoresis as antigens. Serum samples from male and female patients agreed with each other, and no difference was observed between sexes (Carlsson et al. 2004b). This reflects a high conformity taking into consideration that the ASA assay was carried out on fresh sperm cells, while the prostasomal antigen represented frozen material that had been thawed before testing. Twenty sera from the 116 suspected immunoinfertile patients with highest titres of antiprostasome autoantibodies were selected for the identification of the proteins corresponding to the prostasomal antigens. Two-dimensional, silver-stained SDS-PAGE gel of separated prostasomal proteins revealed over 200 protein spots in the molecular mass range of 8–180 kDa, and approximately 70 % of the spots displayed isoelectric points (IPs) below 7, i.e. reflecting anionic proteins (Carlsson et al. 2004c). On immunoblotting, most of the ASA-positive sera recognised 3–10 spots in the molecular mass range between 17 and 70 kDa. The size distribution of all protein spots recognised by patient sera after immunoblotting revealed that 19 out of 20 (95 %) of the serum samples discerned protein(s) with a molecular mass range of approximately 17 kDa. Another protein spot with a molecular mass of approximately 40 kDa was found in 17 out of 20 (85 %) of the serum samples. Proteins in 12 spots were identified and are summarised with their theoretical IPs, molecular masses and accession numbers in Table 9.2. The two most frequent antigens are prolactin-inducible protein (PIP), recognised by 19 out of 20 (95 %) patient sera, and corresponding figures for clusterin are 17 out of 20 (85 %) patient sera (Table 9.2). The remaining ten prostasomal antigens were sporadically occurring and identified as given in Table 9.2. A comparison of this study is feasible with that of Bohring et al. (2001) who investigated seminal plasma samples from 20 infertile patients. It should be noted though, that we used blood sera to avoid the problem of antigen excess, due to the huge representation of prostasomes in seminal plasma. The former authors detected 18 proteins associated with sperm membrane antigens, and six proteins were biochemically identified (Bohring et al. 2001), but none of these proteins associated with sperm membranes and recognised by ASA were identical with the 12 proteins associated with prostasomal membranes and yet recognised by ASA. The clearly disproportionate occurrence of only two immunodominant prostasomal antigens (PIP and clusterin) at the expense of the remaining ten sporadically occurring prostasomal proteins deserves careful consideration. It should be mentioned though that Thacker et al. (2011) identified in a pilot study

Table 9.2 Twelve prostasome immunogens recognised by antisperm antisera of 20 patients with suspected immunological infertility in two-dimensional gel electrophoresis. Identification was carried out by MALDI analyses

No.	pI	Mass (Da)	Accession No.	Protein name
1	7.0	46,800	Q13584	Isocitrate dehydrogenase [NADP]
2	5.9	86,600	Q99728	BRCA1-associated RING domain protein 1
3	6.3	36,500	P10909	Clusterin
4	7.0	36,600	P14550	Alcohol dehydrogenase [NADP(+)]
5	5.7	31,100	094760	N(G),N(G)-dimethylarginine dimethylaminohydrolase 1
6	6.8	38,700	P04083	Annexin A1
7	6.0	36,400	P12429	Annexin A3
8	7.4	32,400	000560	Syntenin-1
9	6.5	22,800	P04792	Heat shock protein beta-1 (HSP27)
10	5.2	20,800	Q04760	Lactoylglutathione lyase
11	5.4	21,900	P32119	Peroxiredoxin-2
12	5.3	16,600	P12273	Prolactin-inducible protein

From Carlsson et al. (2004c)
pI = isoelectric point

the presence of four major proteins in human seminal plasma that were unique and different in fertile and infertile men. These were PIP, clusterin isoform, prostate-specific antigen (PSA) isoform 1 preprotein and semenogelin II precursor. However, immunoblot analyses of individual serum samples containing ASA from infertile couples generally reveal a certain degree of heterogeneity of antigens recognised by human sperm auto- and isoantibodies (Mathur et al. 1988). Shetty et al. (1999) adopted high-resolution two-dimensional electrophoresis with separation of sperm cell antigens in the first dimension by either isoelectric focusing or nonequilibrium pH gradient electrophoresis to screen a range of proteins with different isoelectric points, followed by SDS-PAGE and a sensitive Western blotting method. They found that a number of antigenic spots among both anionic and cationic proteins were reactive with sera from infertile but not fertile individuals reflecting a considerable diversity in the ASA composition of infertile males and females.

The two immunodominant prostasomal antigens (PIP and clusterin) are widely expressed proteins. PIP is a 17 kDa glycoprotein (gp17). It exerts multiple important functions in biological systems and is involved in fertility, immunoregulation, antimicrobial activity, apoptosis and tumour progression. PIP has been identified in prostatic secretion and therewith in seminal plasma (Autiero et al. 1995) which is a rich source for its isolation and characterisation (Chiu and Chamley 2003). The binding ability of PIP to the Fc fragment of IgG has improved the understanding of the functional role of PIP in seminal plasma (Witkin et al. 1983). The reduced level of PIP might be associated with infertility, especially in men with ASA (Bronson 1999). Secreted PIP in seminal plasma has been found to have its localisation on the sperm cell surface (Bergamo et al. 1997). PIP is able to bind to $CD4^+$ T-cell receptor and to block CD4-mediated T-cell programmed death which means that PIP may act as a modulator in an immune response reaction (Gaubin et al. 1999). PIP was among some candidate galectin-3 ligands in prostasomes that were identified by tandem mass spectrometry of proteins that co-purified with galectin-3 during lactose affinity chromatography of the membrane fraction containing solubilised human prostasomes (Block et al. 2011). Galectin-3 is a beta-galactoside-binding protein involved in immunomodulation, cell interactions and cancer progression. The intact galectin-3 molecule contains a carbohydrate recognition domain and a non-lectin domain that interacts with protein and lipid moieties. It has a firm association to human prostasomes and more precisely to the prostasome surface (Block et al. 2011), meaning that also PIP may have a surface localisation on human prostasomes. Hence, the multiple functions of galectin-3 are exerted through ligand binding with specific galectin-3 ligands involved (Ochieng et al. 2004). Galectin-3 ligands in human prostasomes were purified, identified and characterised by Kovak et al. (2013), and using a proteomic approach, clusterin as well was among the candidate galectin-3 binding ligands.

Clusterin has been found in all body fluids and is a major heterodimeric glycoprotein in mammalian semen. It is synthesised and secreted by a wide variety of cells in different species. Clusterin was named for its ability to elicit clustering among Sertoli cells (Blaschuk et al. 1983). Two forms of clusterin have been identified, viz. the secretory form and the nuclear form. The secretory form is synthesised as a 50–60 kDa protein precursor that is glycosylated and proteolytically cleaved into two subunits (alpha and beta chains) (Leskov et al. 2003). The nuclear form is expressed as a 49 kDa inactive protein precursor that does not undergo a proteolytic cleavage (Reddy et al. 1996). The soluble form of clusterin is present in normal human sperm cells (Thacker et al. 2011), and it has also been observed on the surfaces of immature, low-motile and morphologically aberrant sperm cells (O'Bryan et al. 1994). Clusterin participates in many biological processes such as cell–cell interactions, sperm maturation, agglutination of abnormal sperm cells, membrane recycling, apoptosis and lipid transportation and controls complement-induced sperm cell lysis (Silkensen et al. 1999; Wong et al. 1993). Using a panel of polyclonal and monoclonal antibodies against different parts of the clusterin molecule, Lakins et al. (1998) inferred that in normal rat prostate, clusterin has at least five different glyco/isoforms: fully glycosylated mature pro-protein (76 kDa), cleaved fully glycosylated alpha and beta chains (32 and 48 kDa), two intermediate uncleaved processing forms of pro-protein [presumably the high-mannose (64 kDa) and low-mannose species of clusterin (56 kDa) and full length unglycosylated holoprotein (50 kDa)]. It means that prostate-derived clusterin may act in different immunological shapes herewith including its probable presence at the surface of both sperm cells and prostasomes.

Acknowledgement The author apologises to those colleagues whose work was not cited due to space limitations.

References

Aalberts M et al (2012) Identification of distinct populations of prostasomes that differentially express prostate stem cell antigen, annexin A1, and GLIPR2 in humans. Biol Reprod 86(3):82

Aalberts M et al (2013) Spermatozoa recruit prostasomes in response to capacitation induction. Biochim Biophys Acta 1834(11):2326–2335

Aalberts M, Stout TA, Stoorvogel W (2014) Prostasomes: extracellular vesicles from the prostate. Reproduction 147(1):R1–R14

Agerberth B et al (2000) The human antimicrobial and chemotactic peptides LL-37 and alpha-defensins are expressed by specific lymphocyte and monocyte populations. Blood 96(9): 3086–3093

Aitken RJ et al (1985) Relationship between the movement characteristics of human spermatozoa and their ability to penetrate cervical mucus and zona-free hamster oocytes. J Reprod Fertil 73(2):441–449

Allegrucci C et al (2001) Circulating human antisperm antibodies recognize prostasomes. Am J Reprod Immunol 46(3):211–219

Anderson DJ, Michaelson JS, Johnson PM (1989) Trophoblast/leukocyte-common antigen is expressed by human testicular germ cells and appears on the surface of acrosome-reacted sperm. Biol Reprod 41(2):285–293

Andersson E et al (2002) Isolation of human cationic antimicrobial protein-18 from seminal plasma and its association with prostasomes. Hum Reprod 17(10):2529–2534

Arienti G et al (2001) Progesterone-induced increase of sperm cytosolic calcium is enhanced by previous fusion of spermatozoa to prostasomes. Cell Calcium 30(3):222–227

Arienti G et al (2004) Role of human prostasomes in the activation of spermatozoa. J Cell Mol Med 8(1):77–84

Arvidson G et al (1989) Human prostasome membranes exhibit very high cholesterol/phospholipid ratios yielding high molecular ordering. Biochim Biophys Acta 984(2):167–173

Autiero M et al (1995) A 17-kDa CD4-binding glycoprotein present in human seminal plasma and in breast tumor cells. Eur J Immunol 25(5):1461–1464

Bergamo P et al (1997) CD4-mediated anchoring of the seminal antigen gp17 onto the spermatozoon surface. Hum Immunol 58(1):30–41

Blaschuk O, Burdzy K, Fritz IB (1983) Purification and characterization of a cell-aggregating factor (clusterin), the major glycoprotein in ram rete testis fluid. J Biol Chem 258(12): 7714–7720

Block AS et al (2011) Co-purification of Mac-2 binding protein with galectin-3 and association with prostasomes in human semen. Prostate 71(7):711–721

Bohring C, Krause W (2003) Immune infertility: towards a better understanding of sperm (auto)-immunity. The value of proteomic analysis. Hum Reprod 18(5):915–924

Bohring C et al (2001) Isolation and identification of sperm membrane antigens recognized by antisperm antibodies, and their possible role in immunological infertility disease. Mol Hum Reprod 7(2):113–118

Bronson RA (1999) Antisperm antibodies: a critical evaluation and clinical guidelines. J Reprod Immunol 45(2):159–183

Carlsson L et al (2000) Antibacterial activity of human prostasomes. Prostate 44(4):279–286

Carlsson L et al (2004a) Prostasome antigens as targets for sperm agglutinating antibodies demonstrated by 1-D gel electrophoresis and immunoblottings. Int J Androl 27(6):360–367

Carlsson L et al (2004b) A new test for immunological infertility: an ELISA based on prostasomes. Int J Androl 27(3):130–133

Carlsson L et al (2004c) Dominant prostasome immunogens for sperm-agglutinating autoantibodies of infertile men. J Androl 25(5):699–705

Chamley LW, Clarke GN (2007) Antisperm antibodies and conception. Semin Immunopathol 29(2):169–184

Chiu WW, Chamley LW (2003) Human seminal plasma prolactin-inducible protein is an immunoglobulin G-binding protein. J Reprod Immunol 60(2):97–111
Doekes G, van Es LA, Daha MR (1984) Binding and activation of the first complement component by soluble immune complexes: effect of complex size and composition. Scand J Immunol 19(2):99–110
Fabiani R et al (1994) Enhanced recruitment of motile spermatozoa by prostasome inclusion in swim-up medium. Hum Reprod 9(8):1485–1489
Fusi F, Bronson RA (1990) Effects of incubation time in serum and capacitation on spermatozoal reactivity with antisperm antibodies. Fertil Steril 54(5):887–893
Gaubin M et al (1999) Potent inhibition of CD4/TCR-mediated T cell apoptosis by a CD4-binding glycoprotein secreted from breast tumor and seminal vesicle cells. J Immunol 162(5):2631–2638
Ho K, Wolff CA, Suarez SS (2009) CatSper-null mutant spermatozoa are unable to ascend beyond the oviductal reservoir. Reprod Fertil Dev 21(2):345–350
Hoover P, Naz RK (2012) Do men with prostate abnormalities (prostatitis/benign prostatic hyperplasia/prostate cancer) develop immunity to spermatozoa or seminal plasma? Int J Androl 35(4):608–615
Iwasaki A, Gagnon C (1992) Formation of reactive oxygen species in spermatozoa of infertile patients. Fertil Steril 57(2):409–416
Jin M, Larsson A, Nilsson BO (1991) A functionally active complement system is present in uterine secretion of the mouse prior to implantation. Am J Reprod Immunol 26(2):53–57
Kelly RW et al (1991) Extracellular organelles (prostasomes) are immunosuppressive components of human semen. Clin Exp Immunol 86(3):550–556
Kovak MR et al (2013) Proteomic identification of galectin-3 binding ligands and characterization of galectin-3 proteolytic cleavage in human prostasomes. Andrology 1(5):682–691
Lakins J et al (1998) Clusterin biogenesis is altered during apoptosis in the regressing rat ventral prostate. J Biol Chem 273(43):27887–27895
Larrick JW et al (1994) A novel granulocyte-derived peptide with lipopolysaccharide-neutralizing activity. J Immunol 152(1):231–240
Lee HK, Ha TY (1989) Human seminal plasma suppresses delayed-type hypersensitivity responses to intravaginally deposited sheep red blood cells and sperm: separation of immunosuppressive factors. Int Arch Allergy Appl Immunol 88(4):412–419
Leskov KS et al (2003) Synthesis and functional analyses of nuclear clusterin, a cell death protein. J Biol Chem 278(13):11590–11600
Lishko PV, Botchkina IL, Kirichok Y (2011) Progesterone activates the principal Ca2+ channel of human sperm. Nature 471(7338):387–391
Luria A et al (2002) Extracellular adenosine triphosphate stimulates acrosomal exocytosis in bovine spermatozoa via P2 purinoceptor. Biol Reprod 66(2):429–437
Malm J et al (2000) The human cationic antimicrobial protein (hCAP-18) is expressed in the epithelium of human epididymis, is present in seminal plasma at high concentrations, and is attached to spermatozoa. Infect Immun 68(7):4297–4302
Marchiani S et al (2007) Characterization of M540 bodies in human semen: evidence that they are apoptotic bodies. Mol Hum Reprod 13(9):621–631
Mathur S et al (1988) Special antigens on sperm from autoimmune infertile men. Am J Reprod Immunol Microbiol 17(1):5–13
Metz-Boutigue MH et al (1998) Antibacterial peptides are present in chromaffin cell secretory granules. Cell Mol Neurobiol 18(2):249–266
Minelli A et al (1999) CD26 and adenosine deaminase interaction: its role in the fusion between horse membrane vesicles and spermatozoa. Biol Reprod 61(3):802–808
Naz RK, Butler TS (2013) Antibodies to prostate-specific antigen in immunoinfertile women and men. J Reprod Immunol 97(2):217–222
O'Bryan MK et al (1994) The use of anticlusterin monoclonal antibodies for the combined assessment of human sperm morphology and acrosome integrity. Hum Reprod 9(8):1490–1496

Ochieng J, Furtak V, Lukyanov P (2004) Extracellular functions of galectin-3. Glycoconj J 19(7–9):527–535
Omu AE et al (1999) Relationship between unexplained infertility and human leukocyte antigens and expression of circulating autogeneic and allogeneic antisperm antibodies. Clin Exp Obstet Gynecol 26(3–4):199–202
Palmerini CA et al (1999) Increase of human spermatozoa intracellular Ca^{2+} concentration after fusion with prostasomes. Cell Calcium 25(4):291–296
Palmerini CA et al (2003) Fusion of prostasomes to human spermatozoa stimulates the acrosome reaction. Fertil Steril 80(5):1181–1184
Park KH et al (2011) Ca^{2+} signaling tools acquired from prostasomes are required for progesterone-induced sperm motility. Sci Signal 4(173):ra31
Publicover S, Harper CV, Barratt C (2007) [Ca2+]i signalling in sperm–making the most of what you've got. Nat Cell Biol 9(3):235–242
Purcell DF et al (1990) The human cell-surface glycoproteins HuLy-m5, membrane co-factor protein (MCP) of the complement system, and trophoblast leucocyte-common (TLX) antigen, are CD46. Immunology 70(2):155–161
Raposo G, Stoorvogel W (2013) Extracellular vesicles: exosomes, microvesicles, and friends. J Cell Biol 200(4):373–383
Reddy KB et al (1996) Transforming growth factor beta (TGF beta)-induced nuclear localization of apolipoprotein J/clusterin in epithelial cells. Biochemistry 35(19):6157–6163
Renneberg H et al (2001) Identification and characterization of neutral endopeptidase (EC 3. 4. 24. 11) from human prostasomes–localization in prostatic tissue and cell lines. Prostate 46(3):173–183
Ronquist G (1987) Effect of modulators on prostasome membrane-bound ATPase in human seminal plasma. Eur J Clin Invest 17(3):231–236
Ronquist G (2012) Prostasomes are mediators of intercellular communication: from basic research to clinical implications. J Intern Med 271(4):400–413
Ronquist G, Brody I (1985) The prostasome: its secretion and function in man. Biochim Biophys Acta 822(2):203–218
Ronquist G, Hedstrom M (1977) Restoration of detergent-inactivated adenosine triphosphatase activity of human prostatic fluid with concanavalin A. Biochim Biophys Acta 483(2):483–486
Ronquist G, Nilsson BO, Hjerten S (1990) Interaction between prostasomes and spermatozoa from human semen. Arch Androl 24(2):147–157
Ronquist KG et al (2009) Human prostasomes contain chromosomal DNA. Prostate 69(7):737–743
Ronquist GK et al (2011) Prostasomal DNA characterization and transfer into human sperm. Mol Reprod Dev 78(7):467–476
Ronquist GK et al (2012) Prostasomes are heterogeneous regarding size and appearance but affiliated to one DNA-containing exosome family. Prostate 72(16):1736–1745
Ronquist KG et al (2013) Prostasomes from four different species are able to produce extracellular adenosine triphosphate (ATP). Biochim Biophys Acta 1830(10):4604–4610
Rooney IA et al (1993) Physiologic relevance of the membrane attack complex inhibitory protein CD59 in human seminal plasma: CD59 is present on extracellular organelles (prostasomes), binds cell membranes, and inhibits complement-mediated lysis. J Exp Med 177(5):1409–1420
Rooney IA, Heuser JE, Atkinson JP (1996) GPI-anchored complement regulatory proteins in seminal plasma. An analysis of their physical condition and the mechanisms of their binding to exogenous cells. J Clin Invest 97(7):1675–1686
Saez F et al (1998) Antioxidant capacity of prostasomes in human semen. Mol Hum Reprod 4(7):667–672
Saez F et al (2000) Prostasomes inhibit the NADPH oxidase activity of human neutrophils. Mol Hum Reprod 6(10):883–891
Sahlen GE et al (2002) Ultrastructure of the secretion of prostasomes from benign and malignant epithelial cells in the prostate. Prostate 53(3):192–199
Shetty J et al (1999) Human sperm proteome: immunodominant sperm surface antigens identified with sera from infertile men and women. Biol Reprod 61(1):61–69

Silkensen JR et al (1999) Identification of clusterin sequences mediating renal tubular cell interactions. J Pept Res 54(5):449–457
Simons K, Ikonen E (1997) Functional rafts in cell membranes. Nature 387(6633):569–572
Skibinski G et al (1992) Immunosuppression by human seminal plasma–extracellular organelles (prostasomes) modulate activity of phagocytic cells. Am J Reprod Immunol 28(2):97–103
Skibinski G, Kelly RW, James K (1994) Expression of a common secretory granule specific protein as a marker for the extracellular organelles (prostasomes) in human semen. Fertil Steril 61(4):755–759
Stegmayr B, Ronquist G (1982) Promotive effect on human sperm progressive motility by prostasomes. Urol Res 10(5):253–257
Stegmayr B et al (1982) Calcium, magnesium, and zinc contents in organelles of prostatic origin in human seminal plasma. Scand J Urol Nephrol 16(3):199–203
Stridsberg M et al (1996) Prostasomes are neuroendocrine-like vesicles in human semen. Prostate 29(5):287–295
Tarazona R et al (2011) Human prostasomes express CD48 and interfere with NK cell function. Immunobiology 216(1–2):41–46
Tarter TH, Alexander NJ (1984) Complement-inhibiting activity of seminal plasma. Am J Reprod Immunol 6(1):28–32
Teves ME et al (2006) Progesterone at the picomolar range is a chemoattractant for mammalian spermatozoa. Fertil Steril 86(3):745–749
Thacker S et al (2011) Evaluation of sperm proteins in infertile men: a proteomic approach. Fertil Steril 95(8):2745–2748
Trajkovic K et al (2008) Ceramide triggers budding of exosome vesicles into multivesicular endosomes. Science 319(5867):1244–1247
Vanhoof G et al (1992) Distribution of proline-specific aminopeptidases in human tissues and body fluids. Eur J Clin Chem Clin Biochem 30(6):333–338
Varki A (1993) Biological roles of oligosaccharides: all of the theories are correct. Glycobiology 3(2):97–130
Vijayaraghavan S, Hoskins D (1989) Quantitation of bovine sperm cytoplasmic calcium with Quin-2 and Fura-2: evidence that external calcium does not have direct access to the sperm cytoplasm. Cell Calcium 10(4):241–253
Visconti PE (2009) Understanding the molecular basis of sperm capacitation through kinase design. Proc Natl Acad Sci USA 106(3):667–668
Vojtech L et al (2014) Exosomes in human semen carry a distinctive repertoire of small non-coding RNAs with potential regulatory functions. Nucleic Acids Res 42(11):7290–7304
Wilson MJ et al (1998) Prostate specific origin of dipeptidylpeptidase IV (CD-26) in human seminal plasma. J Urol 160(5):1905–1909
Witkin SS et al (1983) An IgG-Fc binding protein in seminal fluid. Am J Reprod Immunol 3(1):23–27
Wong P et al (1993) Genomic organization and expression of the rat TRPM-2 (clusterin) gene, a gene implicated in apoptosis. J Biol Chem 268(7):5021–5031

Chapter 10
The Paternal Contribution to Fetal Tolerance

Anne Schumacher and Ana Claudia Zenclussen

Abstract Recognition of foreign paternal antigens expressed in the semi-allogeneic fetus by maternal immune cells is a requirement for successful pregnancy. However, despite intensive research activity during the last decades, the precise mechanisms contributing to the acceptance of the paternal alloantigens are still puzzling and pregnancy remains a fascinating phenomenon. Moreover, most studies focused on the maternal and fetal contribution to pregnancy success, and relatively little is known about the paternal involvement. In the current review, we address the contribution of paternal-derived factors to fetal-tolerance induction. First, we discuss data suggesting that in both humans and mice, the female body gets prepared for a pregnancy in every cycle, also in regard to male alloantigens delivered at coitus. Then, we provide an overview about factors present in seminal fluid and how these factors influence immune responses in the female reproductive tract. We further discuss ways of paternal alloantigen presentation and identify the immune modulatory properties of seminal fluid-derived factors with a special focus on Treg biology. Finally, we highlight the therapeutic potential of seminal fluid in different clinical applications.

Keywords Alloantigens • Fetal tolerance • Regulatory T cells • Seminal fluid • Seminal plasma • Pregnancy

10.1 Introduction

Awareness of the foreign paternal/fetal alloantigens and their tolerance is a prerequisite for the survival of the semi-allogeneic fetus within the maternal uterus. There is plenty of published evidence that, to achieve this tolerance, both mother and fetus release immune-modulating factors that among other effects augment the number of regulatory immune cells to suppress alloreactive immune responses. By contrast, the contribution of the father to fetal tolerance is still underexplored. When a couple

A. Schumacher • A.C. Zenclussen (✉)
Department of Experimental Obstetrics and Gynecology, Medical Faculty,
Otto-von-Guericke University, Gerhart-Hauptmann Straße 35, 39108 Magdeburg, Germany
e-mail: ana.zenclussen@med.ovgu.de

R. Bronson (ed.), *The Male Role in Pregnancy Loss and Embryo Implantation Failure*, Advances in Experimental Medicine and Biology 868,
DOI 10.1007/978-3-319-18881-2_10

plans to have a baby, they should have sexual intercourse without contraception. Within this time, even if pregnancy does not occur, maternal immune cells present in the vagina and the uterus are repeatedly challenged by paternal alloantigens present in the semen at coitus. Several studies indicate that there is an accumulation of various immune cell populations in the vagina and the uterus at or immediately after ovulation, and these cellular changes are needed to establish fetal tolerance from the very beginning of pregnancy. Early recognition and acceptance of paternal alloantigens has been shown to ensure successful embryo implantation and fetal growth. Preeclampsia (PE) is a serious pregnancy complication associated with poor placentation as a consequence of shallow trophoblast invasion. The risk of developing PE is diminished if repeated contact with paternal alloantigens has taken place before pregnancy arises (Saftlas et al. 2014). Dekker and colleagues discuss in their review that PE is a disease of first pregnancies. The protective effect of multiparity is however lost with change of partner. Artificial donor insemination and oocyte donation result in a substantial increase of PE risk (Dekker et al. 1998). These observations underlie the meaningfulness of early alloantigen awareness for a successful progress of pregnancy as immune maladaptation is related with poor pregnancy outcome. We need to further study mechanisms underlying establishment of early fetal allotolerance. Here, we discuss available data from the literature on the immune-modulating properties of seminal fluid components with a specific focus on regulatory T cells (Treg).

10.2 Immune Cell Fluctuations During the Reproductive Cycle

Fluctuations in various immune cell populations have been observed in uterine tissue during the reproductive cycle in humans and mice. These cellular changes are proposed to prepare the endometrium for the appearance of fetal alloantigens when the embryo is implanting. Hormonal variations and other factors have been held responsible for the observed fluctuations. For instance, uterine natural killer (NK) cells dramatically increase in number in the late secretory phase of the human menstrual cycle finally representing the main immune cell population of all leukocytes in human decidual tissue (Bulmer et al. 1991). Moreover, human and murine uterine mast cells (MCs) oscillate during the reproductive cycle (Padilla et al. 1990; Mori et al. 1997) reaching their maximum number in the receptive phase (estrus) of the murine estrous cycle (Woidacki et al. 2013). In line, we demonstrated that dendritic cells (DCs) and Treg accumulate in the estrus phase in mice (Zenclussen et al. 2013; Teles et al. 2013a). For the latter, cycle-dependent fluctuations have been confirmed also by others in human and mice (Arruvito et al. 2007; Kallikourdis and Betz 2007), and it was proposed that Treg changes are hormone driven (Weinberg et al. 2011; Schumacher et al. 2014). Altogether, it can be assumed that every time a female becomes receptive, her immune system prepares itself for the contact with the foreign paternal antigens delivered with semen at coitus. This ensures the possibility of a very early tolerance initiation toward the paternal antigens that will be presented by fetal tissue at implantation and afterward.

10.3 Composition of Seminal Fluid

Seminal fluid (semen) consists of spermatozoa suspended in seminal plasma. Components of seminal plasma are secreted from rete testis, epididymis, and accessory sex glands including the seminal vesicle, the prostate, and the bulbourethral glands (Juyena and Stelletta 2012). Analysis of mammalian seminal plasma revealed a variety of factors including ions, energy substrates, organic compounds, and nitrogenous compounds (Juyena and Stelletta 2012; Milardi et al. 2013). Additionally, seminal plasma contains high concentrations of factors proven to posses immune regulatory properties. One of these factors is tumor growth factor-β (TGF-β), whereby TGF-β1 and TGF-β3 can be found in high amounts and TGF-β2 is present in lower amounts (Nocera and Chu 1995; Lokeshwar and Block 1992; O'Leary et al. 2013; Srivastava et al. 1996; Tremellen et al. 1998). Therefore, the majority of TGF-β existing in its latent form has to be activated to be fully functional. TGF-β activation was proposed to take place after insemination in the female tract. Here, several enzymes delivered with the seminal plasma or being present in the female tract are suggested to contribute to the activation process. More precisely, plasmin, subtilisin-like endoproteases, tissue- and urokinase-type plasminogen activator (Chu and Kawinski 1998), thrombospondin 1 (Slater and Murphy 1999), and αvβ6 integrin (Breuss et al. 1993) were implicated in TGF-β activation. Moreover, the acidic environment of the human vagina may drive TGF-β activation to some extent. Besides TGF-β, human seminal plasma contains other immune-modulating factors such as interleukin (IL)-8 and soluble IL-2 receptor (Srivastava et al. 1996) prostaglandin E_2 (PGE_2) and 19-hydroxyprostaglandin E (19-hydroxy PGE) (Denison et al. 1999), soluble tumor necrosis factor (TNF) receptors (Liabakk et al. 1993), receptors for the Fc portion of γ-globulin, spermine (Evans et al. 1995), and complement inhibitors (Kelly 1995).

10.4 Recognition and Presentation of Seminal Fluid-Derived Alloantigens

Immediately after insemination, the secretion of growth factors, cytokines, and chemokines from cervical and endometrial tissue is induced, resulting in local pro-inflammatory environment. This provokes a rapid and dramatic influx of immune cells and additionally increases the number of various immune cell populations in situ. In the mouse, release of the granulocyte-macrophage colony-stimulating factor (GM-CSF), IL-6, and several chemokines leads to the infiltration of macrophages, DCs, and granulocytes in subepithelial stromal tissue (De et al. 1991; McMaster et al. 1992; Robertson et al. 1996). In humans, sexual intercourse results in neutrophil infiltration into the superficial epithelium of the cervical tissues (Pandya and Cohen 1985) followed by a recruitment of macrophages, DCs, and lymphocytes into the epithelial layers and deeper stromal tissues (Sharkey et al. 2007). This remarkable recruitment of immune cells and especially of those capable of efficiently present alloantigens suggests that presentation of paternal antigens can take place even before implantation and be involved in pregnancy success or failure already at this early time point. Evidence for this was provided by two studies showing that depletion of Treg before mating or depletion of uterine DCs in the preimplantation period impaired the implantation process drastically (Plaks et al. 2008; Teles et al. 2013a). We confirmed the presence of paternal alloantigens immediately after fecundation in a mouse model. By mating wild-type females with GFP^+ males, we demonstrated that paternal antigens (GFP^+ cells) can be found in vaginal lumen as early as day 0.5 of pregnancy. Moreover, at the same time, we detected GFP^+ cells in decidual tissue and lymph nodes. At later pregnancy stages, we found GFP^+ cells in lymphoid as well as nonlymphoid tissue. However, after day 5 of murine pregnancy (implantation), we could not distinguish whether GFP^+ structures were of paternal or fetal origin (Zenclussen et al. 2010). The appearance of fetal cells in maternal organs as a result of transplacental cell migration is well documented in humans and mice and results in a phenomenon called fetal microchimerism (Yan et al. 2005; Khosrotehrani et al. 2005; Tan et al. 2005). Accordingly, maternal cells migrate to the fetus ending up in maternal microchimerism (Loubière et al. 2006). In more detailed analysis, we then showed that some GFP^+ cells also expressed MHC class II molecules and were positive for the DC marker CD11c suggesting that semen contains DCs capable to present paternal antigens immediately after fecundation. In line with these findings, Witkin and colleagues confirmed the presence of several immune cell populations in semen (Witkin and Goldstein 1988). Together with the observation that maternal antigen-presenting cells (APCs) are enriched in the vagina and uterus after insemination, it can be assumed that paternal alloantigens are presented directly (via paternal APCs) and/or indirectly (via maternal APCs). Moldenhauer and colleagues analyzed the involvement of seminal fluid antigens in maternal T-cell activation and defined the underlying antigen-presenting pathway in a transgenic mouse model. The authors transferred T cells specific for the ovalbumin antigen to female mice that were mated to male mice ubiquitously expressing membrane-bound ovalbumin. They confirmed the presence of the ovalbumin antigen in the seminal plasma and detected activated ovalbumin-specific $CD4^+$ and $CD8^+$ T cells locally in the para-aortic lymph nodes displaying a high proliferative capacity. Furthermore, they proved that seminal plasma, but not sperm, was necessary to induce T-cell proliferation. Interestingly, they showed that paternal ovalbumin presented by maternal APCs was essential for $CD8^+$ T-cell proliferation, but responses were not elicited when ovalbumin was presented by paternal APCs (Moldenhauer et al. 2009) suggesting that indirect rather that direct pathways are involved. However, two other groups proposed antigen presentation via the direct pathway (Root-Bernstein and DeWitt 1995; Clark et al. 2013). It was suggested that the direct presentation of paternal antigens may also determine the occurrence of spontaneous abortion events (Clark et al. 2013). More studies are necessary to clarify the pathways of paternal antigens presentation and their relevance for pregnancy success.

10.5 Immune Regulatory Properties of Seminal Plasma

Sarah Robertson and colleagues discussed four different effector functions mediated by seminal plasma on reproductive processes (Robertson 2005). These functions include the clearance of sperms and microbes delivered with semen at mating, the activation of tissue remodeling processes essential for embryo implantation, the induction of cytokines and growth factors important for preimplantation embryo development, and the induction of tolerance mechanisms toward the foreign paternal alloantigens. Additionally, very recently, the same research group provided important evidence that seminal plasma can also affect the health of male offsprings. By ablation of the plasma fraction from seminal fluid, the authors observed male offsprings exhibiting obesity, distorted metabolic hormones, diminished glucose tolerance, and hypertension suggesting that seminal plasma has long-lasting effects on the health of the male progenies (Bromfield et al. 2014).

Here, we will focus on the immune-modulating properties of seminal plasma components. As one of the major components, TGF-β is able to negatively influence growth activity and function of lymphocytes (Nocera and Chu 1993). Moreover, together with IL-8, TGF-β has been shown to induce IL-1β, IL-6, and LIF expression in endometrial epithelial cells (Gutsche et al. 2003). LIF itself was reported to play an essential role for embryo implantation and development, and its lack has been associated with pregnancy loss in the mouse (Stewart et al. 1992). Whether this is relevant for human pregnancies, it is still a matter of debate. Furthermore, other factors present in seminal plasma were reported to influence immune responses, for instance, the soluble p55 TNF-α receptor known to inhibit TNF-mediated cytotoxicity (Liabakk et al. 1993), spermine which impairs proliferation of NK cells and T lymphocytes (Evans et al. 1995), and inhibitors of the complement system (Tarter and Alexander 1984; Chowdhury et al. 1996). Finally, prostaglandins were suggested to prevent lymphocyte proliferation, NK cell activity, and secretion of pro-inflammatory cytokines (Kelly 1995; Kelly et al. 1997).

10.6 Regulatory T Cells in Human and Murine Pregnancy

Treg represent a unique T-cell subpopulation best known for their function in suppressing autoreactive and alloreactive immune responses, thereby preventing autoimmune diseases and allograft rejection (Sakaguchi et al. 1995). In addition, a crucial role for Treg in the establishment and maintenance of fetal tolerance has widely been reported in both humans and mice (Aluvihare et al. 2004; Zenclussen et al. 2005; Heikkinen et al. 2004; Saito et al. 2005). Most studies observed an augmentation of Treg in peripheral blood and decidua during the first and second trimester in normal pregnant women (Heikkinen et al. 2004; Xiong et al. 2013; Tilburgs et al. 2006). In the third trimester, Treg levels begin to decrease (Seol et al. 2008) and further decline with successive stages of labor (Xiong et al. 2010). Furthermore, Treg suppressive capacity is reduced in term labor and preterm labor suggesting that changes in Treg function may contribute to initiation of labor (Kisielewicz et al. 2010). In mice, normal pregnancy is associated with an augmentation in Treg numbers at very early pregnancy stages, a reduction around implantation time followed by a second increment on day 10 of pregnancy (Teles et al. 2013b; Thuere et al. 2007). Uterine enrichment of Treg can be achieved by a selective recruitment of these cells from the periphery or by conversion from conventional T cells directly at the fetal-maternal interface (Ramhorst et al. 2012; Tilburgs et al. 2008). Cytokines, chemokines, and hormones may serve as attractors for Treg migration into uterine tissue (Kallikourdis et al. 2007; Schumacher et al. 2009; Teles et al. 2013a). We recently suggested that CCR7 is involved in murine Treg homing to nonpregnant uterus (Teles et al. 2013a), while the pregnancy-hormone human chorionic gonadotropin attracts human Treg to trophoblasts (Schumacher et al. 2009) and murine Treg to the fetal-maternal interface when injected at peri-implantation (Schumacher et al. 2013). LH also attracted Treg to the fetal-maternal interface in a mouse model (Schumacher et al. 2014).

The importance of Treg for pregnancy success was confirmed by several studies showing that in both humans and mice, spontaneous abortion is associated with a diminished number and activity of Treg (Aluvihare et al. 2004; Zenclussen et al. 2005; Heikkinen et al. 2004; Somerset et al. 2004; Sasaki et al. 2004; Yang et al. 2008; Jin et al. 2009; Mei et al. 2010; Inada et al. 2013). Additionally, reduced Treg levels and impaired functionality were reported in other pregnancy complications such as extrauterine pregnancies, endometriosis, and preeclampsia (Schumacher et al. 2009; Basta et al. 2010; Sasaki et al. 2007; Toldi et al. 2008; Prins et al. 2009; Santner-Nanan et al. 2009; Quinn et al. 2011; Darmochwal-Kolarz et al. 2012), clearly indicating that only fully functional Treg guarantee successful pregnancy. The adoptive transfer of Treg was shown to diminish the occurrence of spontaneous abortion in the mouse (Zenclussen et al. 2005; Schumacher et al. 2007; Yin et al. 2012). Immunization with paternal antigens in early human pregnancies was associated with an increase of Treg (Wu et al. 2014).

10.7 Antigen Specificity of Treg and Mechanisms of Their Action

There is a general consent that recognition of fetal alloantigens by maternal immune cells is a prerequisite for the induction of an active suppression of anti-fetal immune responses. However, there is still a lively debate whether Treg mediate their protection in an antigen-specific fashion. Although several studies indicated that the recognition of foreign paternal/fetal alloantigens is critically for Treg development and function (Darrasse-Jèze et al. 2006; Mjösberg et al. 2007; Kallikourdis et al. 2007; Tilburgs et al. 2009; Kahn and Baltimore 2010; Schumacher et al. 2007; Liu et al. 2013), there is also evidence arguing against a "pure" alloantigen-driven expansion of Treg during pregnancy (Chen et al. 2013). Based on these findings, it seems presumable that both alloantigens and self-antigens support Treg augmentation and protective activity during pregnancy.

In contrast to the discussed antigen specificity of Treg function, the research community agrees on the indispensable role for Treg in the prevention of overwhelming inflammatory immune responses in various clinical disciplines. Here, Treg have been reported to modulate the number and activity of every other immune cell population. For instance, Treg suppress the proliferation and function of T cells, DCs, and macrophages (Piccirillo and Shevach 2001; Mempel et al. 2006; Cederbom et al. 2000; Misra et al. 2004; Taams et al. 2005) and hamper the proliferation of B cells as well as their antibody secretion (Lim et al. 2005). Treg function is thereby realized either by direct cell-cell contact or via the secretion of immune-suppressive cytokines such as IL-10 and TGF-β (Hara et al. 2001; Wahl et al. 2004; Friedline et al. 2009). In pregnancy, the precise mechanisms underlying Treg protection are still not completely understood. We found that in a murine model of disturbed fetal tolerance, Treg rather function through PD-1, IL-10, and HO-1 than through CTLA-4 and TGF-β (Verdijk et al. 2004; Schumacher et al. 2007, 2012). However, Jin and colleagues proposed a role for CTLA-1 in Treg function in humans (Jin et al. 2009), suggesting that the way of Treg-mediated suppression may differ between experimental models and species. Additionally, we recently showed that Treg regulate the accumulation of conventional $CD8^+$ T cells and the production of pro-inflammatory molecules in the uterus and draining lymph nodes in the preimplantation period. Therefore, Treg dampen local inflammatory processes occurring during the time of implantation and support successful embryo nidation (Teles et al. 2013a).

10.8 Influence of Seminal Fluid on Treg Biology

There is accumulating evidence that seminal fluid-derived antigens and other factors present in seminal fluid play a pivotal role in Treg generation, expansion, migration, and function. In the preimplantation period, seminal fluid causes an expansion of the $CD4^+CD25^+Foxp3^+$ Treg population in uterine-draining lymph nodes and the uterus itself. The need for seminal fluid for uterine Treg expansion was proven by matings with seminal vesicle-deficient and vasectomized males (Guerin et al. 2011; Robertson et al. 2009) suggesting that both seminal plasma and sperms are necessary for Treg induction. We confirmed the need for seminal plasma for in vivo Treg expansion in draining lymph nodes in the preimplantation period by matings with seminal-deficient male mice (Teles et al. 2013a). In addition, in vitro co-culture of Treg in the presence of different concentrations of seminal plasma provoked a significant proliferation of Treg. However, this was not observed for conventional T cells (Teles et al. 2013a). Moreover, pseudopregnancy induced by mechanical stimulation did not result in Treg augmentation again underlying the importance of seminal plasma for Treg elevation (Schumacher et al. 2007). In contrast to observations obtained by Robertson and colleagues, in our mouse model, matings with vasectomized males did not significantly impair Treg increase in lymph nodes (Schumacher et al. 2007; Teles et al. 2013a).

TGF-β, present in high amounts in seminal plasma, might be one of the factors responsible for seminal plasma-driven Treg expansion. In vitro co-cultures of seminal plasma with Treg in the presence of a TGF-β antibody abrogated Treg proliferation (Teles et al. 2013a). In line, Clark and colleagues confirmed an effect of pure TGF-β3 on Treg induction and pregnancy success. They applied pharmaceutical-grade bioactive TGF-β3 into the vaginal tract of abortion-prone females at mating. Application of intravaginal TGF-β3 reduced the abortion rate and increased the numbers of Treg in the vagina (Clark et al. 2008). PGE, known to have the capability to induce Treg number and activity (Baratelli et al. 2005), may also contribute to Treg elevation. Furthermore, seminal fluid induces the expression of uterine CCL19, a chemokine that acts through the CCR7 receptor, and may therefore be involved in Treg recruitment to the uterus (Guerin et al. 2011; Teles et al. 2013a). Moreover, Robertson and colleagues proved that seminal fluid-induced Treg are fully functional. In a mouse model, they nicely showed that Treg induced by seminal fluid-derived paternal antigens efficiently prevented the rejection of engrafted tumor cells expressing the same paternal antigens (Robertson et al. 2009). In agreement with the murine data, human seminal plasma was also shown to increase the proportion of $CD127^{low}$ $CD49d^{low}$ Treg. However, Balandya and colleagues revealed that increased Treg numbers were a result of an increased conversion rate from $CD4^{+}$ non-Treg into $Foxp3^{-}$ Treg and not due to proliferation of preexisting Treg (Balandya et al. 2012).

Altogether, Treg expansion in the preimplantation period may occur due to the presence of paternal alloantigens and other factors in seminal fluid. Paternal alloantigens might be presented by APCs of paternal or maternal origin. After implantation, the Treg pool is then further maintained by the continuous release of fetal antigens from the placenta supporting fetal survival until birth. Interestingly, postpartum fetal-specific Treg may persist in the mother creating a memory to paternal antigens (Schober et al. 2012) and rapidly re-accumulate during subsequent pregnancies (Rowe et al. 2012).

10.9 Therapeutic Potential of Seminal Fluid

The findings discussed above provide evidence that seminal fluid might possess some therapeutic potential in the treatment of infertility and miscarriage. A recently published review compared data on the outcome of in vitro fertilization (IVF) treatments in patients exposed to seminal plasma around the time of oocyte retrieval or embryo transfer with placebo controls or controls with no exposure to seminal plasma. The authors compared the clinical pregnancy and live birth/ongoing pregnancy rate and found a statistically significant improvement in clinical pregnancy rate but no significant improvement in terms of ongoing pregnancy/live birth rates. However, they admitted that available data for the ongoing pregnancy/live birth rates were very limited, and the methodology and quality of the analyzed studies were variable (Crawford et al. 2014). The positive effect of seminal plasma exposure on the clinical pregnancy rate in IVF patients might be explained by an increase in the number of Treg that has been associated with improved pregnancy rates in IVF patients (Zhou et al. 2012). Moreover, seminal fluid-driven Treg protection may also have beneficial effects in other clinical applications. In this regard, exposure of seminal fluid may improve disease activity in autoimmune disorders associated with reduced Treg numbers and function (Pakravan et al. 2014). Altogether, it can be assumed that administration of seminal fluid is a promising tool to modulate undesired immune responses and provoke tolerance in different clinical disciplines.

10.10 Conclusions

Despite intensive research work investigating the factors and mechanisms allowing fetal survival within the hostile uterine environment, there remain several open questions to be answered. One of these questions addresses the father's contribution to fetal tolerance. Here, we pointed out that the father not only provides the genetics to create new life but also contributes to its survival from the very beginning by delivering immune-modulating factors with the semen. These factors allow the establishment and maintenance of a fetal-friendly environment resulting in a successful embryo implantation and fetal growth. Moreover, the health of the progeny after birth seems also to be influenced by paternal-derived factors. Thus, the contribution of the father for the survival of its own child should not be underestimated. In Fig. 10.1, we propose an hypothetical scenario as to how maternal immune responses are regulated upon contact with paternal antigens.

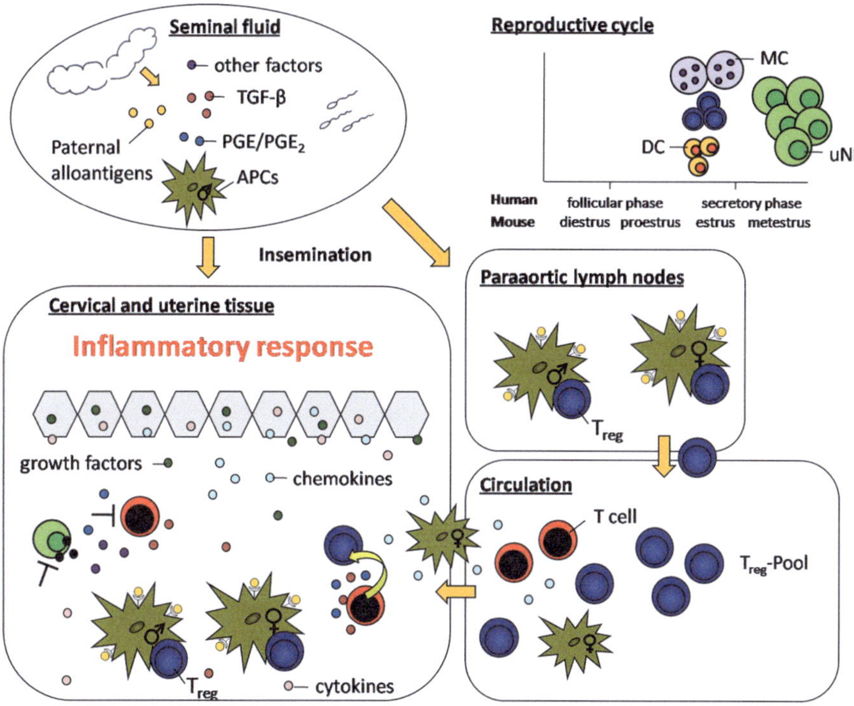

Fig. 10.1 Hypothetical scenario on the immune modulatory capacity of seminal fluid. After contact with paternal antigens a rather tolerant immune response is generated. The graphic illustrates the interactions between immune cells at different compartments

References

Aluvihare VR, Kallikourdis M, Betz AG (2004) Regulatory T cells mediate maternal tolerance to the fetus. Nat Immunol 5(3):266–271. doi:10.1038/ni1037

Arruvito L, Sanz M, Banham AH, Fainboim L (2007) Expansion of CD4+CD25+ and FOXP3+ regulatory T cells during the follicular phase of the menstrual cycle: implications for human reproduction. J Immunol 178(4):2572–2578

Balandya E, Wieland-Alter W, Sanders K, Lahey T (2012) Human seminal plasma fosters CD4+ regulatory T-cell phenotype and transforming growth factor-β1 expression. Am J Reprod Immunol 68(4):322–330. doi:10.1111/j.1600-0897.2012.01176.x

Baratelli F, Lin Y, Zhu L, Yang S, Heuzé-Vourc'h N, Zeng G, Reckamp K, Dohadwala M, Sharma S, Dubinett SM (2005) Prostaglandin E2 induces FOXP3 gene expression and T regulatory cell function in human CD4+ T cells. J Immunol 175(3):1483–1490

Basta P, Majka M, Jozwicki W, Lukaszewska E, Knafel A, Grabiec M, Stasienko E, Wicherek L (2010) The frequency of CD25+CD4+ and FOXP3+ regulatory T cells in ectopic endometrium and ectopic decidua. Reprod Biol Endocrinol 8:116. doi:10.1186/1477-7827-8-116

Breuss JM, Gillett N, Lu L, Sheppard D, Pytela R (1993) Restricted distribution of integrin beta 6 mRNA in primate epithelial tissues. J Histochem Cytochem 41(10):1521–1527

Bromfield JJ, Schjenken JE, Chin PY, Care AS, Jasper MJ, Robertson SA (2014) Maternal tract factors contribute to paternal seminal fluid impact on metabolic phenotype in offspring. Proc Natl Acad Sci 111(6):2200–2205. doi:10.1073/pnas.1305609111

Bulmer JN, Morrison L, Longfellow M, Ritson A, Pace D (1991) Granulated lymphocytes in human endometrium: histochemical and immunohistochemical studies. Hum Reprod 6(6):791–798

Cederbom L, Hall H, Ivars F (2000) CD4+CD25+ regulatory T cells down-regulate co-stimulatory molecules on antigen-presenting cells. Eur J Immunol 30(6):1538–1543. doi:10.1002/1521-4141(200006)30:6<1538:AID-IMMU1538>3.0.CO;2-X

Chen T, Darrasse-Jeze G, Bergot A, Courau T, Churlaud G, Valdivia K, Strominger JL, Ruocco MG, Chaouat G, Klatzmann D (2013) Self-specific memory regulatory T cells protect embryos at implantation in mice. J Immunol 191(5):2273–2281. doi:10.4049/jimmunol.1202413

Chowdhury NA, Kamada M, Takikawa M, Mori H, Gima H, Aono T (1996) Complement-inhibiting activity of human seminal plasma and semen quality. Arch Androl 36(2):109–118

Chu TM, Kawinski E (1998) Plasmin, subtilisin-like endoproteases, tissue plasminogen activator, and urokinase plasminogen activator are involved in activation of latent TGF-beta 1 in human seminal plasma. Biochem Biophys Res Commun 253(1):128–134. doi:10.1006/bbrc.1998.9760

Clark DA, Fernandes J, Fernandez J, Banwatt D (2008) Prevention of spontaneous abortion in the CBA x DBA/2 mouse model by intravaginal TGF-beta and local recruitment of CD4+8+ FOXP3+ cells. Am J Reprod Immunol 59(6):525–534. doi:10.1111/j.1600-0897.2008.00582.x

Clark DA, Rahmati M, Gohner C, Bensussan A, Markert UR, Chaouat G (2013) Seminal plasma peptides may determine maternal immune response that alters success or failure of pregnancy in the abortion-prone CBAxDBA/2 model. J Reprod Immunol 99(1–2):46–53. doi:10.1016/j.jri.2013.03.006

Crawford G, Ray A, Gudi A, Shah A, Homburg R (2014) The role of seminal plasma for improved outcomes during in vitro fertilization treatment: review of the literature and meta-analysis. Hum Reprod Update. doi:10.1093/humupd/dmu052

Darmochwal-Kolarz D, Kludka-Sternik M, Tabarkiewicz J, Kolarz B, Rolinski J, Leszczynska-Gorzelak B, Oleszczuk J (2012) The predominance of Th17 lymphocytes and decreased number and function of Treg cells in preeclampsia. J Reprod Immunol 93(2):75–81. doi:10.1016/j.jri.2012.01.006

Darrasse-Jèze G, Darasse-Jèze G, Klatzmann D, Charlotte F, Salomon BL, Cohen JL (2006) CD4+CD25+ regulatory/suppressor T cells prevent allogeneic fetus rejection in mice. Immunol Lett 102(1):106–109. doi:10.1016/j.imlet.2005.07.002

De M, Choudhuri R, Wood GW (1991) Determination of the number and distribution of macrophages, lymphocytes, and granulocytes in the mouse uterus from mating through implantation. J Leukoc Biol 50(3):252–262

Dekker GA, Robillard PY, Hulsey TC (1998) Immune maladaptation in the etiology of preeclampsia: a review of corroborative epidemiologic studies. Obstet Gynecol Surv 53(6):377–382

Denison FC, Grant VE, Calder AA, Kelly RW (1999) Seminal plasma components stimulate interleukin-8 and interleukin-10 release. Mol Hum Reprod 5(3):220–226

Evans CH, Lee TS, Flugelman AA (1995) Spermine-directed immunosuppression of cervical carcinoma cell sensitivity to a majority of lymphokine-activated killer lymphocyte cytotoxicity. Nat Immun 14(3):157–163

Friedline RH, Brown DS, Nguyen H, Kornfeld H, Lee J, Zhang Y, Appleby M, Der SD, Kang J, Chambers CA (2009) CD4+ regulatory T cells require CTLA-4 for the maintenance of systemic tolerance. J Exp Med 206(2):421–434. doi:10.1084/jem.20081811

Guerin LR, Moldenhauer LM, Prins JR, Bromfield JJ, Hayball JD, Robertson SA (2011) Seminal fluid regulates accumulation of FOXP3+ regulatory T cells in the preimplantation mouse uterus through expanding the FOXP3+ cell pool and CCL19-mediated recruitment. Biol Reprod 85(2):397–408. doi:10.1095/biolreprod.110.088591

Gutsche S, von Wolff M, Strowitzki T, Thaler CJ (2003) Seminal plasma induces mRNA expression of IL-1beta, IL-6 and LIF in endometrial epithelial cells in vitro. Mol Hum Reprod 9(12):785–791

Hara M, Kingsley CI, Niimi M, Read S, Turvey SE, Bushell AR, Morris PJ, Powrie F, Wood KJ (2001) IL-10 is required for regulatory T cells to mediate tolerance to alloantigens in vivo. J Immunol 166(6):3789–3796

Heikkinen J, Möttönen M, Alanen A, Lassila O (2004) Phenotypic characterization of regulatory T cells in the human decidua. Clin Exp Immunol 136(2):373–378. doi:10.1111/j.1365-2249.2004.02441.x

Inada K, Shima T, Nakashima A, Aoki K, Ito M, Saito S (2013) Characterization of regulatory T cells in decidua of miscarriage cases with abnormal or normal fetal chromosomal content. J Reprod Immunol 97(1):104–111. doi:10.1016/j.jri.2012.12.001

Jin L, Chen Q, Zhang T, Guo P, Li D (2009) The CD4+CD25 bright regulatory T cells and CTLA-4 expression in peripheral and decidual lymphocytes are down-regulated in human miscarriage. Clin Immunol 133(3):402–410. doi:10.1016/j.clim.2009.08.009

Juyena NS, Stelletta C (2012) Seminal plasma: an essential attribute to spermatozoa. J Androl 33(4):536–551. doi:10.2164/jandrol.110.012583

Kahn DA, Baltimore D (2010) Pregnancy induces a fetal antigen-specific maternal T regulatory cell response that contributes to tolerance. Proc Natl Acad Sci 107(20):9299–9304. doi:10.1073/pnas.1003909107

Kallikourdis M, Betz AG (2007) Periodic accumulation of regulatory T cells in the uterus: preparation for the implantation of a semi-allogeneic fetus? PLoS One 2(4), e382. doi:10.1371/journal.pone.0000382

Kallikourdis M, Andersen KG, Welch KA, Betz AG (2007) Alloantigen-enhanced accumulation of CCR5+ 'effector' regulatory T cells in the gravid uterus. Proc Natl Acad Sci USA 104(2):594–599. doi:10.1073/pnas.0604268104

Kelly RW (1995) Immunosuppressive mechanisms in semen: implications for contraception. Hum Reprod 10(7):1686–1693

Kelly RW, Carr GG, Critchley HO (1997) A cytokine switch induced by human seminal plasma: an immune modulation with implications for sexually transmitted disease. Hum Reprod 12(4):677–681

Khosrotehrani K, Johnson KL, Guégan S, Stroh H, Bianchi DW (2005) Natural history of fetal cell microchimerism during and following murine pregnancy. J Reprod Immunol 66(1):1–12. doi:10.1016/j.jri.2005.02.001

Kisielewicz A, Schaier M, Schmitt E, Hug F, Haensch GM, Meuer S, Zeier M, Sohn C, Steinborn A (2010) A distinct subset of HLA-DR+-regulatory T cells is involved in the induction of preterm labor during pregnancy and in the induction of organ rejection after transplantation. Clin Immunol 137(2):209–220. doi:10.1016/j.clim.2010.07.008

Liabakk NB, Lien E, Sundan A, Sunde A, Austgulen R, Espevik T (1993) High concentrations of the soluble p55 tumour necrosis factor receptor in human seminal plasma. Hum Reprod 8(11):1837–1842

Lim HW, Hillsamer P, Banham AH, Kim CH (2005) Cutting edge: direct suppression of B cells by CD4+ CD25+ regulatory T cells. J Immunol 175(7):4180–4183
Liu C, Wang X, Sun X (2013) Assessment of sperm antigen specific T regulatory cells in women with recurrent miscarriage. Early Hum Dev 89(2):95–100. doi:10.1016/j.earlhumdev.2012.08.003
Lokeshwar BL, Block NL (1992) Isolation of a prostate carcinoma cell proliferation-inhibiting factor from human seminal plasma and its similarity to transforming growth factor beta. Cancer Res 52(20):5821–5825
Loubière LS, Lambert NC, Flinn LJ, Erickson TD, Yan Z, Guthrie KA, Vickers KT, Nelson JL (2006) Maternal microchimerism in healthy adults in lymphocytes, monocyte/macrophages and NK cells. Lab Invest 86(11):1185–1192. doi:10.1038/labinvest.3700471
McMaster MT, Newton RC, Dey SK, Andrews GK (1992) Activation and distribution of inflammatory cells in the mouse uterus during the preimplantation period. J Immunol 148(6): 1699–1705
Mei S, Tan J, Chen H, Chen Y, Zhang J (2010) Changes of CD4+CD25high regulatory T cells and FOXP3 expression in unexplained recurrent spontaneous abortion patients. Fertil Steril 94(6):2244–2247. doi:10.1016/j.fertnstert.2009.11.020
Mempel TR, Pittet MJ, Khazaie K, Weninger W, Weissleder R, von Boehmer H, von Andrian UH (2006) Regulatory T cells reversibly suppress cytotoxic T cell function independent of effector differentiation. Immunity 25(1):129–141. doi:10.1016/j.immuni.2006.04.015
Milardi D, Grande G, Vincenzoni F, Castagnola M, Marana R (2013) Proteomics of human seminal plasma: identification of biomarker candidates for fertility and infertility and the evolution of technology. Mol Reprod Dev 80(5):350–357. doi:10.1002/mrd.22178
Misra N, Bayry J, Lacroix-Desmazes S, Kazatchkine MD, Kaveri SV (2004) Cutting edge: human CD4+CD25+ T cells restrain the maturation and antigen-presenting function of dendritic cells. J Immunol 172(8):4676–4680
Mjösberg J, Berg G, Ernerudh J, Ekerfelt C (2007) CD4+ CD25+ regulatory T cells in human pregnancy: development of a Treg-MLC-ELISPOT suppression assay and indications of paternal specific Tregs. Immunology 120(4):456–466. doi:10.1111/j.1365-2567.2006.02529.x
Moldenhauer LM, Diener KR, Thring DM, Brown MP, Hayball JD, Robertson SA (2009) Cross-presentation of male seminal fluid antigens elicits T cell activation to initiate the female immune response to pregnancy. J Immunol 182(12):8080–8093. doi:10.4049/jimmunol.0804018
Mori A, Zhai YL, Toki T, Nikaido T, Fujii S (1997) Distribution and heterogeneity of mast cells in the human uterus. Hum Reprod 12(2):368–372
Nocera M, Chu TM (1993) Transforming growth factor beta as an immunosuppressive protein in human seminal plasma. Am J Reprod Immunol 30(1):1–8
Nocera M, Chu TM (1995) Characterization of latent transforming growth factor-beta from human seminal plasma. Am J Reprod Immunol 33(4):282–291
O'Leary S, Armstrong DT, Robertson SA (2013) Transforming growth factor-? (TGF?) in porcine seminal plasma. Reprod Fertil Dev 23(6):748. doi:10.1071/RD11001
Padilla L, Reinicke K, Montesino H, Villena F, Asencio H, Cruz M, Rudolph MI (1990) Histamine content and mast cells distribution in mouse uterus: the effect of sexual hormones, gestation and labor. Cell Mol Biol 36(1):93–100
Pakravan N, Ghaffarinia A, Jalili C, Riazi-Rad F, Tajedini M, Mostafaie A (2014) Seminal vesicle fluid ameliorates experimental autoimmune encephalomyelitis by increasing FoxP3+ regulatory T-cell numbers. Cell Mol Immunol. doi:10.1038/cmi.2014.88
Pandya IJ, Cohen J (1985) The leukocytic reaction of the human uterine cervix to spermatozoa. Fertil Steril 43(3):417–421
Piccirillo CA, Shevach EM (2001) Cutting edge: control of CD8+ T cell activation by CD4+CD25+ immunoregulatory cells. J Immunol 167(3):1137–1140
Plaks V, Birnberg T, Berkutzki T, Sela S, BenYashar A, Kalchenko V, Mor G, Keshet E, Dekel N, Neeman M, Jung S (2008) Uterine DCs are crucial for decidua formation during embryo implantation in mice. J Clin Invest. doi:10.1172/JCI36682
Prins JR, Boelens HM, Heimweg J, van der Heide S, Dubois AE, van Oosterhout AJ, Erwich JJHM (2009) Preeclampsia is associated with lower percentages of regulatory T cells in maternal blood. Hypertens Pregnancy 28(3):300–311. doi:10.1080/10641950802601237

Quinn KH, Lacoursiere DY, Cui L, Bui J, Parast MM (2011) The unique pathophysiology of early-onset severe preeclampsia: role of decidual T regulatory cells. J Reprod Immunol 91(1–2):76–82. doi:10.1016/j.jri.2011.05.006

Ramhorst R, Fraccaroli L, Aldo P, Alvero AB, Cardenas I, Leirós CP, Mor G (2012) Modulation and recruitment of inducible regulatory T cells by first trimester trophoblast cells. Am J Reprod Immunol 67(1):17–27. doi:10.1111/j.1600-0897.2011.01056.x

Robertson SA (2005) Seminal plasma and male factor signalling in the female reproductive tract. Cell Tissue Res 322(1):43–52. doi:10.1007/s00441-005-1127-3

Robertson SA, Mau VJ, Tremellen KP, Seamark RF (1996) Role of high molecular weight seminal vesicle proteins in eliciting the uterine inflammatory response to semen in mice. J Reprod Fertil 107(2):265–277

Robertson SA, Guerin LR, Bromfield JJ, Branson KM, Ahlström AC, Care AS (2009) Seminal fluid drives expansion of the CD4+CD25+ T regulatory cell pool and induces tolerance to paternal alloantigens in mice. Biol Reprod 80(5):1036–1045. doi:10.1095/biolreprod.108.074658

Root-Bernstein RS, DeWitt SH (1995) Semen alloantigens and lymphocytotoxic antibodies in AIDS and ICL. Genetica 95(1–3):133–156

Rowe JH, Ertelt JM, Xin L, Way SS (2012) Pregnancy imprints regulatory memory that sustains anergy to fetal antigen. Nature 490(7418):102–106. doi:10.1038/nature11462

Saftlas AF, Rubenstein L, Prater K, Harland KK, Field E, Triche EW (2014) Cumulative exposure to paternal seminal fluid prior to conception and subsequent risk of preeclampsia. J Reprod Immunol 101–102:104–110. doi:10.1016/j.jri.2013.07.006

Saito S, Sasaki Y, Sakai M (2005) CD4(+)CD25high regulatory T cells in human pregnancy. J Reprod Immunol 65(2):111–120. doi:10.1016/j.jri.2005.01.004

Sakaguchi S, Sakaguchi N, Asano M, Itoh M, Toda M (1995) Immunologic self-tolerance maintained by activated T cells expressing IL-2 receptor alpha-chains (CD25). Breakdown of a single mechanism of self-tolerance causes various autoimmune diseases. J Immunol 155(3):1151–1164

Santner-Nanan B, Peek MJ, Khanam R, Richarts L, Zhu E, de St F, Groth B, Nanan R (2009) Systemic increase in the ratio between Foxp3+ and IL-17-producing CD4+ T cells in healthy pregnancy but not in preeclampsia. J Immunol 183(11):7023–7030. doi:10.4049/jimmunol.0901154

Sasaki Y, Sakai M, Miyazaki S, Higuma S, Shiozaki A, Saito S (2004) Decidual and peripheral blood CD4+CD25+ regulatory T cells in early pregnancy subjects and spontaneous abortion cases. Mol Hum Reprod 10(5):347–353. doi:10.1093/molehr/gah044

Sasaki Y, Darmochwal-Kolarz D, Suzuki D, Sakai M, Ito M, Shima T, Shiozaki A, Rolinski J, Saito S (2007) Proportion of peripheral blood and decidual CD4(+) CD25(bright) regulatory T cells in pre-eclampsia. Clin Exp Immunol 149(1):139–145. doi:10.1111/j.1365-2249.2007.03397.x

Schober L, Radnai D, Schmitt E, Mahnke K, Sohn C, Steinborn A (2012) Term and preterm labor: decreased suppressive activity and changes in composition of the regulatory T-cell pool. Immunol Cell Biol 90(10):935–944. doi:10.1038/icb.2012.33

Schumacher A, Wafula PO, Bertoja AZ, Sollwedel A, Thuere C, Wollenberg I, Yagita H, Volk H, Zenclussen AC (2007) Mechanisms of action of regulatory T cells specific for paternal antigens during pregnancy. Obstet Gynecol 110(5):1137–1145. doi:10.1097/01.AOG.0000284625.10175.31

Schumacher A, Brachwitz N, Sohr S, Engeland K, Langwisch S, Dolaptchieva M, Alexander T, Taran A, Malfertheiner SF, Costa S, Zimmermann G, Nitschke C, Volk H, Alexander H, Gunzer M, Zenclussen AC (2009) Human chorionic gonadotropin attracts regulatory T cells into the fetal-maternal interface during early human pregnancy. J Immunol 182(9):5488–5497. doi:10.4049/jimmunol.0803177

Schumacher A, Wafula PO, Teles A, El-Mousleh T, Linzke N, Zenclussen ML, Langwisch S, Heinze K, Wollenberg I, Casalis PA, Volk H, Fest S, Zenclussen AC, Kassiotis G (2012) Blockage of heme oxygenase-1 abrogates the protective effect of regulatory T cells on murine pregnancy and promotes the maturation of dendritic cells. PLoS One 7(8), e42301. doi:10.1371/journal.pone.0042301

Schumacher A, Heinze K, Witte J, Poloski E, Linzke N, Woidacki K, Zenclussen AC (2013) Human chorionic gonadotropin as a central regulator of pregnancy immune tolerance. J Immunol 190(6):2650–2658. doi:10.4049/jimmunol.1202698

Schumacher A, Poloski E, Spörke D, Zenclussen AC (2014) Luteinizing hormone contributes to fetal tolerance by regulating adaptive immune responses. Am J Reprod Immunol 71(5):434–440. doi:10.1111/aji.12215

Seol H, Oh M, Lim J, Jung N, Yoon S, Kim H (2008) The role of CD4+CD25bright regulatory T cells in the maintenance of pregnancy, premature rupture of membranes, and labor. Yonsei Med J 49(3):366. doi:10.3349/ymj.2008.49.3.366

Sharkey DJ, Macpherson AM, Tremellen KP, Robertson SA (2007) Seminal plasma differentially regulates inflammatory cytokine gene expression in human cervical and vaginal epithelial cells. Mol Hum Reprod 13(7):491–501. doi:10.1093/molehr/gam028

Slater M, Murphy CR (1999) Thrombospondin is sequentially expressed and then de-expressed during early pregnancy in the rat uterus. Histochem J 31(7):471–475

Somerset DA, Zheng Y, Kilby MD, Sansom DM, Drayson MT (2004) Normal human pregnancy is associated with an elevation in the immune suppressive CD25+ CD4+ regulatory T-cell subset. Immunology 112(1):38–43. doi:10.1111/j.1365-2567.2004.01869.x

Srivastava MD, Lippes J, Srivastava BI (1996) Cytokines of the human reproductive tract. Am J Reprod Immunol 36(3):157–166

Stewart CL, Kaspar P, Brunet LJ, Bhatt H, Gadi I, Köntgen F, Abbondanzo SJ (1992) Blastocyst implantation depends on maternal expression of leukaemia inhibitory factor. Nature 359(6390):76–79. doi:10.1038/359076a0

Taams LS, van Amelsfort JMR, Tiemessen MM, Jacobs KMG, de Jong EC, Akbar AN, Bijlsma JWJ, Lafeber FPJG (2005) Modulation of monocyte/macrophage function by human CD4+CD25+ regulatory T cells. Hum Immunol 66(3):222–230. doi:10.1016/j.humimm.2004.12.006

Tan X, Liao H, Sun L, Okabe M, Xiao Z, Dawe GS (2005) Fetal microchimerism in the maternal mouse brain: a novel population of fetal progenitor or stem cells able to cross the blood-brain barrier? Stem Cells 23(10):1443–1452. doi:10.1634/stemcells.2004-0169

Tarter TH, Alexander NJ (1984) Complement-inhibiting activity of seminal plasma. Am J Reprod Immunol 6(1):28–32

Teles A, Schumacher A, Kühnle M, Linzke N, Thuere C, Reichardt P, Tadokoro CE, Hämmerling GJ, Zenclussen AC (2013a) Control of uterine microenvironment by Foxp3+ cells facilitates embryo implantation. Front Immunol 4:158. doi:10.3389/fimmu.2013.00158

Teles A, Thuere C, Wafula PO, El-Mousleh T, Zenclussen ML, Zenclussen AC (2013b) Origin of Foxp3(+) cells during pregnancy. Am J Clin Exp Immunol 2(3):222–233

Thuere C, Zenclussen ML, Schumacher A, Langwisch S, Schulte-Wrede U, Teles A, Paeschke S, Volk H, Zenclussen AC (2007) Kinetics of regulatory T cells during murine pregnancy. Am J Reprod Immunol 58(6):514–523. doi:10.1111/j.1600-0897.2007.00538.x

Tilburgs T, Roelen DL, van der Mast BJ, van Schip JJ, Kleijburg C, de Groot-Swings GM, Kanhai HHH, Claas FHJ, Scherjon SA (2006) Differential distribution of CD4(+)CD25(bright) and CD8(+)CD28(-) T-cells in decidua and maternal blood during human pregnancy. Placenta 27:S47–S53. doi:10.1016/j.placenta.2005.11.008

Tilburgs T, Roelen DL, van der Mast BJ, de Groot-Swings GM, Kleijburg C, Scherjon SA, Claas FH (2008) Evidence for a selective migration of fetus-specific CD4+CD25bright regulatory T cells from the peripheral blood to the decidua in human pregnancy. J Immunol 180(8):5737–5745

Tilburgs T, Scherjon SA, van der Mast BJ, Haasnoot GW, Versteeg-V D, Voort-Maarschalk M, Roelen DL, van Rood JJ, Claas FHJ (2009) Fetal-maternal HLA-C mismatch is associated with decidual T cell activation and induction of functional T regulatory cells. J Reprod Immunol 82(2):148–157. doi:10.1016/j.jri.2009.05.003

Toldi G, Švec P, Vásárhelyi B, Mészáros G, Rigó J, Tulassay T, Treszl A (2008) Decreased number of FoxP3+ regulatory T cells in preeclampsia. Acta Obstet Gynecol Scand 87(11):1229–1233. doi:10.1080/00016340802389470

Tremellen KP, Seamark RF, Robertson SA (1998) Seminal transforming growth factor beta1 stimulates granulocyte-macrophage colony-stimulating factor production and inflammatory cell recruitment in the murine uterus. Biol Reprod 58(5):1217–1225

Verdijk RM, Kloosterman A, Pool J, van de Keur M, Naipal AMIH, van Halteren AGS, Brand A, Mutis T, Goulmy E (2004) Pregnancy induces minor histocompatibility antigen-specific cytotoxic T cells: implications for stem cell transplantation and immunotherapy. Blood 103(5):1961–1964. doi:10.1182/blood-2003-05-1625

Wahl SM, Swisher J, McCartney-Francis N, Chen W (2004) TGF-beta: the perpetrator of immune suppression by regulatory T cells and suicidal T cells. J Leukoc Biol 76(1):15–24. doi:10.1189/jlb.1103539

Weinberg A, Enomoto L, Marcus R, Canniff J (2011) Effect of menstrual cycle variation in female sex hormones on cellular immunity and regulation. J Reprod Immunol 89(1):70–77. doi:10.1016/j.jri.2010.11.009

Witkin SS, Goldstein M (1988) Reduced levels of T suppressor/cytotoxic lymphocytes in semen from vasovasostomized men: relationship to sperm autoantibodies. J Reprod Immunol 14(3):283–290

Woidacki K, Popovic M, Metz M, Schumacher A, Linzke N, Teles A, Poirier F, Fest S, Jensen F, Rabinovich GA, Maurer M, Zenclussen AC (2013) Mast cells rescue implantation defects caused by c-kit deficiency. Cell Death Dis 4(1):e462. doi:10.1038/cddis.2012.214

Wu L, Luo LH, Zhang YX, Li Q, Xu B, Zhou GX, Luan HB, Liu YS (2014) Alteration of Th17 and Treg cells in patients with unexplained recurrent spontaneous abortion before and after lymphocyte immunization therapy. Reprod Biol Endocrinol 12:74. doi:10.1186/1477-7827-12-74

Xiong H, Zhou C, Qi G (2010) Proportional changes of CD4+CD25+Foxp3+ regulatory T cells in maternal peripheral blood during pregnancy and labor at term and preterm. Clin Invest Med 33(6):E422

Xiong Y, Yuan Z, He L (2013) Effects of estrogen on CD4(+) CD25(+) regulatory T cell in peripheral blood during pregnancy. Asian Pac J Trop Med 6(9):748–752. doi:10.1016/S1995-7645(13)60131-5

Yan Z, Lambert NC, Guthrie KA, Porter AJ, Loubiere LS, Madeleine MM, Stevens AM, Hermes HM, Nelson JL (2005) Male microchimerism in women without sons: quantitative assessment and correlation with pregnancy history. Am J Med 118(8):899–906. doi:10.1016/j.amjmed.2005.03.037

Yang H, Qiu L, Chen G, Ye Z, Lü C, Lin Q (2008) Proportional change of CD4+CD25+ regulatory T cells in decidua and peripheral blood in unexplained recurrent spontaneous abortion patients. Fertil Steril 89(3):656–661. doi:10.1016/j.fertnstert.2007.03.037

Yin Y, Han X, Shi Q, Zhao Y, He Y (2012) Adoptive transfer of CD4+CD25+ regulatory T cells for prevention and treatment of spontaneous abortion. Eur J Obstet Gynecol Reprod Biol 161(2):177–181. doi:10.1016/j.ejogrb.2011.12.023

Zenclussen AC, Gerlof K, Zenclussen ML, Sollwedel A, Bertoja AZ, Ritter T, Kotsch K, Leber J, Volk H (2005) Abnormal T-cell reactivity against paternal antigens in spontaneous abortion: adoptive transfer of pregnancy-induced CD4+CD25+ T regulatory cells prevents fetal rejection in a murine abortion model. Am J Pathol 166(3):811–822. doi:10.1016/S0002-9440(10)62302-4

Zenclussen ML, Thuere C, Ahmad N, Wafula PO, Fest S, Teles A, Leber A, Casalis PA, Bechmann I, Priller J, Volk H, Zenclussen AC (2010) The persistence of paternal antigens in the maternal body is involved in regulatory T-cell expansion and fetal-maternal tolerance in murine pregnancy. Am J Reprod Immunol 63(3):200–208. doi:10.1111/j.1600-0897.2009.00793.x

Zenclussen AC, Olivieri DN, Dustin ML, Tadokoro CE (2013) In vivo multiphoton microscopy technique to reveal the physiology of the mouse uterus. Am J Reprod Immunol 69(3):281–289. doi:10.1111/aji.12066

Zhou J, Wang Z, Zhao X, Wang J, Sun H, Hu Y (2012) An increase of Treg cells in the peripheral blood is associated with a better in vitro fertilization treatment outcome. Am J Reprod Immunol 68(2):100–106. doi:10.1111/j.1600-0897.2012.01153.x

Index

R. Bronson (ed.), *The Male Role in Pregnancy Loss and Embryo Implantation Failure*, Advances in Experimental Medicine and Biology 868,
DOI 10.1007/978-3-319-18881-2